DAB DIGITAL RADIO: LICENSED TO FAIL

GRANT GODDARD

Radio Books

DAB Digital Radio: Licensed To Fail
by **Grant Goddard**

Published in 2010 by Radio Books

www.radiobooks.org

British Library Cataloguing in Publication Data:
A catalogue record for this book is available from the British Library

Library of Congress Cataloging in Publication Data:
A catalog record for this book is available from the Library of Congress

ISBN: 978-0-9564963-0-0 paperback

Cover photography: Rebecca Lewis

CONTENTS

FOREWORD

There is nothing wrong with DAB (Digital Audio Broadcasting) radio. The technology, the motivation to widen listener choice and the opportunity to improve broadcast audio quality are all fine objectives. The only thing wrong with DAB radio is its specific implementation in the United Kingdom.

A large part of this problem was the timing. When DAB was first demonstrated at the Radio Festival in 1993, the majority of us did not have mobile phones, did not have computers, and compact discs were in their ascendancy. Since then, the technological world around us has changed almost beyond recognition. Nearly two decades later, the very fact that the industry is still debating how to implement DAB radio successfully demonstrates that something went badly wrong with the timing. So badly wrong that, in my opinion, the window of opportunity for DAB has now passed. Marketing DAB radio in 2010 is like trying to persuade the consumer that a *Sinclair ZX80* computer is 'cutting edge'.

Another part of the problem was the initial motivation. The commercial industry stakeholders seemed driven to invest in DAB infrastructure, rather than content, because it created an opportunity to control this new broadcast platform. Radio stations might be good at radio, but that does not mean they will necessarily be good at running other businesses. *Capital Radio's* diversification into restaurants served as a salutary lesson. Because the radio industry's motivation to invest in DAB did not focus on the consumer, the outcome was that the consumer became lost in the execution strategy.

Finally, commercial radio's lack of competitive get-up-and-go let it down. Control of DAB was handed to a cartel of the biggest broadcasters. There are good reasons why the law forbids cartels, and one of them is that they may not produce outcomes that are in the consumer's interest. This was the case with DAB. Being king of the DAB castle might have been great for radio sector egos, but it proved a disaster for radio sector balance sheets. Power was exercised ruthlessly and the regulator turned a blind eye. A digital station start-up like *PrimeTime* managed to attract DAB radio's biggest audience but was thrown on the scrapheap because it was not cartel-owned.

Now it looks as if the radio industry has been sat in a DAB waiting room for the last three years. There has been lots of talk but no action. When you are a monopoly or a cartel, you expect success to come knocking on your door, rather than having to go out and make it happen yourself. The new Digital Economy Act changes nothing. The government will not create automatically successful radio stations and hand out 'licences to print money' on a plate ….

Or maybe it will, as the government's renewal of the *Classic FM* national licence in the Digital Economy Act was just that. As long as such practices persist, parts of the commercial radio industry will never learn to grow up and compete for licences, listeners, revenues and profits. Therein lies the problem with DAB.

Grant Goddard
July 2010

1.

10 December 2008

DAB: fiddling while Rome burns?

The planned migration of radio broadcasting from analogue to digital platforms in the UK currently sits on a knife-edge. After a decade of existence, the DAB platform is still struggling. Only 9.2% of commercial radio hours listened are via DAB;[1] while 79% of new radios sold in the UK are still old-fashioned analogue rather than DAB.[2] The financial pressures on commercial radio owners are already immense, and the burden of continuing to simulcast on both analogue and digital terrestrial transmitters cannot be borne much longer.

When I wrote about this dire situation in October 2008, I noted that "*Ofcom [is] threatening to revoke the analogue licence of any [simulcasting] station giving up on DAB*" and I asserted that the regulator's "*once carrot-and-stick approach to digital regulation now looks like a hostage situation.*"[3] If stations who had accepted an automatic analogue licence renewal are still forced to continue simulcasting on DAB (some at a cost of many times their analogue transmission) by the regulator, many will simply go out of business.

My attention was drawn this morning to a speech made by Ofcom's Director of Radio, Peter Davies, at the recent Voice of the Listener & Viewer Conference in London, as quoted in The Radio Magazine (headline: 'Ofcom: hundreds more DAB transmitters needed'):

"*We need to build a lot more transmitters than we currently have. The BBC currently has around 100 DAB transmitters. It may need four or five times that number in order to achieve the equivalent coverage of analogue. But, in the end, if it builds those transmitters, the DAB network would probably still be cheaper to run than today's FM network. It's just too early to set a [switchover] date and far more needs to be done to improve the service before that can become a reality.*"[4]

However, the costs of such a DAB build-out programme are significant. The BBC's existing single national DAB multiplex network of 96 transmitters covering 86% of the population costs £6m per annum. To extend that multiplex to 230 transmitters covering 90% of the population would cost an additional £5m per annum. To extend the existing multiplex to the 1,000 transmitters necessary to cover 99% of the population would cost an additional £34m per annum. Now remember that the BBC only has one single national DAB network, whereas the commercial radio sector has one national DAB network, plus a separate layer of local DAB multiplexes that cover most of the UK, plus a further layer of regional DAB multiplexes in the most populous areas. Now imagine what the costs to the commercial sector might be to extend and improve coverage in all these areas.

Although Peter was talking explicitly about the BBC situation, the implication is that the commercial sector too should invest even further in DAB transmission infrastructure, and yet Ofcom must be aware that station owners can barely afford the present network of DAB multiplexes that already cover 90% of the population. It might appear that Ofcom is preoccupied with burdening the commercial radio sector with even more transmission costs, at a time when the industry is already fighting for its life as a result of falling audiences and declining revenues (even before the advertising downturn).

I am reminded of Peter's speech about DAB to The Radio Festival in July 2008:

"Increased coverage of DAB will be absolutely essential if it is ever to become a full replacement for FM for most services…… That brings us to the tricky part – defining what existing coverage is and how we improve it. This is still work in progress but we are approaching it in three stages. Firstly, we need to define what existing FM coverage is. That's not nearly as simple as it might sound. Radio is not like television where you stick an aerial on the roof and you get reception or you don't. Radio is used in every room in the house, usually with a portable aerial. It's used outdoors on a wide variety of devices and it's listened to in cars. So we need to look at geographic coverage as well as population coverage, and we need to look at indoor coverage in different parts of the house. FM coverage gradually fades as you move around, so we need to decide how strong the signal needs to be to be usable. And, surprisingly, this work has never really been done in any kind of consistent manner for the UK as a whole, so it has taken a little while to agree a framework and calculate the numbers.

Having done that, we then have to do the same for existing DAB coverage. Now DAB has all the same issues as FM, but it also has different characteristics. It doesn't fade in the same way – you either get it or you don't – so we need a different set of definitions here. Once we have defined what existing DAB coverage is, we then have to work out what it would take to get existing DAB coverage up to the level of existing FM coverage. Now, we have already done a lot of work on this, and certainly enough to inform the interim report, and the whole thing will be finalised in time for the [DCMS] Digital Radio Working Group final report later this year."[5]

Undoubtedly, these are all important DAB technical issues that (belatedly) demand attention. However, in the grand scheme of things, with the commercial radio sector poised on a precipice of viability, how exactly will this work by Ofcom do anything but add to the sector's existing financial problems?

[PS: Just a reminder that Ofcom's own research in 2007 found that 50% of UK commercial radio licensees either made a loss or an annual profit of less than £100,000.]

---

[1] source: RAJAR Q3 2008
[2] source: DRDB/GfK Q2 2008 four-quarter moving average
[3] http://www.broadcastnow.co.uk/news/multi-platform/comment/talking-radio-grant-goddard/1909761.article
[4] source: The Radio Magazine
[5] author's recording

2.

11 December 2008

DAB v internet: the tortoise and the hare

On Wednesday 10 December 2008, Lord Carter told the Parliamentary Culture, Media & Sport Committee:

"Radio can be received on mobile phones and through the television. Could you have digital radio without DAB? Yes, you probably could. If we do want DAB, we need to push it along a bit or technology will drive it out."[6]

"*Push it along a bit*" probably means state intervention and/or state subsidy.

"*Technology will drive it out*" probably means technologies such as IP-delivered radio via the internet, Wi-Fi, Wi-Max, 3G and 4G, as well as broadcast radio delivered via Freeview, Freesat, Sky and cable.

The same day, evidence was published that demonstrates how one of these platforms – internet-delivered radio – is already poised to eclipse DAB radio. "*With broadband internet access rising from 51% of UK households in 2007 to 56% in 2008 and the high profile launch of the BBC iPlayer, listening to the radio online has never been easier or more popular*", said the new RAJAR internet radio listening report.[7] Its definition of internet listening is:

- listening live via the internet
- listening again via the internet
- personalised online radio
- podcasts.

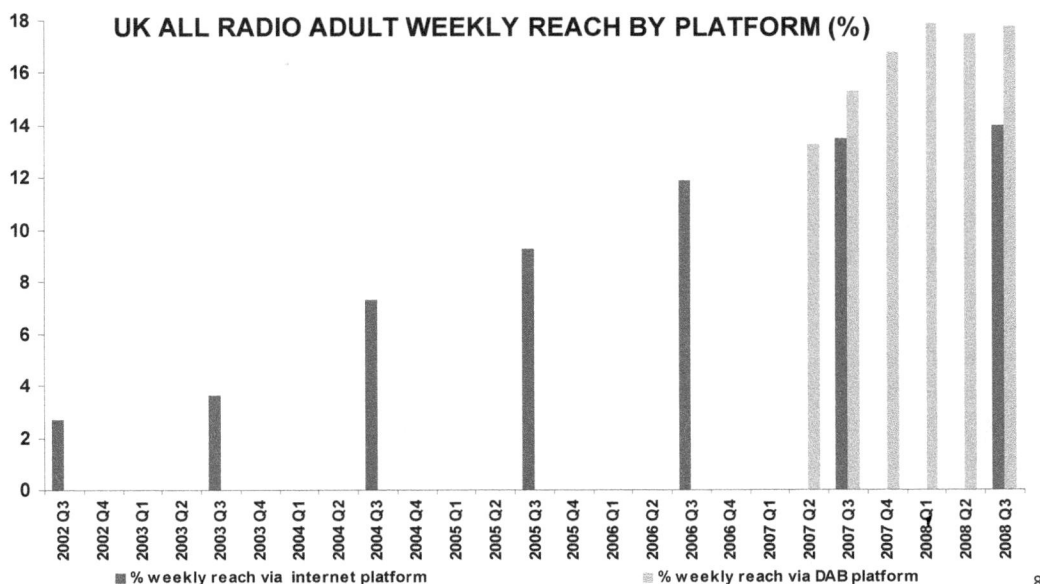

UK ALL RADIO ADULT WEEKLY REACH BY PLATFORM (%)

■ % weekly reach via internet platform   ▨ % weekly reach via DAB platform   8

11

The most informative graph in the RAJAR report was the one that wasn't there ... the one that compares the weekly adult (15+) reach of the DAB platform with the internet platform.

Unfortunately, usage data for the DAB platform are not available on a comparable basis prior to Q2 2007. Suffice to say that commercial radio launched its national Digital One DAB multiplex on 15 November 1999, which could be considered 'Year Zero' for DAB (although it was some time before DAB receivers filtered into shops). What is startling is that the reach of internet radio is so close behind that of DAB. If you were to add up the market value of all the marketing spots promoting the DAB platform that have run on BBC TV and radio and commercial radio over the last decade, their total would run into £m. Add the cost of the sterling efforts of the Digital Radio Development Bureau, jointly funded by the BBC and commercial radio, since 2001 to convince us of the value of DAB radio.

Now compare this with the marketing cost to date spent persuading us to listen to radio via the internet (lots of mentions within BBC radio programmes, but fewer on commercial radio), and it pales by comparison. And yet, listening via the internet is way up there, just behind DAB, driven largely by consumer demand rather than by public intervention.

The other interesting statistic in the RAJAR report was the glaring difference between the online impact of the BBC and the commercial radio sector. Of those who listen to radio via the internet,

- 71% listen via a BBC radio website
- 25% listen via a UK commercial radio website
- 13% listen via a non-BBC, non-UK commercial radio website.[9]

This merely confirms something that was evident already – in the 1990's, the biggest players in the UK commercial radio industry decided to put all their 'future of radio' eggs in the 'DAB' basket and, as a result, neglected to make a comparable investment in the online platform. The BBC has been much more careful (and, admittedly, has the immense resources available) to develop content across a number of platforms simultaneously, and is now reaping the return. Commercial radio could have developed its own 'last.fm' but chose instead to invest huge sums in the DAB platform infrastructure, rather than content, and is now paying the price.

Lord Carter will have to make a difficult (and potentially expensive) political recommendation between now and January 2009 about the future of the DAB platform:[10]

OPTION 1 – Massive state financial intervention to prop up the expensive DAB transmission infrastructure. Who benefits? UK industry. The end result is a closed, almost UK-exclusive system (just like right-hand drive cars). UK radio set manufacturers sell lots of DAB radios in the UK because it is not worthwhile for the global consumer electronics groups to manufacture DAB radios for such a small addressable market. The large UK commercial radio groups and the BBC benefit because they already own both the entire DAB multiplex infrastructure and most of the content broadcast on it, ensuring that most radio listening in the UK remains under their control. Who loses? The consumer. They get a marginally increased choice of radio content that, so far, has failed to propel the DAB platform to mass take-up.

OPTION 2 – No state intervention to support the DAB platform. Who loses? UK industry. DAB remains economically unviable (just as it has been for a decade), forcing commercial radio groups to withdraw from the platform (with substantial balance sheet write-downs). DAB becomes the province of the BBC to offer minority interest services. UK radio set manufacturers lose most of their promised UK market for DAB radios. 7m DAB radio owners complain to Ofcom that all they can receive now on DAB are BBC stations. End result. The UK joins the rest of the world in accepting that IP-delivered radio is an emerging global platform from which the UK benefits from economies of scale (cheap receivers, evolution and innovation). The UK has to admit that DAB seemed like a promising technology in the pre-broadband late 1980s, but its slow implementation was overtaken by technological developments elsewhere and the globalisation of content.

As recently as 2004, The Guardian reported:

"The DTI hopes digital radio will become a rare British industry success story; Ofcom thinks it could get some juicy spectrum to sell off; manufacturers and retailers see rich pickings ('the flat-screen TV of tomorrow', as the man from Dixons told me). Everyone, that is, except the British consumer, who is showing worrying signs of being dazzled by the new technology. According to Stephen Carter, Ofcom chief executive and digital radio owner, most Britons would be on my wife's side – pretty sure that DAB is a good thing, but not quite sure what it is. Last Thursday, in a drum-beating speech to the Social Market Foundation, Carter described the radio industry's foray into digital platforms as at a tipping point between a Sky-style digital success story and an industry-wide egg-on-face scenario."[11]

Four years later, are we any more certain about DAB? There may be a lesson to be learnt from Taiwan:

"The development of DAB in Taiwan passed through three stages: planning, preparation and a final stage characterized by setbacks. It now looks like it may disappear altogether...... After two years of trials, DAB experienced problems, partly because of a lack of promotion, inadequate public knowledge of the technology and high-priced DAB radios that few were willing to purchase. As a result there were too few consumers to keep DAB up and running. In July this year, Taiwan Mobile announced that Tai Yi would be dissolved, and the outlook for other DAB providers is not very bright. The biggest problem for Taiwan's DAB industry was a lack of forward-looking policies......"[12]

[6]http://www.telegraph.co.uk/finance/newsbysector/mediatechnologyandtelecoms/3702966/Future-of-radio-set-for-January.html
[7] http://www.rajar.co.uk/docs/news/MIDAS3_report.pdf
[8] source: RAJAR
[9] http://www.rajar.co.uk/docs/news/MIDAS3_report.pdf
[10] http://www.guardian.co.uk/media/2008/dec/10/bbc-ofcom-licence-fee-lord-carter
[11] http://www.guardian.co.uk/media/2004/sep/20/mondaymediasection.digitaltv
[12] http://www.taipeitimes.com/News/editorials/archives/2008/11/30/2003429926

3.

15 December 2008

Nokia: a 'Man Friday' for radio?

Radio has a problem. Young people are listening less to radio in aggregate. This is the result of two main factors: their declining numbers within the population (there will be fewer than 8m 15-24 year olds in the UK by 2014, compared to 8.2m in 2008); and the increasing competition for young people's leisure time. Radio as a whole is losing listening amongst 15-44 year olds, but commercial radio is losing proportionately more than the BBC. This is disastrous for the commercial sector, which defines 15-44 year olds as its 'heartland audience' for advertisers.

CHANGE IN AGGREGATE HOURS LISTENED TO RADIO (% change Q3 2007 to Q3 2008)		
age group	all radio	commercial radio
15-24	-2.9	-3.3
25-34	-3.3	-4.6
35-44	-5.5	-4.9
45-54	0.7	2.2
55-64	0.0	1.5
65+	0.5	-2.5 [13]

So who is working the hardest to enable and promote the notion of radio listening amongst young people? Could it be Nokia?

Nokia had a 38% market share last quarter of mobile devices globally. In Q3 2008, Nokia sold a staggering 118m mobile devices worldwide, 27.4m of which were in Europe. In the UK, of the 94 Nokia models available, 70 include FM radios and 24 include Wi-Fi capability (19 have both FM radio and Wi-Fi). As a result, the overwhelming majority of new Nokia devices sold in the UK offer consumers listening to either FM broadcast radio and/or IP-delivered radio connected via Wi-Fi. Does this make Nokia the biggest selling brand of radio receivers in the world?

Would not a generic campaign to promote radio listening on mobile phones prove a worthwhile marketing project to be funded jointly by commercial radio and the BBC? The mobile phone hardware is (literally) already sitting in millions of people's pockets, offering them the capability to listen to radio. Of course, mobile phone operators are never going to promote the radio listening function on the handsets they sell, for the simple reason that it earns them no revenues, and every quarter-hour spent listening to the radio is a quarter-hour lost of phone usage.

Is the UK radio industry capitalising on this huge volume of FM receivers incorporated into phones with which Nokia and its competitors are flooding the market, but whose radio function seems to sit mostly unused in people's pockets and handbags? Seemingly, no.

Instead, the industry is wedded to the notion of spending millions of marketing pounds trying to convince consumers to purchase yet another piece of hardware that enables them to receive the 'DAB' digital radio platform. The hurdle is that the average retail price of a DAB radio receiver is still £90+.

In this converged world, is there a mobile phone available in the UK that incorporates the DAB platform? No. Why not? Because 'FM radio' is a long established, global broadcast platform used in almost every country, whereas the 'DAB radio' system is only commercially underway in the UK, Denmark, Norway and, imminently, Australia and China. Will 'DAB radio' ever become a global system that replaces 'FM radio'? No, because the US (the biggest consumer electronics market in the world) has already adopted a completely different digital radio standard. Would Nokia make a phone that includes DAB? Despite a recent report, it would seem highly unlikely.[14] The consumer market for DAB simply isn't big enough for a global player like Nokia.

Which is precisely why UK receiver manufacturers, such as Pure and Roberts, continue to dominate our domestic market for DAB radio hardware – the addressable market is simply not big enough for most global brands to be interested in 'DAB radio'. But neither Pure nor Roberts will ever make mobile phones or cool-design i-Pods that include DAB radio and which might appeal to fashion-conscious, brand-obsessed, young people. As a result, the DAB platform is condemned to be largely the province of older demographics who listen at home on DAB 'kitchen radios'. And, importantly, they are mostly listening to their same, favourite analogue stations via DAB platform simulcasts that they used to listen to on FM/AM. New, digital-only radio stations barely get a look-in in the radio ratings.

Neither do the UK sales figures of DAB hardware look particularly impressive, compared to Nokia's success in pumping FM radios into the market. In the decade since DAB was introduced, more than 7m DAB receivers have been sold. But, during the last year alone, more than 8m analogue radios were sold in the UK. Amazingly, 79% of new radios sold in the UK during the last year were analogue, rather than DAB. Despite a landmark pronouncement in 2006 by online electronics retailer Dixons that it would no longer sell analogue radios, consumers have continued to demonstrate their interest in purchasing inexpensive AM/FM radios. Dixons has been forced to eat humble pie and now stocks four models of analogue portable radio, the cheapest of which is £8.79, alongside 38 models of DAB radio, the cheapest of which is three times that price.

In the face of consumer reticence, the UK commercial radio industry, supported by Ofcom and the government, has been busy the last decade desperately trying to persuade the public to migrate its radio listening to the DAB platform. The sticking point here is the pre-requisite for consumers to spend considerable sums replacing all the analogue radios they own with more expensive, new digital ones. Meanwhile, global heavyweights like Nokia, pursuing their own strategy to satisfy consumer needs, continue to supply the market with millions of analogue FM radios incorporated into a myriad of converged, portable devices. Could the UK government 'persuade' Nokia not to push its FM radios in the UK market? Er, probably not. In which case, its proposed analogue radio 'switch-off' remains a completely lost cause.[15]

Perhaps, instead of viewing Nokia and its ilk as an irrelevancy to its long-held digital migration plans, the UK radio industry needs to simply capitalise on the massive penetration of FM-enabled (and now IP-enabled) phones already within the consumer electronics market. These phones are the 'sleeping giant' that could potentially reinvigorate radio listening, particularly amongst the young demographics. All their owners need is a 'call to action' – a marketing campaign to make them realise that they already have the world's most immediate, live, portable broadcast medium in their pocket.

---

[13] source: RAJAR

[14] http://www.pocket-lint.com/news/19643/nokia-dab-radio-phones-possible

[15] http://www.guardian.co.uk/media/2008/dec/12/digital-radio-radio

4.

19 December 2008

Digital Radio Working Group: it must be 'Numberwang'!

The Final Report of the Digital Radio Working Group published today includes an 'Aspirational Timetable' which, it says, will "*act as a useful guide for those working towards digital migration in the coming months and years.*" The projected dates in the timetable include:

- End 2010 – "*DAB sales to exceed sales of analogue radios*"
- 2014 – "***All** new cars to be fitted with digital radios*"
- 2015 (approx) – "*Migration criteria met.*"

One of three specified "*migration criteria*" is:

- "*that at least 50% of total radio listening is to digital platforms*"

which would look like this by (year-end) 2015:

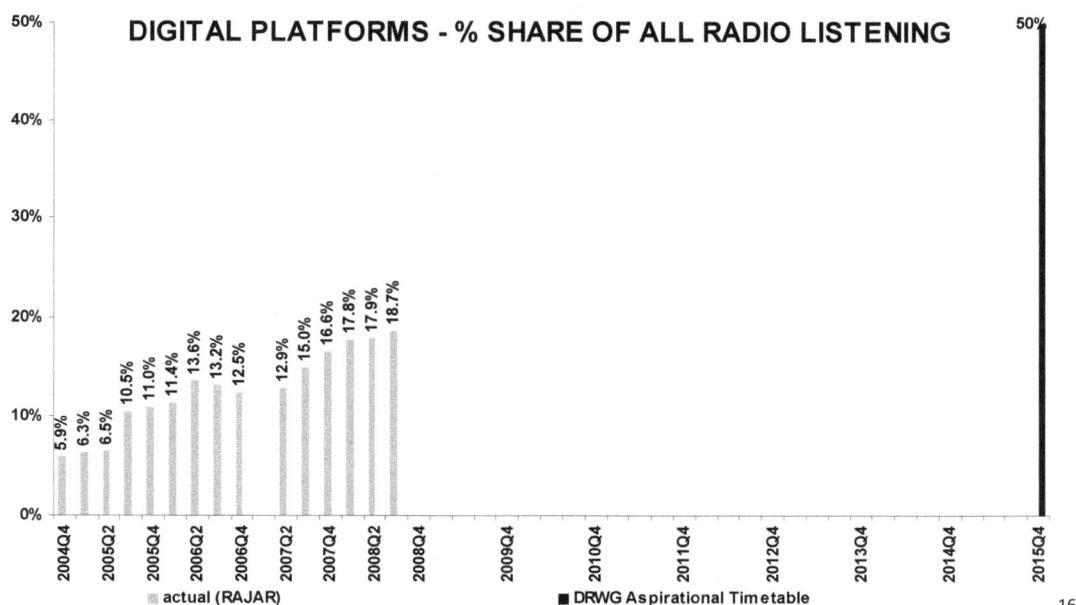

How likely is this outcome???

It might prove instructive to re-examine earlier forecasts for digital radio take-up published by three leading stakeholders – Ofcom, RadioCentre and the Digital Radio Development Bureau:

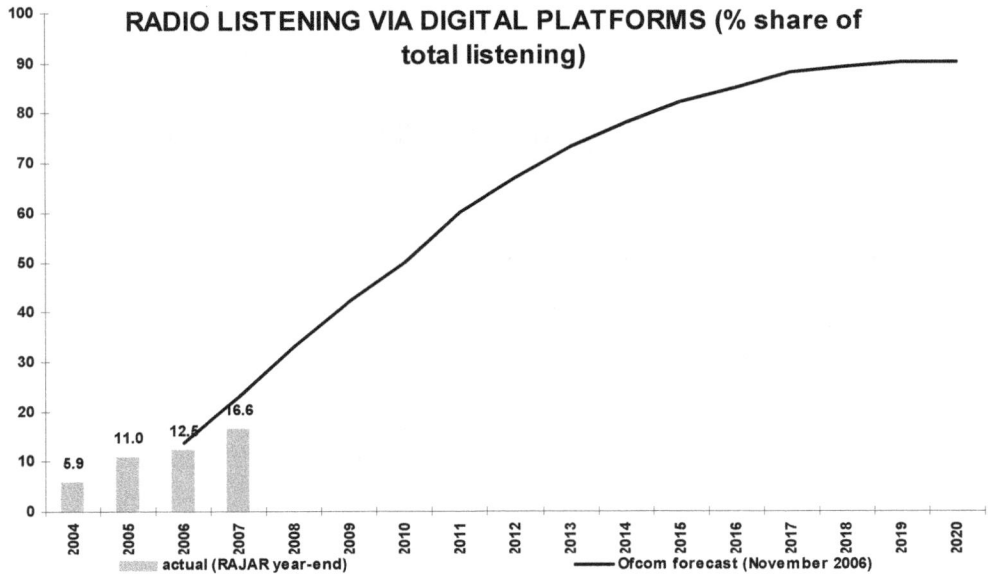

RADIO LISTENING VIA DIGITAL PLATFORMS (% share of total listening)

5.9 | 11.0 | 12.5 | 16.6

actual (RAJAR year-end) — Ofcom forecast (November 2006)

17

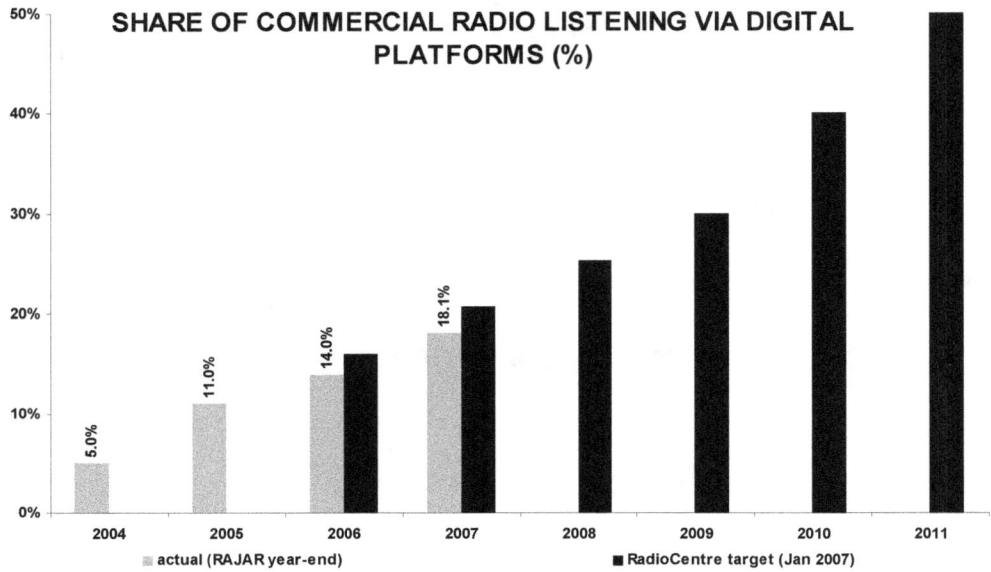

SHARE OF COMMERCIAL RADIO LISTENING VIA DIGITAL PLATFORMS (%)

5.0% | 11.0% | 14.0% | 18.1%

actual (RAJAR year-end) ■ RadioCentre target (Jan 2007)

18

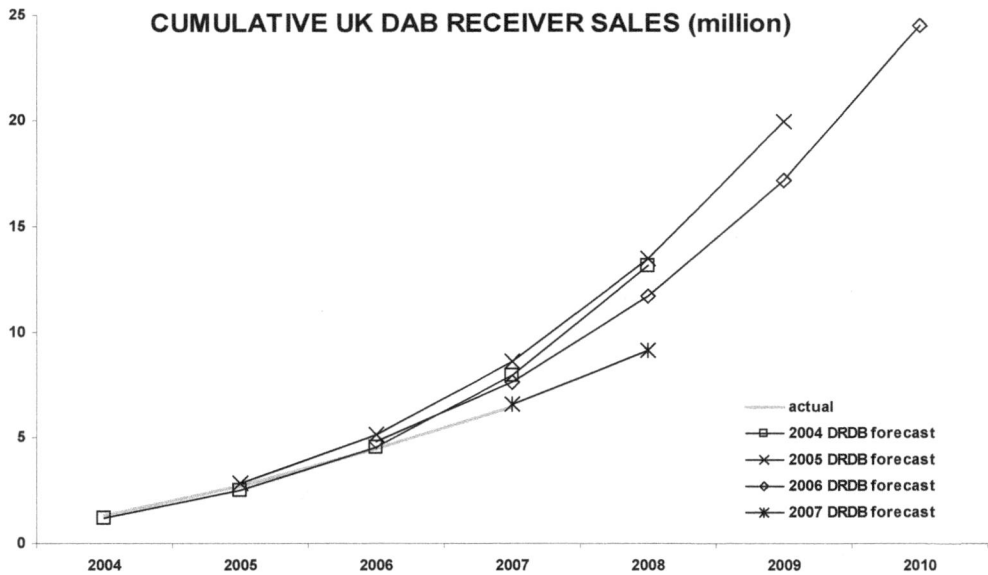

CUMULATIVE UK DAB RECEIVER SALES (million)

actual
2004 DRDB forecast
2005 DRDB forecast
2006 DRDB forecast
2007 DRDB forecast

19

This last graph is interesting because the Digital Radio Development Bureau published progressively less optimistic annual forecasts for DAB set sales in 2004, 2005, 2006 and 2007. Its 2007 forecast only projected figures to 2008. When I enquired in September 2007 why the forecast horizon had been reduced by three years, the DRDB told me:

"The problem with forecasting a cumulative to 2011 is that there are too many variables. If we based it on what there is available now in the traditional radio market, we could certainly come up with a figure. But if, as suggested in the forecast, DAB moves into other form factors, such as mobile phones, docking stations, MP3, MP4 etc, then that 'traditional' figure would be selling the market short and would not be indicative of the potential cumulative market for DAB."[20]

Fifteen months on, DAB has made slow progress moving into these other 'form factors', with mobile phones and cars still on the starting blocks. Notably, DRDB has yet to publish a 2008 forecast.

None of this statistical evidence offers confidence that the Digital Radio Working Group's *"Aspirational Timetable"* is anything more than 'pie in the sky'.

---

[16] source: RAJAR & Digital Radio Working Group
[17] source: RAJAR & Ofcom
[18] source: RAJAR & RadioCentre
[19] source: Digital Radio Development Bureau & GfK
[20] e-mail to author

5.

6 January 2009

Shipwrecked on desert island DAB

One important question was sidestepped by the Digital Radio Working Group in its enthusiasm for the DAB platform in the Final Report: if DAB only comes to be adopted in a handful of countries, what are the 'opportunity costs' for UK consumers? In other words, if UK consumers are forced by government policy to purchase DAB receivers to replace their analogue radios, what other consumer hardware will they not purchase, either because it does not incorporate DAB radio, or because they have already spent their allocated budget replacing all five or six analogue receivers in their household with DAB radios?

The answer might be provided by the annual International Consumer Electronics Show [CES] taking place this week in Las Vegas, which describes itself as *"the world's largest consumer technology tradeshow"* with 2,700 exhibiting companies, 500 expert speakers and 200 conference sessions.[21]

The Digital Radio Working Group had written in its Final Report that:

".... the DAB standard used in the UK and all three variants will be receivable on [radio] sets which manufacturers will be producing from [2009], so creating a European-wide market for digital radio."[22]

You might imagine that such innovations in DAB radio hardware would be reflected at this week's CES event? Apparently not. Only 6 out of the 2,700 exhibiting companies list 'DAB' in their descriptions – the UK's Frontier Silicon (*"the leading supplier of audio processors for digital radios powering over 70% of all DAB radio products"*); Germany's Fraunhofer Institute (*"audio/video compression technologies"*); Taiwan's Joycell (broadcast antenna manufacturers); China's Blue Tinum and Shenzhen Baoan Fenda which manufacture DAB/FM/internet radios; and Hong Kong's Kenwin Industrial which makes plastic injection moulds for electronics products.[23] Additionally, not one of the 200 conference sessions at CES is about DAB. The reality is that, for most of the 130,000 people attending the event, DAB will simply not exist.

But, if you do a search for 'internet radio' at CES, you find a list of 393 exhibitors, 320 products and 32 conference sessions.[24] Now compare that with 'DAB': 6 exhibitors, 8 products, 0 sessions. Furthermore, the newly formed Internet Media Device Alliance, a group of companies significantly involved in internet radio, will be launching at CES.[25] One of its steering committee members is Anthony Sethill, CEO of Frontier Silicon, who said: *"Frontier's role in the formation of the IMDA affirms our position as the leading supplier of Internet radio connected audio products to the global consumer electronics market."*[26]

The significant word there is 'global'. Despite its current dominance of the largely UK market for DAB, Frontier needs a global market for its product lines .... something that DAB's limited take-up will never offer it.

So why does the Digital Radio Working Group want to shipwreck UK radio listeners on a desert island of DAB (for accuracy, I should add that you can take your DAB radio to Denmark or Norway and it will work there too)? The answer might be in paragraph 3.10 of its Report, which states:

*"We strongly believe that in order for **radio** to preserve the qualities which make it such a valued part of our everyday lives, and to allow it to build a strong future, it must have a space where it can be the master of its own destiny and have the freedom to take risks"* [emphasis added].

If you replace the word 'radio' with 'the BBC and UK commercial radio companies' and then read this sentence again, it becomes perfectly clear that what the Working Group is advocating is protectionism of the British radio broadcasting industry – protectionism from unregulated radio content delivered from non-UK sources via internet radio. Heaven forbid that we UK residents might prefer listening to Ryan Seacrest over Johnny Vaughan, because the government will seemingly do as much as possible to stop such an outrage happening.

If you think this is a fantastical notion, I suggest you read paragraph 3.9 of the same Report, which is unapologetically 'patriotic':

"Radio is an important part of the national discourse and perhaps an even more important voice in local democracy. These principles are the bedrock of radio in the UK and we believe they are something which citizens not only value, but expect."

The fact is that UK radio, much more than television, offers an easy platform for politicians and their policies to be propagated to mass audiences of voters (viz *Radio 4*'s 'Today' programme). Incredibly, the Central Office of Information has long been commercial radio's biggest advertiser! The best way to preserve this cosy relationship is to build a wall around it.

For the mandarins, it might look like a nice walled garden to play in. For us consumers, it has all the hallmarks of a content prison.

---

[21] http://cesweb.org/news/releaseDetail.asp?id=11646
[22] http://www.culture.gov.uk/images/publications/DRWG_Final_Report.pdf
[23] http://myces.bdmetrics.com/CompanySearch.aspx?configId=2&keyword=dab&sctid=3
[24] http://www.cesweb.org/searchResults.asp?keyword=%22internet+radio%22&configid=2
[25] http://www.imdalliance.org/
[26] http://www.frontier-silicon.com/media/releases/08/1219_IMDA.htm

6.

11 January 2009

Classic FM: always check the expiry date before purchase

When Global Radio paid £375 million for GCap Media in 2008, the portfolio of stations it acquired included *Classic FM*, the most listened to and most profitable of the UK's three national commercial radio stations, and the only one of the three on FM. *Classic FM* was almost the only jewel remaining in GCap's tarnished crown, after its management had destroyed the audiences/revenues of *Capital FM* and its other city FM stations by implementing disastrous content and commercial strategies. *Classic FM* presently has an 11% reach, a 3.8% share, 66% of its adult hours listened derive from the desirable ABC1 demographic, whilst 85% derive from 'housewives'. Its only competitor in the classical music format is national *BBC Radio Three*, which has only a 4% reach and a 1.2% share but, of course, carries no commercials. *Classic FM* is a cash cow.[27]

There is only one problem for Global Radio. *Classic FM*'s licence expires on 30 September 2011 and it cannot be automatically renewed. This is a big problem. Whereas local commercial radio licences are still awarded (and re-awarded) by Ofcom under a 'beauty contest' system, national commercial radio licences are not. The system for national commercial radio licences is simple. Sealed bids are placed in envelopes. Ofcom opens the envelopes. The bidder willing to pay the highest price wins the licence. That's it. This system is enshrined in legislation. Even if Ofcom wants a different system, it cannot change it without legislation.

As *Classic FM*'s new owner, Global Radio definitely wants a different system that will enable it to hang on to this most valuable asset. Global has been busy bending the ears of anybody and everybody who it might be able to persuade to interpret the broadcasting rules in a way that lets it keep *Classic FM* after 2011. Even Ofcom has had its lawyers busy examining the legislation to see what flexibility it has to interpret the rules in a way that might maintain the status quo.

Unfortunately, the legislation in the Broadcasting Act 1990 is quite specific:

"[Ofcom] shall, after considering all the cash bids submitted by the applicants for a national licence, award the licence to the applicant who submitted the highest bid."[28]

There is one, and only one, caveat in the legislation:

*"[Ofcom] may disregard the requirement imposed by subsection (1) [above] and award the licence to an applicant who has not submitted the highest bid if it appears to them that there are **exceptional circumstances** which make it appropriate for them to award the licence to that applicant; and where it appears to [Ofcom], in the context of the licence, that any circumstances are to be regarded as **exceptional circumstances** for the purposes of this subsection, those circumstances may be so regarded by them despite the fact that similar*

circumstances have been so regarded by them in the context of any other licence or licences" [emphasis added].

Nothing more explicit is mentioned in the legislation about these possibly *"exceptional circumstances."* The problem facing Ofcom is that, if it were to award the licence to Global Radio in a hypothetical situation where it had not been the highest bidder, whoever was the highest bidder would be likely to seek a judicial review, forcing Ofcom to explain in front of a set of judges the precise nature of the *"exceptional circumstances"* it had invoked. This would not be a pretty sight. There are no precedents because this part of the legislation has never been used before.

So what is the precise meaning of the 'cash bid' that has to be submitted to Ofcom in a sealed envelope? It is an amount to be paid annually by the winner throughout the licence period (increased annually by the rate of inflation). When *Classic FM* won the licence in 1991, it agreed to pay £670,000 per annum, plus 4% of its revenues as demanded by the regulator.

Later on, the Broadcasting Act 1996 allowed the regulator to extend *Classic FM's* licence once, but on new terms, if the station agreed to simulcast its output on DAB. The regulator set *Classic FM's* new licence payment as £1 million per annum plus 14% of its revenues from 1999. This new licence would have expired in 2007.

Then, the Communications Act 2003 allowed Ofcom to extend *Classic FM's* licence again for a further four years but, once again, it could re-set the terms. Ofcom reduced *Classic FM's* licence payment to £50,000 plus 6% of its revenues from 2007. This is the licence that expires in 2011.

| | RAJAR Q3 2008 | | LICENCE PAYMENTS | | | |
| | | | to 2007/8 | | 2007/8 to 2011/12 | |
	% reach	% share	£/annum	% revenue	£/annum	% revenue
Classic FM	11	3.8	1,161,000	14%	50,000	6%
TalkSport	5	1.9	563,000	6%	100,000	0%
Absolute Radio	5	1.4	1,125,000	12%	100,000	0% [29]

Why did Ofcom decide to reduce the payments so substantially in its 2006 decision? It argued that the growth of listening via digital platforms was *"leading to a decline in the scarcity value of the analogue spectrum."* Additionally, it argued that the licensee's *"share of advertising, derived as a result of access to the analogue spectrum, is likely to fall."*[30]

Ofcom had forecast in November 2006 that digital platforms would account for 33% of radio listening by 2008, and 50% by 2010. By the time the *Classic FM* licence was due to expire in 2011, Ofcom anticipated that digital platforms would be responsible for 60% of radio listening overall. In other words, the FM licence would, by 2011, be accountable for only the minority of listening to *Classic FM*.

Ofcom's forecast proved to be extremely wide of the mark. By Q3 2008, only 18.7% of radio listening accrued from digital platforms, little more than half of what Ofcom anticipated. The 50% threshold is unlikely to be reached even by 2015, and certainly not by Ofcom's target of 2010. As a result of these forecasting failures, *Classic FM* (along with the other two national commercial stations) is now paying Ofcom an amazingly discounted rate for the licence fee to use analogue spectrum. The combined licence fees of the three national licensees would have been £7 million per annum under the previous regime, whereas these were reduced by Ofcom to less than £1.5 million (by Ofcom's own estimate).[31]

The net result of these changes is that Global Radio has a bargain licence on its books. *Classic FM* probably generates more than £20 million revenues per annum, but Global now pays only £1.3 million for its licence. The bad news is that Global Radio's cash cow will end in September 2011. If Global does not win the re-advertised national FM licence, the value of its balance sheet could be up to halved. On the other hand, to keep this prize asset, it will have to bid significantly more than the £50,000 annual licence fee it is paying now, so that *Classic FM*'s future profitability would be impacted anyway, even if Global managed to keep the licence.

RADIO LISTENING VIA DIGITAL PLATFORMS (% share of total listening)

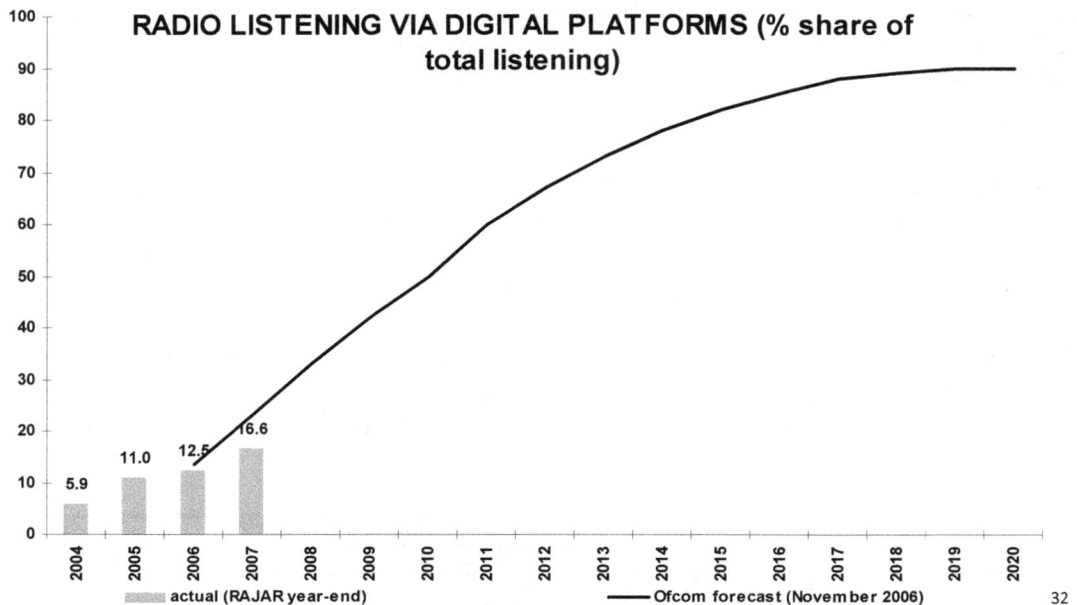

actual (RAJAR year-end)　　　　　Ofcom forecast (November 2006)　　　32

However, there are plenty of other media owners out there who would like to have the UK's only national commercial radio FM licence in their portfolio. The fact that the DAB platform has not grown anywhere near as quickly as anticipated in the UK simply makes this FM licence more valuable. The last time the licence was advertised in 1991, bids were only open to European Union applicants. Since then, legislation has opened up the bidding process worldwide. The licence format does not have to be classical music – the licensee can operate any format of its choice, apart from pop music (this caveat is in the legislation).

The fly in the ointment is that Ofcom adopted a new policy in 2007 that all its analogue local and national radio licences would be scheduled to expire on 31 December 2015, or five years from their commencement, whichever is longer.[33] For *Classic FM*, this means that its next licence period would theoretically run only from 1 October 2011 to 1 October 2016. If a new bidder won the licence by offering the highest cash bid, five years is hardly enough time for a nascent business to establish itself and become profitable, particularly if it were to adopt a format other than classical music. The Ofcom policy seems unworkable in practice, and also seems biased in the incumbent's favour.

Now, with an understanding of Global Radio's desperation to hang on to its *Classic FM* licence almost at any cost, it is useful to re-read Paragraph 2.3 of the Final Report of the Digital Radio Working Group. Remember that Global Radio owns about 50% of the UK's commercial radio DAB transmission capacity, and Global Radio accounts for 39% of commercial radio listening. The Report said:

*"In exchange for its ongoing and future commitment to DAB, we believe **the radio industry** must have greater certainty and control of its future. Therefore, we propose that the government must relax some of the existing legislative and regulatory burdens placed on **the radio industry**, which will require parliamentary time, as outlined below and Ofcom should consider how to reduce some of the existing regulatory burdens.*

*First, **the commercial radio industry** must be granted a further renewal of its analogue services which are carried on DAB, and of DAB multiplex licences."* [34] [emphasis added]

Now read this quote once more, but replace the phrase '**the radio industry**' or '**the commercial radio industry**' with '**Global Radio**'. Aha! Wouldn't it be great for Global Radio if the government could be persuaded to step in and somehow automatically renew its *"analogue service"* Classic FM licence, thus avoiding a licence auction in 2010? Even more so if Global could be allowed to continue paying only £50,000 per annum (plus 6% of revenues) for the FM spectrum it uses? If you were Global, would you not be eager to offer the government a deal whereby you maintain your costly DAB infrastructure (and maybe even extend it) as the price you have to pay for securing the future of your most significant balance sheet asset?

From reading its Final Report, it certainly looks as if the Digital Radio Working Group bought into this argument. The next hurdle for Global Radio is to persuade Lord Carter and his Digital Britain team to buy into the same deal, which is: we promise to keep the DAB platform alive, despite it losing us a small fortune, if you 'arrange' legislation that enables us to keep the *Classic FM* licence for another decade. Thus, the government avoids the embarrassment of the DAB platform failing in the UK, and Global Radio might stand a better chance of staying in business.

To date, the other commercial radio owners have seemed happy to go along with this plan. They, like Global, would get to renew their radio licences automatically too (although none of their licences are as individually valuable as *Classic FM*'s). On the other hand, they too will be burdened with the continued costs of simulcasting their services on the DAB platform, with almost no financial return. However, despite most radio owners' private dislike of the whole DAB 'fiasco', publicly they continue to stress their continuing support. Nobody turns down a 'free lunch', and a free licence renewal is an enticing offer for a radio industry still built upon oligopoly power rather than open competition.

The only question now is whether the government considers it politically worthwhile to 'help' the commercial radio sector with new legislation that would extend the licence status quo, in return for forcing onto consumers a 'new' DAB radio technology that is more than a decade old and has long been superseded by innovation.

Lord Carter's pronouncements during the next fortnight might give us an idea of how important/unimportant it is to the government to:
- bale out privately held Global Radio;
- force further investment in improving/developing the DAB platform.

---

[27] source: RAJAR
[28] http://www.opsi.gov.uk/acts/acts1990/ukpga_19900042_en_10

[29] source: Ofcom & Radio Authority
[30] http://www.ofcom.org.uk/consult/condocs/methodology/financialterms/financialterms.pdf
[31] http://www.ofcom.org.uk/consult/condocs/methodology/financialterms/financialterms.pdf
[32] source: RAJAR & Ofcom
[33] http://www.ofcom.org.uk/consult/condocs/futureradio07/summary/
[34] http://www.culture.gov.uk/images/publications/DRWG_Final_Report.pdf

7.

19 January 2009

The Digital One DAB radio multiplex: fixing 'market failure' ten years too late

Digital One is the owner of the UK's first and only national commercial radio DAB multiplex. If you produce commercial radio content that you wish to make available nationally on the DAB platform, you have to go to Digital One and agree a price and a contract. That price is set by Digital One, not by Ofcom or any other regulatory body. Digital One is the national DAB 'gatekeeper' and it decides what commercial radio brands we hear and what we don't hear on DAB. It would be hard not to consider Digital One's operation monopolistic.

Furthermore, Digital One is part of a vertically integrated business. Its controlling shareholder is Global Radio (formerly GCap Media, formerly GWR Group), the UK's largest commercial radio group. In this way, Digital One/Global Radio's business is an end-to-end operation that includes: generating radio content ('stations'), some of which are carried on the DAB platform; selling advertising space around that content, some of which is carried on the DAB platform; owning the national DAB platform in the UK; and owning the 'gatekeeper' role for other radio content providers wanting access to that national DAB platform. (This 'gatekeeper' role was bestowed upon the DAB multiplex owner, rather than Ofcom, by the 1996 Broadcasting Act.)

Does Digital One's business work in the interests of a competitive broadcasting sector or the listening public? Is this not a case where some kind of intervention by the regulator is appropriate? Within Ofcom's own definition of 'market failure in traditional broadcasting', one of the main six reasons it uses to justify regulatory intervention is where:

"Restricted access to spectrum makes entry impossible on market grounds and, without competition, the ability of the market to deliver the most efficient solution is impaired."[35]

Ofcom then explains this issue in more detail:

"A tendency towards monopoly/oligopoly. Economies of scope and scale are inherent in broadcasting and will tend to encourage the concentration of ownership in large, often vertically-integrated companies. The result of an unregulated market might therefore be reduced competition, less choice for viewers and either higher prices or lower quality than would be available in a competitive market."[36]

Is this not exactly what has happened with the national commercial radio DAB platform? Digital One seems to have operated its 'gatekeeper' monopoly over the platform in a way that that has reduced competition, offered less choice to listeners, and maintained high carriage prices. The end result? After a decade of operation, there is only one radio station that has elected to contract with Digital One to be carried on its DAB platform of its own volition. There is enough bandwidth on the multiplex for a clutch more national stations, but that capacity remains unused.

Digital One was awarded a 12-year DAB licence in June 1998 to operate the *"first and only national commercial digital multiplex licence."*[37] It promised to pay the regulator a licence fee of £10,000 per annum. However, until very recently, if you had approached Digital One and asked the cost of putting a radio station on its multiplex, you would have been expected to pay more than £1 million per annum. Furthermore, if your proposed content competed directly with that of Digital One/Global Radio's own digital radio stations, carriage might not have been offered, even at that price.

Therefore, it proves somewhat surprising to see today that Digital One issued a press release and published an advertisement inviting *"expressions of interest from companies ready to contract and launch digital radio stations in 2009"* on its DAB multiplex.[38] It is even more surprising to learn that *"capacity is available for mainstream stations, as well as more specialist channels appealing to a diversity of tastes and interests."* And it is shocking to read that *"Digital One is reviewing its charges for capacity"* and that *"it is anticipated that prices will initially be set below Digital One's 2008 rate card, in order to provide an incentive for approved applicants to invest in high quality services...."*

The appropriate time to have published such a 'call for content' was June 1998, immediately after Digital One was awarded the DAB multiplex licence by the regulator. Perhaps then the sad story of the DAB platform's slow development in the UK would have turned out differently. By now, Digital One might have fostered a broad range of audio content on the national DAB platform provided by a variety of producers, creating a 'compelling consumer proposition' that could have motivated the public to purchase DAB radios in significant numbers. But, unfortunately, it did not turn out that way and now, after a decade, DAB remains barely off the starting blocks.

Instead, for a decade, Digital One has clung on to the notions that:

- its monopoly over the DAB infrastructure is valuable in itself, even if the capacity is mostly unused (is a rail network valuable without trains?)
- its 'gatekeeper' role enables it to push its own digital services to listeners, at the expense of competitors and potential competitors
- high carriage fees for external users will quickly put them out of business
- listeners will lap up its own controlling shareholder's content on the DAB platform, however little is invested in its production (one computer + 100 CDs = digital radio station)
- 'control' of a broadcast platform is alone sufficient to create a profitable monopolistic business platform .

It hardly inspires confidence in the Digital One DAB platform that Global Radio's predecessor, GCap Media, closed three of its own digital-only stations carried on its platform last year, and sold *Planet Rock* to an entrepreneur with no other radio interests. Neither is it a good advertisement for Digital One that its platform *"providing coverage to 90% of the population of Great Britain"* only succeeds in securing a peak half-hour audience of 79,000 adults for its last remaining digital-only audio contractor, *Planet Rock.*[39]

Digital One's licence for the *"first and only national commercial digital multiplex licence"* will expire on 14 November 2011. Would I sign a contract with a company that has unashamedly hogged the UK DAB national multiplex for its own selfish ambitions since 1998, but now suddenly wants to offer me capacity on its multiplex, just as its own life is expiring? My attitude would be: so you've screwed up almost a decade of your 12-year monopoly and lost everything but your shirt in the process, but now, on your deathbed, you want me to pay you good money for carriage on a platform that you yourself have helped ruin?

DIGITAL-ONLY RADIO STATIONS ON DIGITAL ONE PLATFORM										
launched	owner	station	digital platforms					RAJAR Q3 2008		
			national DAB	local DAB	Freeview	FreeSat	internet	% share of total radio listening	peak half-hour audience (adults)
1999	Planet Rock	Planet Rock	■			■		0.50	79,000
2006	GCap Media	TheJazz	closed 2008						
2000	GCap Media	Capital Life	closed 2008						
2000	C4/UBC	Oneword Radio	closed 2008						
1999	GCap Media	Core	closed 2008						
2000	Saga	PrimeTime Radio	closed 2006						
2000	Bloomberg	Talkmoney	closed 2003						
2000	ITN	ITN News	closed 2002						[40]

Digital One's announcement today reminds me of those grocery stores that put cans of food in a 10p bargain bin that are not only damaged, but are also only a few days away from their expiry date. You expect me to buy these? I guess we will see if there is somebody out there desperate enough to take the bait. I can think of many radio formats unavailable on AM/FM that should have a national platform in the UK. Would any of them work on DAB? Ten years ago, yes, they might have done. Now, no. The DAB platform has proven to be a failure with consumers, and Digital One has played a very large part in making it so. And yet, Digital One has decided now to advertise its newfound enthusiasm for *"enhanced choice, variety and innovation"* on its DAB platform.

A case of: too much, too little, too late…….

[35] http://www.ofcom.org.uk/consult/condocs/psb2_1/annex11.pdf

[36] http://www.ofcom.org.uk/consult/condocs/psb/psb/forward/

[37] http://www.ofcom.org.uk/static/archive/rau/newsroom/news-release/98/pr56.htm

[38] http://www.ukdigitalradio.com/news/display.asp?searchnews=&year=&id=310
http://www.ukdigitalradio.com/news/1585605.pdf

[39] http://www.ukdigitalradio.com/news/1585605.pdf & source: RAJAR

[40] source: Grant Goddard & RAJAR

8.

21 January 2009

DAB radio: now you hear it (in-store), now you don't (in-home)

The Digital Radio Development Bureau [DRDB] announced yesterday that, after a one-year trial, Ofcom *"has agreed to put in place a permanent licensing regime for all retailers across the country"* to install DAB repeaters that will boost the signal in-store. According to DRDB:

"Many electrical retailers suffer from poor analogue and DAB signal strength due to the steel framed infrastructure of the building or their basement location. Installing a DAB repeater on the roof of the store means a signal can be boosted in-store and DAB radios can more easily be demonstrated, thus increasing sales potential."[41]

Currys owner DSGi's Trading Manager Amanda Cottrell said:

"We know from experience that demonstrating DAB radio in-store is the best way to show consumers the benefits of more station choice, ease of tuning and clean, digital quality sound. Consumers like to get hands-on with new technology and these DAB repeaters will help us to maximise sales in areas where demonstration was a problem."

I understand the retail sales floor problem, but am I the only one worried that the solution implemented here might not be quite appropriate? I admit it is a very long time since I studied consumer law (1981, Durham New College), but my thinking is that these actions could potentially lead to consumer redress under UK legislation. Have the legal eagles at Ofcom considered this fully?

Under Section 15 of the Sale of Goods Act 1979, when goods are sold by 'sample' (ie: consumer sees in-store demonstration sample of DAB radio receiver, but store supplies consumer with sealed, boxed good), *"the goods must correspond to the sample in quality."*[42] The law requires *"that the goods will be free from any defect, making their quality unsatisfactory, which would **not be apparent on reasonable examination of the sample.**"* [my emphasis]

Under the new 'repeater' system, when the consumer examines the in-store sample of the DAB receiver, the receiver will be capable of offering 'perfect' reception of DAB radio stations. This is due to the installation of special in-store equipment. A fixed antenna has been installed on the roof of the building, pointed directly to the nearest DAB transmitter mast, and its received signal supplies a relay transmitter (transmitting the same stations) placed on the shop floor adjacent to the DAB radio receiver demonstration area.

When the consumer takes the sealed, boxed DAB radio home, they may open it and find that reception of radio stations on their hardware is not as good as it was in-store. This is because their radio is not receiving the DAB signal from a relay transmitter only metres away from the receiver, as it was in-store. Instead, it is receiving signals from the nearest DAB transmitter, probably miles away, and that signal may or may not penetrate the building in which they are using the radio.

The consumer could theoretically apply to Ofcom to install a relay transmitter in their home, in order to replicate the precise conditions in which the sample DAB receiver was demonstrated in-store. Ofcom's response to the consumer's application would certainly be 'no'. Thus, the in-store 'sample' DAB receiver was purposefully demonstrated to the consumer under an artificially created environment that cannot ever be reproduced within the consumer's home.

This would not be the first time that the marketing of DAB radio in the UK has come under legal scrutiny for potentially misleading consumers. In 2004, Ofcom banned an advertisement broadcast on London station Jazz FM which had claimed falsely that DAB radio offers consumers *"CD-quality sound."*[43] In 2005, the Advertising Standards Authority upheld a complaint against DAB multiplex owner Switchdigital for a misleading radio advert which had claimed that DAB radio was *"distortion free"* and *"crystal clear."*[44] In its verdict, the ASA said it had *"received no evidence to show that DAB digital radio was superior to analogue radio in terms of audio quality."*

The problems concerning the paucity of DAB reception in some circumstances (basements, steel buildings, built-up areas) have been known to the broadcast industry for a long time. At the 2006 TechCon event, Grae Allen, then manager of digital distribution at EMAP Radio, had explained that:

"[the] Wiesbaden 1995 [radio conference] and all the other DAB planning dealt with mobile reception – in-car and portable outdoors. It made assumptions about aerial heights being just above ground level and, to provide good service to 99% of locations, the conclusion was that it required 58dbμV per metre to maintain that quality of service, and it made some assumptions about the performance of receivers and aerials."[45]

In practice, he said, *"some receivers do not quite live up to expectations – some have lossy aerial systems and suffer from self-noise."* Grae said that 2006's European Regional Radio Conference *"[was] moving DAB to become a truly indoor medium. The new planning model has around 10dbμV higher field strength than was envisaged in the original plan."*

In 2006, BT Movio had been about to launch a mobile TV service using DAB spectrum, and Grae said:

"That raises a question. We are seeing increasing numbers of hand-held receivers, such as the BT Movio receiver, that do not have an aerial of any significant size. So, in some areas, we may have to go to higher field strengths to deliver to handhelds indoors. So how are we going to improve the coverage? Unfortunately, the people who fill in RAJAR diaries don't tend to live in large numbers alongside the sheep in the fields [where DAB transmitters are mostly located]. They live in the cities and the urban sprawl, and that's where we need to deliver the high field strengths that are required for the types of receivers that are becoming popular, and the level of service that is expected. In the future, as I envisage it, we will see a need to put more and more [transmitter] sites inside the cities in areas where we actually need significant power where people are living and working."[46]

Mark Thomas, then head of broadcast technical policy at Ofcom, admitted at the 2006 TechCon event that the original DAB power allocations had proven too low:

"The Radio Authority had no data of how [DAB] receivers performed, so it had to make some very broad-brush assumptions. More recently, now that we have a lot of receivers in the market and we can see how they behave, an industry group has been working under Ofcom's chairmanship for the last two years to look into the issue in more detail and come up with some modus operandi for new transmitter sites."[47]

Mark concluded: *"The Ofcom approach is that the industry co-operates between commercial operators with each other, and with the BBC, in identifying the sites that will improve field strength of DAB services to consumers and will also avoid the issues surrounding Adjacent Channel Interference. ACI also adds to the investment challenge that all of this spectrum development is building."*

Now zoom forward from 2006 to December 2008 and read the Final Report of the Digital Radio Working Group, which said:

"We believe that action is needed to improve the quality and robustness of the existing [DAB] multiplexes' coverage. We recognise that such a request has significant financial implications for multiplex operators..."[48]

So, it would appear that, from 2004 onwards (when Mark acknowledges Ofcom was aware of the problem), the UK radio industry has continued to market and sell millions of DAB radios to UK consumers, in the full knowledge that its DAB transmission infrastructure requires a significant upgrade to provide consumers with sufficiently robust DAB radio reception in built-up areas and in homes.

The latest DRDB 'repeater' sales initiative merely tackles the symptom of poor DAB reception which has existed for years, and the solution is limited entirely to electronics retailers. What is still missing is a solution to the core problem of the *"quality and robustness"* of DAB radio reception ..... for consumers.

---

[41] http://www.drdb.org/article.php?id=747&from=arc~s=2009-1-1~e=2009-1-31
[42] http://www.johnantell.co.uk/SOGA1979.htm
[43] http://www.ofcom.org.uk/accessibility/rtfs/bulletins/archive04/a15.rtf
[44] http://www.asa.org.uk/NR/rdonlyres/A51D8768-EFEF-4D35-AA00-6F859587D989/0/Broadcast_report_5_Oct_05.pdf
[45] author's recording
[46] author's recording
[47] author's recording
[48] http://www.asa.org.uk/NR/rdonlyres/A51D8768-EFEF-4D35-AA00-6F859587D989/0/Broadcast_report_5_Oct_05.pdf

9.

27 January 2009

DAB radio receiver sales suffer negative growth

DAB radio receiver unit sales fell by 10% year-on-year in the final quarter of 2008 in the UK, jeopardising the digital platform's future as a mass market replacement for analogue radio. This is the first quarter to have recorded negative year-on-year growth since DAB sales records began six years ago. It marks a significant setback for DAB stakeholders who had invested in a six-week marketing campaign during the run-up to Christmas which promoted the DAB platform heavily on BBC and commercial radio.

DAB RADIO RECEIVER SALES: QUARTERLY

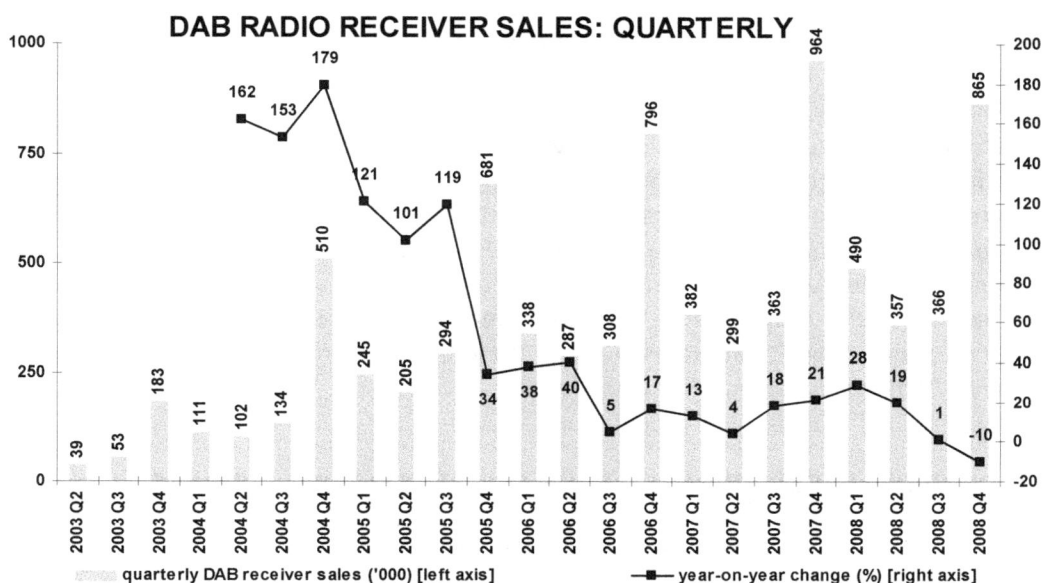

quarterly DAB receiver sales ('000) [left axis]   —■— year-on-year change (%) [right axis]

49

The Digital Radio Development Bureau, the trade body charged with promoting the DAB platform, issued a press release today stating that the *"one ray of sunshine in a gloomy Christmas season for retailers was DAB digital radio."*[50] Its statement failed to mention the negative growth experienced in what is traditionally the most critical quarter of the year for DAB radio sales. Retail data collected by GfK for the DRDB clearly show the declining growth rate of DAB radio sales having started in the second quarter of 2008, a trend that is likely to have been further exacerbated by the 'credit crunch'.

However, this disastrous sales performance has not prevented those UK companies who are pushing the DAB platform from continuing to talk up the success of their technology. Imagination Technologies, the parent company of the Pure Digital brand of DAB radio receivers, today announced *"record export growth for 2008"* and that it *"had more than tripled overseas sales in the year ending 31 December 2008."*[51] Hossein Yassaie, Chief Executive of Imagination Technologies, said: *"Our strong overseas growth is further evidence*

33

that DAB digital radio is gaining traction worldwide, and that the transition to digital radio is inevitable."

However, overseas markets account for only 15% of Pure Digital sales (half-year to end October 2008), so why did Imagination Technologies feel it worthwhile to issue a press release for a relatively insignificant revenue stream?[52] It is probably because Imagination has to convince Lord Carter that the government should back DAB radio technology as part of his recommendations within the forthcoming Digital Britain report. Imagination Technologies has bet the farm on DAB becoming a successful, global technology. If the UK government does not decide to force radio listeners to migrate to DAB technology, Imagination could lose its shirt.

Imagination Technology's interim results, published six weeks ago, admitted that revenues from its Pure Digital DAB radio receivers were up only 2% year-on-year, a result it attributed to "*the downturn*" in the UK market, which still accounts for 85% of its global sales.[53] Chief Executive Hossein Yassaie said there had been a "*UK slow-down*" of DAB radio receiver sales and noted that "*the introduction of lower price radios and the onset of the recession meant that the increase of the UK DAB market was less than 5%*." Pure Digital Marketing Director Colin Crawford said this week: "*Our [DAB] sales at Christmas were good, though a little bit down on last year*."[54]

Disappointing sales figures seem only to have encouraged the DAB protagonists to push the boundaries of their government lobbying beyond the limits of truthfulness. In its latest annual report, Imagination Technologies claimed that "*DAB has reinvigorated the now rapidly growing UK radio market and effectively replaced analogue radio*."[55] The latter statement is untrue. According to industry data, only 21% of radio receivers sold in the UK during the last twelve months were DAB, the remaining 79% being old fashioned analogue. The overwhelming majority of radios in use in the UK remain analogue, and DAB is nowhere near having "*effectively replaced*" them.

UK RADIO RECEIVER SALES: FOUR-QUARTER MOVING AVERAGE

DAB receivers as percentage of total receiver sales

| 8% | 9% | 10% | 12% | 13% | 14% | 14% | 17% | 18% | 18% | 19% | 20% | 20% | 21% | 21% |

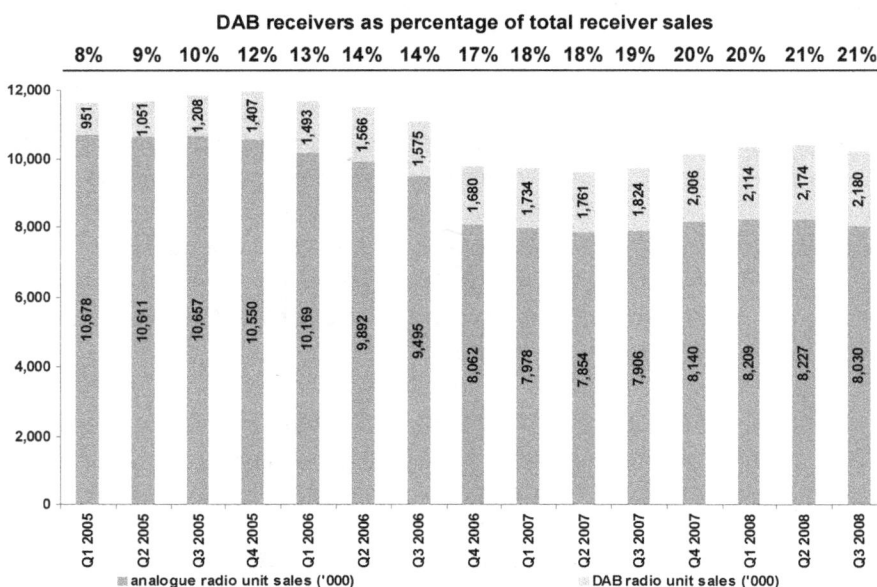

	Q1 2005	Q2 2005	Q3 2005	Q4 2005	Q1 2006	Q2 2006	Q3 2006	Q4 2006	Q1 2007	Q2 2007	Q3 2007	Q4 2007	Q1 2008	Q2 2008	Q3 2008
DAB	951	1,051	1,208	1,407	1,493	1,566	1,575	1,680	1,734	1,761	1,824	2,006	2,114	2,174	2,180
analogue	10,678	10,611	10,657	10,550	10,169	9,892	9,495	8,062	7,978	7,854	7,906	8,140	8,209	8,227	8,030

■ analogue radio unit sales ('000)   □ DAB radio unit sales ('000)

Another corporate victim of over-enthusiastic government lobbying for DAB is Frontier Silicon, whose Chief Executive Anthony Sethill was quoted in a company press release issued in December 2008 as saying: *"Digital radio is here to stay, with DAB sets outselling analogue models by six to one."*[57] Once again, the industry data demonstrate this statement to be a blatant untruth, and simply part of a desperate campaign by a clutch of inter-connected companies to convince the government that DAB technology is already a 'success' in the UK.

Frontier Silicon is a privately owned UK company which describes itself as *"the world's leading supplier of innovative semiconductor, module and software solutions for digital radio and connected audio systems."*[58] Its electronic modules are in 80% of all DAB radios, making it *"the number one supplier to the DAB/DAB+ market."* In 2003, Imagination Technologies took a 17% equity stake and £1.25m of loan stock in Frontier Silicon. Imagination has an 80% share of the worldwide market for the intellectual property on DAB chips, which are then incorporated into Frontier Silicon's modules. However, in 2008, Imagination's stake in Frontier Silicon had to be written down from £7 million to £3.6 million, likely a result of slowing DAB take-up.[59]

Another of Frontier Silicon's ten investors is Digital One, the owner of the UK's only national commercial radio DAB multiplex. Digital One is controlled by Global Radio, the UK's largest commercial radio group, owner of one national station, dozens of local stations and with stakes in the majority of local DAB multiplexes. For Imagination Technologies, Frontier Silicon, Digital One and Global Radio, a decision by the UK government to implement a forced consumer migration to DAB radio would have a hugely beneficial impact on their financial performances. For Imagination, which reported its first profitable year in 2007/8 (£1.88 million pre-tax profits), it might even turn the company's forecast 2010/11 pre-tax profit of £11.84 million into a reality.

More than a decade ago, the idea of a few bright sparks in the government's Department of Trade & Industry was that DAB radio technology could be quickly made a hit in the home market, take-up would then spread globally, and DAB would become a hugely profitable technological export for the UK. This dream continues to be espoused by Intellect, the trade association of the UK technology industry, which told Lord Carter in December 2008:

"The UK is the home of the major chip manufacturer of DAB silicon, as well as two leading receiver manufacturers and, as such, is uniquely positioned to benefit from the potential expansion of DAB not just in the UK, but globally. We believe that this example of high value manufacturing could make a substantial contribution to the UK's future prosperity............"[60]

Unfortunately, the dream is not working out as planned. DAB take-up in the UK market has proven laboriously slow and is in danger of being superseded by newer technologies. Worse, overseas markets have shown little interest in DAB. In Europe, only Denmark has a DAB market as developed as the UK's. Globally, Australia is about to launch DAB but the largest market, the US, has chosen a different digital radio standard. Several countries have experimented with DAB and since abandoned the technology.

With overseas markets looking less likely to prove a source of significant export revenues, the UK technology companies pushing DAB have become increasingly desperate to ensure that their products at least succeed in their home market. Hence, their desperation to persuade the government to force a consumer switchover from FM to DAB. The average household owns six radios, and a government-backed FM switch-off will force all six to be replaced with shiny, new DAB radios. That's a lot of potential revenue for a select number of UK technology companies.

---

[49] Digital Radio Development Bureau/GfK
[50] http://www.drdb.org/article.php?id=751&from=lat
[51] http://www.pure.com/press/release.asp?ID=327
[52] http://www.imgtec.com/corporate/newsdetail.asp?NewsID=419
[53] http://www.imgtec.com/corporate/newsdetail.asp?NewsID=419
[54] http://www.techradar.com/news/audio/interview-pure-digital-outlines-the-future-of-radio-503126
[55] http://www.annualreports.com/HostedData/AnnualReports/PDF/Imagination%20Tech.pdf
[56] Digital Radio Development Bureau/GfK
[57] http://www.frontier-silicon.com/media/releases/08/1219_DRWG.htm
[58] http://www.frontier-silicon.com/about/index.htm
[59] http://www.annualreports.com/HostedData/AnnualReports/PDF/Imagination%20Tech.pdf
[60] http://www.intellectuk.org/component/option,com_docman/task,doc_download/gid,3041/Itemid,102

10.

29 January 2009

Warning! Digital radio objectives may appear closer than they are in reality

By coincidence, the Interim Report[61] of Lord Carter's Digital Britain team was released on the same day as the latest quarterly RAJAR radio ratings data.[62] The former focused optimistically on the inevitability of the UK replacing its existing analogue radio system with the DAB platform. The Digital Britain report stated:

"We are making a clear statement of Government and policy commitment to enabling DAB to be a primary distribution network for radio" and *"we will create a plan for digital migration of radio......."*[63]

This coincidence of timing between Lord Carter and RAJAR offered a perfect PR opportunity for the radio industry to emphasise just how successful its drive towards digital migration has been to date. But where exactly were the stories of dazzling digital radio success?

The RAJAR press release noted that *"digital listening hours [are] up 10% year on year"* and *"DAB ownership [is] up 35% year on year."*[64]

The brief RadioCentre press release avoided mentioning digital radio altogether.[65]

The BBC press release only mentioned digital radio in the context of its digital-only stations, but nothing specifically about the DAB platform.[66]

The Digital Radio Development Bureau [DRDB] press release was headed *"Digital listening and hours up"* and noted that *"radio listening via a digital platform has increased year on year while remaining stable quarter on quarter."*[67]

The Bauer Radio press release avoided all mention of the DAB platform. So, not much evidence today of digital radio's success.

What about the DRDB's statement that digital platform usage is *"stable quarter on quarter"*? Only two months ago, DRDB announced the launch of a joint BBC and commercial radio Christmas marketing campaign *"aimed at driving sales of DAB radios this season."*[68] Although DAB radio hardware sales in the final quarter of the year had subsequently proven disappointing, might not the campaign have also encouraged some consumers to use the DAB platform more, if they already owned the DAB hardware?

Apparently not. Whilst it is true that the latest RAJAR data show increases in listener usage of digital platforms year-on-year, that growth is nowhere near fast enough to make DAB *"a primary distribution network for radio"* anytime soon, as Lord Carter has advocated. The Digital Britain report has simply decided to endorse wholesale the earlier proposals contained in the Digital Radio Working Group's Final Report, as it stated:

"We will create a plan for digital migration of radio, which the Government intends to put in place once….. 50% of radio listening is digital."[69]

Furthermore, the Report pledged the Government to *"work with industry to satisfy the migration criteria by 2015 and, where possible, identify initiatives which could bring forward the migration timetable."*

DIGITAL PLATFORMS - % SHARE OF ALL RADIO LISTENING

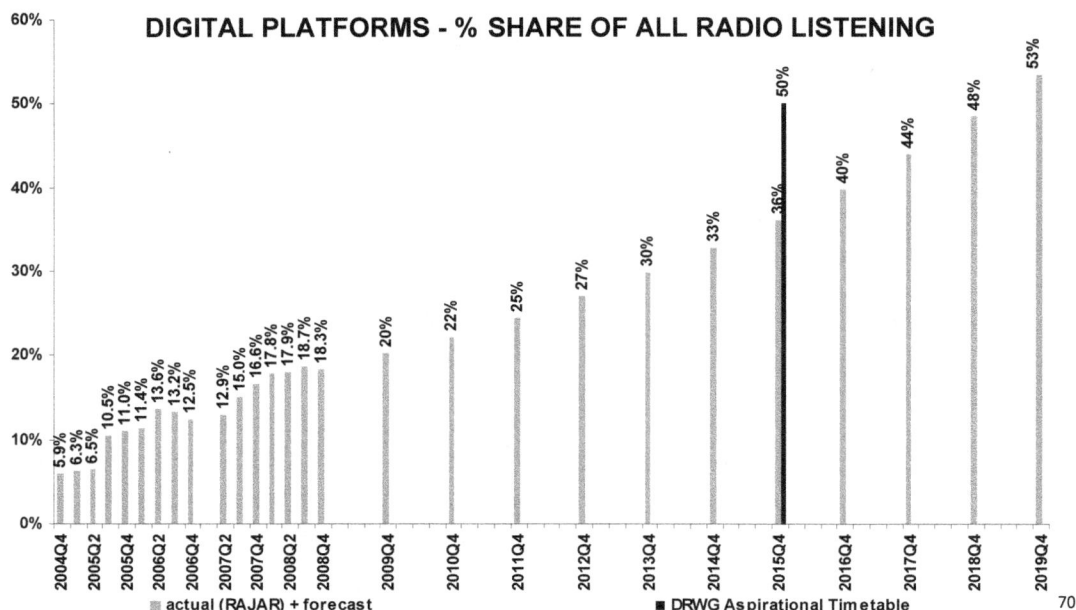

Chart data (actual (RAJAR) + forecast; DRWG Aspirational Timetable):

Quarter	Value
2004Q4	5.9%
2005Q2	6.3%
2005Q4	6.5%
2006Q2	10.5%
2006Q4	11.0%
2007Q2	11.4%
2007Q4	13.6%
2008Q2	13.2%
2008Q4	12.5%
2007Q2	12.9%
2007Q4	15.0%
2008Q2	16.6%
2008Q4	17.8%
	17.9%
	18.7%
	18.3%
2009Q4	20%
2010Q4	22%
2011Q4	25%
2012Q4	27%
2013Q4	30%
2014Q4	33%
2015Q4	36% / 50%
2016Q4	40%
2017Q4	44%
2018Q4	48%
2019Q4	53%

70

These are bold words. The RAJAR data show digital platforms' share of listening was 18.3% in Q4 2008, down from 18.7% the previous quarter, but up from 16.6% year-on-year. If this last year's rate of growth is projected and compounded into the future, the 50% criterion would not be attained until 2019. To achieve the desired outcome by 2015, let alone before 2015, would necessitate a remarkable change in radio listening habits, the likes of which have not been witnessed to date.

DAB is presently, by far and away, the most significant platform for digital radio listening (the others are digital TV, the internet and 'digital' mobile phone). As a result, Lord Carter's anticipated increase in digital radio listening is heavily dependent upon consumer purchase of DAB radio receivers, rather than simply a switch from one available technology to another. However, the disappointing sales of DAB hardware last quarter point to sales growth being unlikely to move into positive territory during 2009.

DAB receiver sales in 2008 did not meet the forecast made by the DRDB in 2007, let alone the more optimistic forecasts of previous years. The DRDB did not publish a sales forecast in 2008, but there is little doubt that the growth trend is beginning to look more linear than exponential. DAB receiver uptake is presently the main pre-requisite for growth in digital radio usage and one that is looking increasingly uncertain.

The other essential factor is consumer usage, not just ownership, of DAB radios. If owners continue to listen on their other analogue radios (the average household has six radios) rather than via DAB, it will still take a long time to reach the 50% threshold. It surely must be the exclusive content available on the DAB platform that will promote its usage (though

CUMULATIVE UK DAB RECEIVER SALES (million)

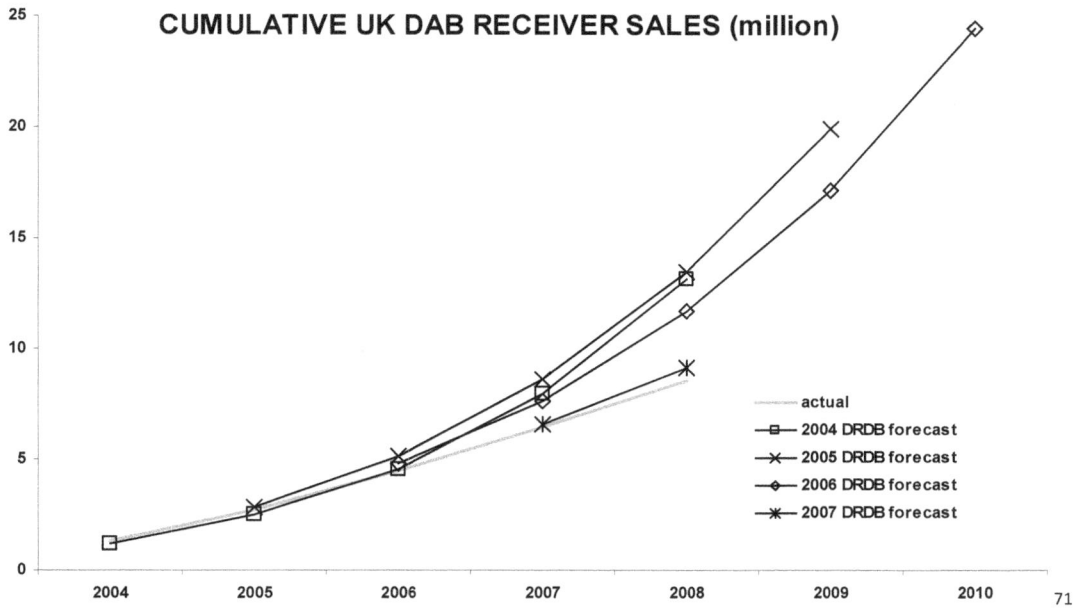

Legend:
- actual
- 2004 DRDB forecast
- 2005 DRDB forecast
- 2006 DRDB forecast
- 2007 DRDB forecast

other factors such as DAB's ease of use and signal strength will play a part). However, 2008 saw a significant reduction in available DAB content, precipitated by GCap Media/Global Radio's decision to withdraw almost entirely from the DAB content market.

COMMERCIAL RADIO: HOURS LISTENED BY PLATFORM (%)

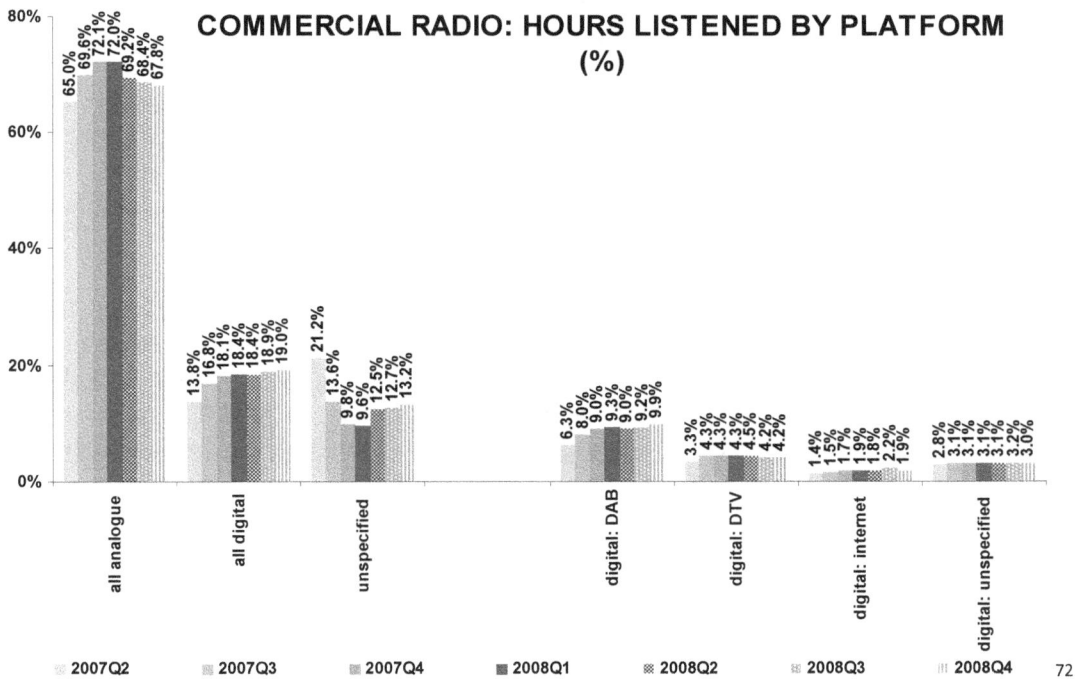

Legend: 2007Q2, 2007Q3, 2007Q4, 2008Q1, 2008Q2, 2008Q3, 2008Q4

Somewhat surprisingly, given this reduction in available content, the DAB platform's share of commercial radio listening showed a significant increase last quarter (to 9.9% from 9.2% the previous quarter) but the aggregate usage of digital platforms has stayed remarkably flat during 2008 at around 19%. Put simply, we are not seeing much, if any, growth in digital platform usage for commercial radio. (NB: much of the apparent growth in the graphs above and below derives from re-distribution of earlier 'unspecified platform' respondent data in recent quarters.)

BBC RADIO: HOURS LISTENED BY PLATFORM (%)

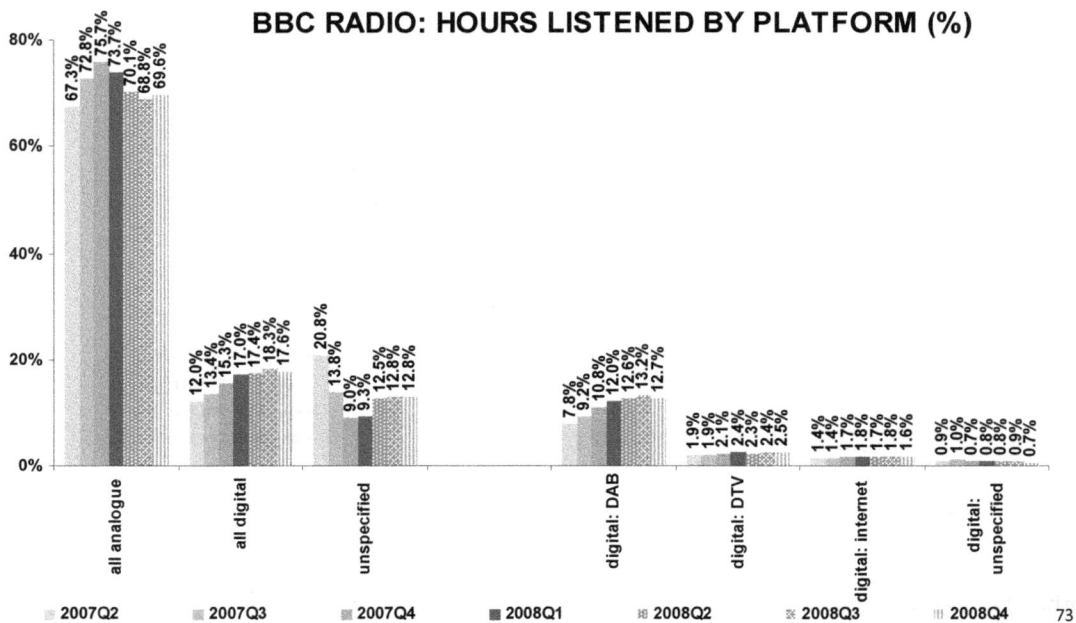

If commercial radio's success with digital platforms seems 'stuck', then the BBC could be in an even worse position. In the last quarter, usage of both the DAB and internet platforms declined, leading to old fashioned analogue radio having accounted for a greater proportion of listening than in the previous quarter (up from 68.8% to 69.6%). This is particularly alarming, given the BBC's much more extensive cross-promotion of its digital platforms across all media, and given the integration of radio into the BBC iPlayer in 2008. It is true that one quarter's data alone might only prove to be a statistical aberration, but it is worrying news to arrive on the very day that Lord Carter chose to pin his colours to the 'radio must be DAB' mast.

Digital-only stations are not proving to be as attractive to listeners as they need to be in order to drive up usage of digital platforms quickly towards the desired 50% criterion. Year-on-year, hours listened to national digital-only stations are down 7%, yet DAB receiver ownership increased by 35% over the same period, according to RAJAR. In aggregate, 16 national digital-only stations accounted for 33 million hours listening per week last quarter, a drop in the ocean compared to radio's total 1 billion hours listened per week.

So, the reason it might have been so quiet today on the digital radio PR front is that there really was not much good news from RAJAR to be shouting about, from either the BBC or commercial radio perspective. And the plan laid out in the Digital Britain document, which might look great in theory, still depends upon:

- increased consumer expenditure on DAB radio hardware
- increased investment in DAB content
- increased investment in DAB transmission infrastructure
- and thus does not appear to be a plan at all steeped in reality, in a time when discretionary expenditure (personal and corporate) is less forthcoming than ever.

NATIONAL DIGITAL-ONLY RADIO STATIONS

launched	group	station	national DAB	local DAB	Freeview	FreeSat	internet	% share of total radio listening Q4 2008	total hours per week ('000) Q4 2007	total hours per week ('000) Q4 2008	year-on-year change in hours listened (%)
2002	BBC	BBC7	■				■	0.53	4,530	**5,380**	19
1999	-	Planet Rock	■		■		■	0.48	3,148	**4,872**	55
2003	Bauer	The Hits		■	■		■	0.45	5,692	**4,557**	-20
2002	BBC	BBC 6 Music	■				■	0.31	2,643	**3,178**	20
2003	Bauer	Smash Hits Radio		■		■	■	0.25	3,096	**2,563**	-17
2002	BBC	BBC 1Xtra	■				■	0.25	2,599	**2,514**	-3
2002	BBC	BBC Asian Network	■				■	0.20	2,600	**1,992**	-23
2002	BBC	BBC Five Live Sports Extra	■			■	■	0.14	1,288	**1,419**	10
2003		Mojo Radio			closed 2008			0.12	717	**1,266**	77
2003	Bauer	Heat		■			■	0.12	1,338	**1,216**	-9
2005	Global	Chill		■				0.10	986	**1,028**	4
2001	Global	The Arrow		■				0.10	920	**1,015**	10
2004	TIML	Absolute Radio Classic Rock					■	0.09	1,045	878	-16
2003	Bauer	Q					■	0.07	830	700	-16
2008	-	NME Radio		■				0.04		356	
2005	TIML	Absolute Radio Xtreme	■					0.03	234	346	48
2005	Folder	Fun Radio		■		■			209		
2006		TheJazz			closed 2008				1,882		
2000		Capital Life			closed 2008				700		
2000		Oneword Radio			closed 2008				583		
2000		Virgin Radio Groove			closed 2008				488		
1999		Core			closed 2008				377		
		TOTAL						**3.28**	**35,905**	**33,280**	**-7**

"national" as defined by RAJAR past and present
BBC Asian Network also on AM in Midlands

[74]

The priority for the radio industry in 2009 must be survival, pure and simple. For commercial radio, it is survival in the worsening struggle against the twin evils of falling listening and declining revenues. For the BBC, it is the struggle to ensure that the commercial radio sector survives. Without a successful commercial radio sector, the BBC's own radio services could be under threat.

Let us hope that the Final Report for Digital Britain incorporates a greater dose of realism and pragmatism, or unfolding events might easily catch up with it even before its publication.

[61] http://www.culture.gov.uk/images/publications/digital_britain_interimreportjan09.pdf
[62] http://www.rajar.co.uk/docs/news/data_release_2008_Q4.pdf
[63] http://www.culture.gov.uk/images/publications/digital_britain_interimreportjan09.pdf
[64] http://www.rajar.co.uk/docs/news/data_release_2008_Q4.pdf
[65] http://www.radiocentre.org/rc2008/showContent.aspx?pubID=307
[66] http://www.bbc.co.uk/pressoffice/pressreleases/stories/2009/01_january/29/rajar.shtml
[67] http://www.drdb.org/article.php?id=755&from=hom
[68] http://www.drdb.org/article.php?id=737&from=arc~s=2008-12-1~e=2008-12-31
[69] http://www.culture.gov.uk/images/publications/DRWG_Final_Report.pdf
[70] source: RAJAR & Digital Radio Working Group
[71] source: Digital Radio Development Bureau
[72] source: RAJAR
[73] source: RAJAR
[74] source: RAJAR

11.

3 February 2009

Digital Britain: the devil is in the indefinite article

Commenting last week on the publication of the government's Digital Britain report, RadioCentre Chief Executive Andrew Harrison said that *"the devil will be in the detail."*[75] Absolutely true because, sometimes, a single word can tell you more about the direction that government policy is taking than a weighty tome. In the case of DAB, the wording of the Digital Britain report raised one such question: does the government want the DAB platform to supplement FM/AM radio, or does it want DAB to supplant it?

The Final Report of the Digital Radio Working Group in December 2008 had recommended:

*"DAB as **the** primary platform for national, regional and large local stations"* [emphasis added].[76]

However, last week's Interim Report of Digital Britain made a commitment:

*"to enabling DAB to be **a** primary distribution network for radio"* [emphasis added].[77]

This may seem like an insignificant detail but, for the radio industry, it certainly is not. If DAB is to be **the** primary platform, the implication is that if your radio station is not available on the DAB platform, your business will be marginalised. It implies that the FM and AM platforms will be closed down, which would be a disastrous outcome for smaller commercial radio stations who may not be able to afford the cost of DAB transmission and/or who cannot find space on their local multiplex (if that multiplex even exists) [see article in The Guardian 'Committed to its listeners'[78]].

On the other hand, if DAB is to be **a** primary platform, the implication is that it will be available to consumers as an adjunct to existing FM/AM radio and to IP-delivered content. In this scenario, the ideal radio receiver of the future will be one which, to the user, is 'platform neutral' but has capabilities to receive DAB, FM/AM and IP. The user would simply select 'Radio 4: live' on the radio's interface, and the radio itself would determine which was the most reliable delivery platform in that location to serve Radio 4 live. Or, the user might select 'Radio 4: The Archers' and it would deliver the most recent episode by IP.

Strangely, the subtle difference between 'a' and 'the' seemed to be ignored by some stakeholders.

Laurence Harrison, director of consumer electronics at Intellect, said:

*"This commitment to DAB as **the** primary distribution network for radio is exactly the sort of strong and decisive leadership we wanted to see from government"* [emphasis added].[79]

Frontier Silicon, in its first press release:

*".....welcomed the Government's commitment to DAB as **the** primary distribution network for future radio broadcasting in the UK"* [emphasis added].[80]

Frontier Silicon, in a second press release:

*".....the Government's endorsement of the digital migration of radio and commitment to DAB as **the** primary distribution network for future radio broadcasting"* [81] [emphasis added]

The precise wording was also reported badly by some media.

The Guardian's Media Monkey wrote:

*"....DAB radio was **the** 'primary distribution network' for radio...."*[82] [emphasis added]

The Sunday Times wrote:

*"....DAB digital technology, set to become **the** 'primary distribution network' for radio...."*[83] [emphasis added]

The Daily Mail wrote:

*"Lord Carter, the Communications Minister, said: 'We are making a public commitment to DAB as **the** primary distribution medium'..."*[84] [emphasis added]

The Telegraph wrote:

*"The Digital Britain report...... gives a firm commitment to digital radio (DAB) as **the** primary way of listening to content in the future."* [85] [emphasis added]

The BBC wrote:

*"The culture secretary said digital audio broadcasting (DAB) will become **the** 'primary distribution network'....."*[86] [emphasis added]

No wonder the public is confused. The potential implication of the Digital Britain report on areas of the UK where DAB reception is presently non-existent is just starting to be realised. *"FM reception in Eden Valley may disappear"* said one local Cumbria newspaper headline yesterday. More coverage like this will inevitably follow.[87]

How many civil servants must have scoured the precise wording of the Digital Britain report before it was published? The change of emphasis from 'the' to 'a' is unlikely to have been accidental. If the DAB platform were to fail (no acceleration in consumer take-up, no increased exclusive content), then the government will find it needs a 'get out of jail free' card. The word 'a' provides it with the perfect caveat, next month, next year, whenever.

---

[75] http://www.brandrepublic.com/News/877280/Digital-Britain-Radio-industry-welcomes-Carter-proposals
[76] http://www.culture.gov.uk/images/publications/DRWG_Final_Report.pdf
[77] http://www.culture.gov.uk/images/publications/digital_britain_interimreportjan09.pdf
[78] http://www.guardian.co.uk/media/2009/feb/02/media-recession-innovation-grassroots
[79] http://www.drdb.org/article.php?id=755&from=arc~s=2009-1-1~e=2009-1-31
[80] http://www.frontier-silicon.com/media/releases/09/0129_digbritirep.htm

[81] http://www.frontier-silicon.com/media/releases/09/0202_chorus3.htm

[82] http://www.guardian.co.uk/media/2009/feb/02/media-monkeys-diary

[83] http://business.timesonline.co.uk/tol/business/industry_sectors/media/article5627243.ece

[84] http://www.dailymail.co.uk/news/article-1132028/Digital-radio-switchover-silence-AM-FM.html

[85] http://www.telegraph.co.uk/finance/newsbysector/mediatechnologyandtelecoms/4386355/Regulators-and-Government-at-odds-over-the-future-of-Channel-4.html

[86] http://news.bbc.co.uk/1/hi/technology/7858498.stm

[87] http://www.newsandstar.co.uk/news/1.507633

12.

11 February 2009

DAB: the medium of consumer choice?

It appears there may be a factual error in the Digital Britain Interim Report. I assume it was an accidental mistake in drafting. Obviously, a government document would not deliberately misrepresent the facts.

The Interim Report states on page 32:

"Dedicated analogue radio sets are no longer part of the retail mainstream: analogue continues to be used in bundled products (e.g. radio alarms). But, in dedicated radio, DAB has become the medium of consumer choice."[88]

There are two distinct assertions here:

- *"dedicated analogue radio sets are no longer part of the retail mainstream"*
- *"DAB has become the medium of consumer choice."*

The second assertion was made by the Interim Report strictly in the context of *"dedicated"* radio hardware, but the statement was quickly abstracted as a standalone fact. The Guardian wrote that the Report *"said DAB had become 'the medium of consumer choice'."*[89] The Telegraph wrote that *"the Report states that DAB digital radio has 'become the platform of choice' for radio listening in the UK...."*[90] and, in a separate article, said that *"Ministers claimed that DAB radio is now 'the medium of consumer choice'"* though it questioned the assertion.[91] Marketing Week wrote that the Report *"says DAB has become 'the medium of consumer choice' in the UK...."*[92] This same assertion was repeated on web sites such as Broadcasting World and Radio-Info.[93]

ASSERTION 1
"dedicated analogue radio sets are no longer part of the retail mainstream"

I have sat through several Powerpoint presentations at radio conferences, both in the UK and overseas, which claimed that analogue radio receivers (AM/FM) have almost disappeared from retail outlets in the UK. The facts tell a very different story.

A survey of electronic consumer goods on sale from the web sites of three of the UK's most prominent consumer electronics retailers reveals that the analogue radio platform is still alive and well. In fact, at Argos and Comet, electronic goods incorporating the analogue radio platform alone continue to outnumber those with digital platforms.

NO. OF MODELS OF ELECTRONICS HARDWARE INCORPORATING RADIO PLATFORMS			
radio platforms	Argos	Currys	Comet
FM and/or AM	113	48	59
FM + DAB	56	55	43
DAB	2	9	10
DAB + FM + internet	4	4	2
DAB + internet	1	0	0
internet	3	2	0
internet + FM	1	2	2

excludes in-car and mobile phone hardware
web sites @ 10Feb2009   [94]

Interestingly, when DAB radio receivers were first introduced, most models were single-platform (such as the 'Pure Evoke 1'). This has changed significantly, so that the vast majority of DAB radio receivers presently on sale are dual-platform (mostly DAB + FM). This change provides a significant 'safety net' at all levels of the value chain, should the DAB platform fail to develop into a mass medium for radio broadcasting.

For consumers, the incorporation of the FM platform into DAB radios should encourage hardware purchase, removing the perceived risk of platform failure (viz ITV Digital). However, the continued availability of the FM platform in 'DAB radios' is likely to impact consumer usage of the DAB platform. If a consumer buys a 'DAB radio', but they continue to use the FM platform incorporated within the hardware for part of their radio listening, they are contributing to the DAB platform's struggle to gain sufficient traction that FM broadcasting can ever be switched off.

Additionally, one wonders how many RAJAR respondents use their 'DAB radio' to listen to stations on the incorporated FM platform, but report this listening in their diaries incorrectly as 'digital' rather than 'analogue'. Surely, if I buy a 'digital radio' which clearly says 'digital radio' on its facia, then all the content I listen to using that radio must be 'digital radio'? No wonder the marketplace is confused.

Examining the other side of the retail marketplace, in terms of consumer purchases of DAB radios, an appendix attached to the Digital Britain Interim Report demonstrates (page 41) clearly that the vast majority of radios purchased in the UK are not DAB, according to data collected by GfK for the Digital Radio Development Bureau.[95] The graph below updates this same data.

The data show that 79% of radio receivers purchased in the UK during the last twelve months were analogue and did not incorporate the DAB platform. The vast majority of radios sold in the UK continue to be analogue, not DAB, which is why, as demonstrated above, electronics retailers continue to stock so much hardware incorporating analogue radio. In some types of hardware, notably personal media players, the market is still almost entirely dominated by analogue radio (in those models that include radio).

UK RADIO RECEIVER SALES: FOUR-QUARTER MOVING AVERAGE

DAB receivers as percentage of total receiver sales

| 8% | 9% | 10% | 12% | 13% | 14% | 14% | 17% | 18% | 18% | 19% | 20% | 20% | 21% | 21% |

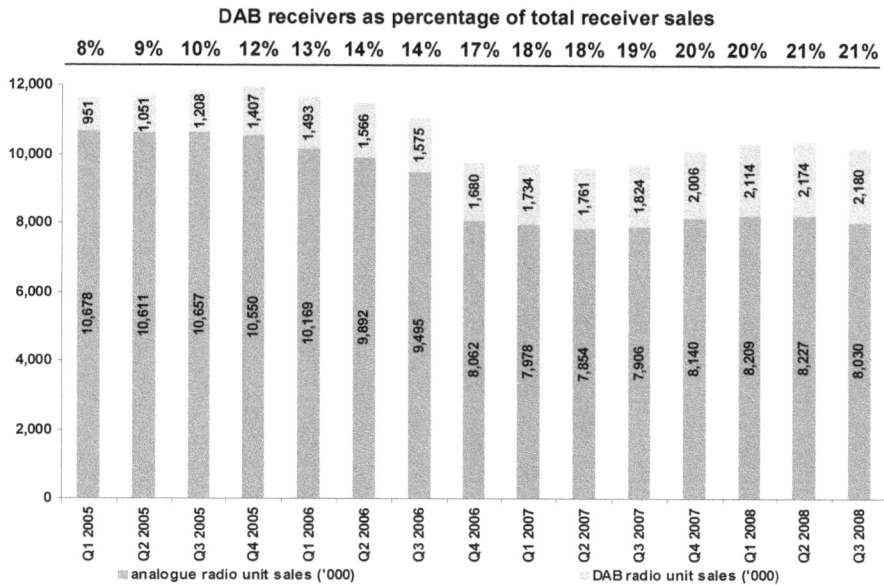

analogue radio unit sales ('000)   DAB radio unit sales ('000)

DAB radio unit sales ('000) per quarter: 951, 1,051, 1,208, 1,407, 1,493, 1,566, 1,575, 1,680, 1,734, 1,761, 1,824, 2,006, 2,114, 2,174, 2,180

analogue radio unit sales ('000) per quarter: 10,678, 10,611, 10,657, 10,550, 10,169, 9,892, 9,495, 8,062, 7,978, 7,854, 7,906, 8,140, 8,209, 8,227, 8,030

Quarters: Q1 2005, Q2 2005, Q3 2005, Q4 2005, Q1 2006, Q2 2006, Q3 2006, Q4 2006, Q1 2007, Q2 2007, Q3 2007, Q4 2007, Q1 2008, Q2 2008, Q3 2008

96

In conclusion, the assertion made in the Digital Britain document that *"dedicated analogue radio sets are no longer part of the retail mainstream"* seems incorrect.

ASSERTION 2
"DAB has become the medium of consumer choice"

The latest RAJAR radio audience data from Q4 2008 demonstrate that the overwhelming majority of radio listening continues to be consumed via the analogue platform, not via DAB.

COMMERCIAL RADIO LISTENING BY PLATFORM (Q4 2008)

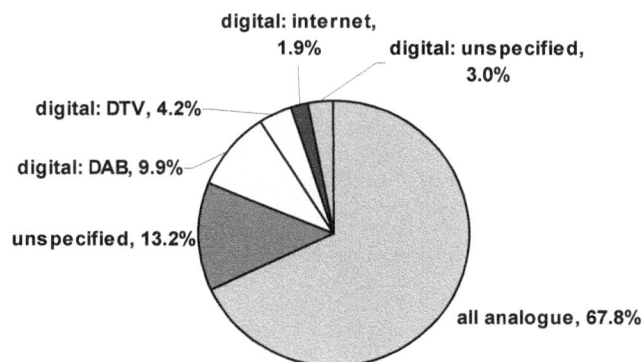

- digital: internet, 1.9%
- digital: unspecified, 3.0%
- digital: DTV, 4.2%
- digital: DAB, 9.9%
- unspecified, 13.2%
- all analogue, 67.8%

97

In the case of commercial radio, 10% of hours listened are via the DAB platform, whereas 68% of hours listened are via analogue.

BBC RADIO LISTENING BY PLATFORM (Q4 2008)

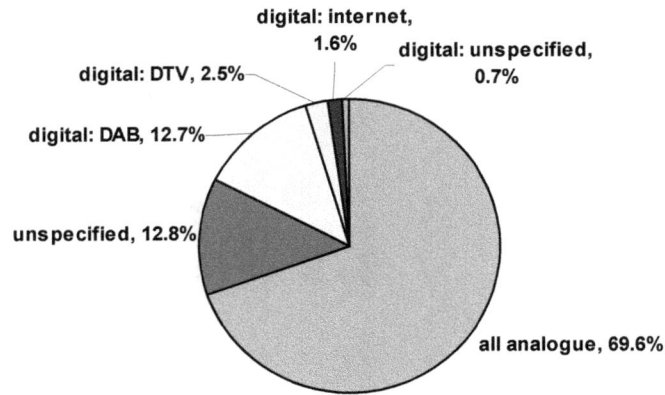

digital: internet, 1.6%

digital: unspecified, 0.7%

digital: DTV, 2.5%

digital: DAB, 12.7%

unspecified, 12.8%

all analogue, 69.6%

98

For BBC radio, 13% of hours listened are via the DAB platform, whereas 70% of hours are consumed via analogue.

In conclusion, the assertion made in some press coverage that "*DAB has become the medium of consumer choice*" is incorrect.

SUMMARY:

Analogue radio is alive and well in the UK because consumers continue to demand and purchase electronic goods that incorporate the analogue radio platform; and because radio listeners are consuming content predominantly delivered via the analogue platform. These are the facts.

---

[88] http://www.culture.gov.uk/images/publications/digital_britain_interimreportjan09.pdf
[89] http://www.guardian.co.uk/media/2009/jan/29/digital-radio-lord-carter-report
[90] http://www.telegraph.co.uk/technology/news/4388722/Digital-Britain-report-is-ambitious-if-imperfect.html
[91] http://www.telegraph.co.uk/technology/news/4387972/Carter-report-AM-and-FM-radio-signals-to-be-switched-off.html
[92] http://www.marketingweek.co.uk/news/government-delays-analogue-radio-switch-off/2064131.article
[93] http://www.radio-info.com/sections/2-breaking-news/news_items/4537-despite-objections-great-britain-may-abandon-am-fm-analog-radio-and-go-digital
[94] source: Grant Goddard
[95] http://www.culture.gov.uk/images/publications/digital_britain_interimreportjan09_annex1.pdf
[96] source: Digital Radio Development Bureau/GfK
[97] source: RAJAR
[98] source: RAJAR

13.

17 February 2009

Localness: please, sir, can I have some less?

The government's announcement that an independent review group will look at the 'localness' issues relating to content on commercial radio could re-ignite a war of words between the stakeholders that a year ago ended in a tense ceasefire.[99] Last time, hostilities between the large radio owners and Ofcom became elevated to such an extent that the regulator's chief executive Ed Richards even used the Annual Ofcom Lecture to chastise the commercial radio industry for its persistent lobbying to loosen its 'localness' obligations:

"Some [radio owners] have called for a huge relaxation in relation to localness, some in the industry even call for a complete removal of all regulation. They believe that localness is either no longer valued or that its value is significantly outweighed by its cost. The problem is that the evidence is to the contrary. What our research tells us is that people continue to want to hear local programming. …. But we are not convinced that the market alone will deliver this if left to its own devices. We recognise very clearly the significant economic challenges faced by the radio sector, but our forthcoming proposals will not involve eliminating the obligation to deliver local programming or its reduction to a negligible level."[100]

Ofcom subsequently published its policy statement on localness in February 2008 and although, on the surface, it might have looked as if a ceasefire had broken out between the two sides, behind the scenes the industry's lobbying for further reductions of its 'localness' obligations continued regardless.[101] Ofcom had estimated that its policy changes would save the radio sector £9.4m to £11.7m per annum from a cost base of around £620m. For the radio industry, these potential savings were simply not enough. Andrew Harrison, chief executive of RadioCentre, argued that *"the heavy burden of the existing localness regulation and legislation [..] is holding back current profitability and future investment in the sector."*[102]

By December 2008, industry lobbying had succeeded in persuading the Digital Radio Working Group to recommend in its Final Report that:

"commercial radio must be given greater freedom to shape its digital future to provide a sustainable future for local radio in a digital world through a relaxation of analogue localness requirements………"[103]

and to comment that:

"…. a model which focuses so heavily on where content is made may not be the best way to deliver either what listeners will most want in the future or allow the industry space to grow. We therefore recommend that the commercial radio sector, Ofcom and the government should look closely at the current localness regime in the coming months…….."[104]

What proved interesting about last week's government announcement of the independent review into 'localness' was that it contained no mention of Ofcom whatsoever.[105] Even though the press release noted that the review would examine *"to what extent are the current requirements for a pre-determined number of hours of local content, and the locality in which content is produced, appropriate and sustainable"*, as implemented by Ofcom, it did not mention the regulator by name. This omission is downright weird. The Communications Act 2003 states clearly that:

"It shall be the duty of OFCOM to carry out their functions in relation to local sound broadcasting services in the manner that they consider is best calculated to secure: (a) that programmes consisting of or including local material are included in such services but, in the case of each service, only if and to the extent (if any) that OFCOM consider appropriate in that case; and (b) that, where such programmes are included in such a service, what appears to OFCOM to be a suitable proportion of them consists of locally-made programmes."[106]

Furthermore, the Act states that *"OFCOM must: (a) draw up guidance…."* and *"OFCOM may revise the guidance from time to time"*, but it *"must consult"* licence holders and stakeholders beforehand.[107] The legislation is crystal clear as to where the responsibility resides. What we are seeing in the government announcement is an intervention at a higher level as a result of perceived dissatisfaction with the way that Ofcom has implemented its responsibilities on this *"particularly contentious"* issue, as Ed Richards described it.[108]

Ofcom's 2007 consultation on 'localness' in radio had elicited 43 responses, from which the regulator *"noted the calls from the commercial radio industry for a reduction of locally-made programming…."*[109] Ofcom stated determinedly: *"We believe that our proposed guidelines already represent a substantial deregulation of locally-made programming in many cases."* However, it looks as if further lobbying has undermined the Ofcom position, and the regulator is now being sidelined by direct government action on this issue, which could lead to new legislation or to new implementation of existing legislation.

So what precisely does the commercial radio industry want changed by Lord Carter in Ofcom's localness requirements?

- local commercial stations required to broadcast no more than 4 hours a day of locally made programming
- regional commercial stations not required to broadcast locally made programming
- local news broadcasts on local stations can be produced in centralised newsrooms
- stations serving populations of less than 750,000 (i.e.: two thirds of the UK's stations) permitted to locate their studios outside the area they serve
- the 4 hours a day of 'local' programming can be simulcast across co-located stations and still count as locally made programming.

And what concessions would the commercial radio industry offer Lord Carter in return for its newly, co-located, networked content, 'local' stations?

- news bulletins (not all local) 13 hours a day on local stations
- news bulletins 24 hours a day (not all regional) on regional stations (13 hours a day on specialist music stations)
- extended news bulletins (of unspecified length)
- a commitment to safeguarding stations' remaining local content (weather, traffic, what's on, charity appeals, community information).

However, these demands and concessions position the 'localness' issue strictly in the context of content regulation. In fact, there is a much bigger game being played out, which concerns the further investment required in the DAB platform to try and make it a success with consumers. Essentially, the commercial radio industry is trying to put a gun to Lord Carter's head and is demanding: 'we won't invest any more money in DAB to make it work, unless you stop Ofcom making us do local things we don't want to do'.

The initial response to the commercial radio lobby was likely to be: 'you acquired all these local radio stations, knowing that they had localness obligations. If you wanted a national radio station, why didn't you buy one of those instead?' It does seem a bit like Stagecoach begging the government to transform its local bus routes into a national coach service. However, Lord Carter is trying to grapple with the issues and forge a compromise whilst still insisting that "*government can not, nor should it, be the main driving force for digital radio.*"

The biggest danger here is that the 'localness' issue becomes a mere sideshow to the much more politically and commercially significant decision over the future of the DAB platform. As such, 'localness' risks becoming a mere pawn in a complex set of negotiations that are essentially designed to maintain the balance sheet valuations of the largest radio groups which have already made significant investments over a decade in costly DAB infrastructure.

Sadly, this is not the first time that the 'localness' issue has been invoked merely as a quid pro quo within a much bigger game. In the original Bill that became the Communications Act 2003, there was no 'localness' clause for local radio, just as there never had been in previous broadcast legislation. It was inserted at the last minute as what the then Minister for Broadcasting, Dr. Kim Howells MP, admitted was "*the quid pro quo for greater liberalisation in the radio market*", allowing more concentrated ownership of local radio than the Bill had originally proposed.[110] In the ensuing House of Commons debate, Michael Fabricant MP successfully stoked the flames of fear:

"*What if Clear Channel – a United States organisation for which I have a considerable respect, but which the [UK commercial radio] industry is rather concerned about – were to acquire a number of radio stations and found that it could pull in large audiences, based in the US, and not be all that local? Its presenters could be based in New York, for example, and it could put in pre-recorded local identifications. Everything could be done on a PC-based system. The stations would sound like local radio, even though they were not; and, because they had a good playlist, they might pull in a big audience. Would we not want back-stop powers in such a case?*"[111]

Six years later, neither Clear Channel nor its competitors have bothered to enter the UK radio market. Instead of the then touted prospect of US-financed global radio, we now have Irish-financed Global Radio wanting to run as much of its UK local radio empire as possible from Leicester Square. At the end of the day, for the listener, does the distinction matter whether a local radio station's studio is in New York or New Bond Street? If I were a listener in NorthEast England, when I choose to listen to local radio, rather than national radio, if it does not fulfil my desire for 'local', then it offers me zero utility. If I am digging my car out of a three-foot snowdrift and the jolly 'local' radio presenter does not mention the inclement weather from her faraway studio, it simply isn't local radio.

Surely, a 'localness' policy for radio should put the citizen/consumer/listener at the heart of its doctrine, something which Ofcom policies to date have failed to do. But neither does the commercial radio industry come out of this smelling of roses. I have yet to see one UK case study backed with evidential data which demonstrate that a decrease in local content on a local radio station has resulted in audience growth. Reduced costs? Yes. Improved profit margins? Yes. But local commercial radio stations have always been gifted scarce analogue radio spectrum for free, in return for their public service content commitments. A local radio station that is not trying to maximise its audience but, instead, aims to maximise profits by reducing costs, cutting local content and knowing full well that its audience will inevitably decline, would seem to be misusing valuable spectrum.

It remains to be seen whether this latest initiative to review radio's localness requirements will result in new regulation that finally puts the listener at the centre of its policies, rather than simply responding to the needs of either the box-ticking regulator or the de-localising, large radio groups.

On a personal note, over several years I researched the issue of 'localness' and 'localism' in local radio, and I wrote an unpublished paper[112] a year ago that examined the issues and suggested a way forward that would reinstate the local radio listener at the heart of localness regulatory policy. If the laws or regulatory regime do have to be changed, my only hope is that they are changed for the better, and not for the worse.

---

[99] http://www.culture.gov.uk/reference_library/media_releases/5881.aspx
[100] http://www.ofcom.org.uk/media/speeches/2007/10/annuallecture
[101] http://www.ofcom.org.uk/consult/condocs/futureradio07/statement/statement.pdf
[102] http://www.guardian.co.uk/media/2009/jan/29/digital-britain-media-industry-reaction?commentpage=1
[103] http://www.culture.gov.uk/images/publications/DRWG_Final_Report.pdf
[104] http://www.culture.gov.uk/images/publications/DRWG_Final_Report.pdf
[105] http://www.culture.gov.uk/reference_library/media_releases/5881.aspx
[106] http://www.opsi.gov.uk/ACTS/acts2003/ukpga_20030021_en_29
[107] http://www.opsi.gov.uk/ACTS/acts2003/ukpga_20030021_en_29
[108] http://www.ofcom.org.uk/media/speeches/2007/10/annuallecture
[109] http://www.ofcom.org.uk/consult/condocs/futureradio07/statement/statement.pdf
[110] Hansard, House of Commons, Standing Committee E, Communications Bill, 28 January 2003 (morning), column 743.
[111] Hansard, House of Commons, Standing Committee E, Communications Bill, 23 January 2003 (afternoon), columns 73-731.
[112] Grant Goddard, "UK Commercial Radio: A New Way To Regulate 'Localness'", unpublished report, November 2007.

14.

19 February 2009

DAB: there is no alternative?

The most startling suggestion in the recent report on 'The Drive to Digital' commissioned by RadioCentre is the part that details the prerequisites for commercial radio to *"forge ahead with DAB"*:

"This requires changes to terms of trade and the active support of the other principal players in radio – the government, Ofcom, the BBC and Arqiva – including **commitment not to pursue alternative technologies to DAB**" [emphasis added].[113]

In other words, commercial radio considers that the way to make the DAB platform a successful technology is to force the remaining stakeholders – notably the BBC – to stop using other alternative digital delivery platforms (the internet, Freeview, Sky, FreeSat, mobile phones) to distribute radio. This would effectively force consumers who want to listen to, for example, digital station *BBC7* to purchase DAB radios whereas, at present, the station can be received on the full range of digital platforms.

This sounds like an extreme solution to a challenging problem, beating consumers with a DAB 'stick'. After almost a decade, the industry has had to reluctantly admit that its 'carrot' approach has failed to convince the public of the value of DAB radio. The RadioCentre report acknowledges that:

"[DAB] has been plagued by a damaging combination of slow take-up, poor coverage, high costs and uncompelling content" and that *"there is not as much DAB-only material as hoped, and very little that's truly compelling – there's no 'must have' content as with sports & movies on Sky [TV]."*[114]

The notion of forcing, rather than persuading, the public to use the DAB platform had been touched upon in the Final Report of the Digital Radio Working Group published in December 2008. It noted that:

"many of the consumer groups believe that, once an announcement [of an AM/FM switch-off date] is made, no equipment should be sold that does not deliver both DAB and FM."[115]

Such a proposal would prove impossible to put into practice. Most consumer electronics hardware is made by global companies whose models benefit from 'universality' and not from having to manufacture a special UK-only version that would incorporate the DAB platform. Right now, there is not a single mobile phone on sale in the UK that includes the DAB platform, and that situation is unlikely to change because Nokia, Samsung, Sony, LG and Motorola understandably consider FM radio to be the universal radio platform.

A similarly unrealistic proposal for DAB surfaced in March 2008, when Channel 4 Radio commissioned an independent report that proposed:

"to distribute one digital (DAB+) radio set [free of charge] to each household – approximately 26 million sets in total – to stimulate mass take up of digital radio. The sets would be provided over a period of three years, starting in 2010, with 80% distributed over the first two. The total cost of the 'switch-on' plan (DAB+ sets, marketing campaign and administration) would be £383m [...]. Preliminary thinking is that distribution would use vouchers that would be redeemed in larger retail outlets or via promotional codes online." [116]

The report anticipated that such a mass consumer giveaway *"could result in 60% digital listening by 2012"* whereas, without it, *"digital listening may not reach 60% until 2017, with analogue switch-off no earlier than 2020."* However, the hypothesis failed to consider that a household given a free DAB radio might not necessarily use it, if there were no radio content of sufficient appeal broadcast on the platform. Given that the average household has six radio receivers, a free distribution such as this might simply result in a glut of unused DAB receivers advertised on E-bay.

Such unrealistic proposals only serve to demonstrate a phenomenon highlighted by a web site that is currently nominating DAB radio for the 'Fiasco Award 2009' in Spain:

"The fact that a technology is possible does not necessarily mean that people is willing to pay for it, and the fact that Institutions and Companies support it does not mean they did the necessary previous research: they were probably just thinking that they didn't want to be left behind.. [sic]" [117]

In the case of the DAB platform, its forced take-up would be the last opportunity remaining for the largest UK commercial radio owners to throw a protectionist cloak around their assets. Through their joint ownership of the DAB platform infrastructure in the UK, this handful of companies hope to limit UK citizens' future radio listening to **their** content broadcast on **their** stations received via **their** DAB platform. To make this scenario work, of course it would be essential to eradicate competing digital radio platforms.

And why are radio owners so desperate? An excellent US article this week by Seeking Alpha's Jeff Jarvis expressed the reasons most eloquently:

"We've been wringing hands over newspapers and magazines, but TV and radio aren't far behind. Broadcast is next. It's a failure of distribution as a business model. Distribution is a scarcity business: 'I control the tower/press/wire and you don't and that's what makes my business.' Not long ago, they said that owning these channels was tantamount to owning a mint. No more. The same was said of content. But it's relationships (read: links) that create value today. Young [Broadcasting, filing for bankruptcy with $1bn debt] tried to build relationships, once upon a time. At WKRN in Nashville, Mike Sechrist did amazing work starting blogs, building relationships with bloggers, training the community in the skills of the TV priesthood. But he left and all that disappeared. Been there, done that, I can imagine executives saying as they try to stuff the hole in the dike with borrowed dollars. Didn't work. The local TV and radio business, once a privilege to be part of, is next to fall. Timber." [118]

As if that was not enough, the credit crunch has exposed the flimsy financial arrangements of recent radio acquisition deals. This was perfectly explained by Jerry Del Colliano's consistently provocative US blog in an entry entitled 'Radio: bankrupt in 6 to 12 months':

"Consolidated radio groups are facing bankruptcy because some will not be able to

restructure their massive debt -- the debt they acquired in the first place when they paid too much for overvalued radio stations. No one worried about it then. But now, it's time to pay the piper. Why else do you think radio people who know better are hunkering down for what they know is coming -- default."

"One reader, a radio executive, claims New York money types are not just talking about the possibility of radio groups defaulting, but the probability. Some think it can happen within six months to a year. Radio groups like Cumulus, Univision, Clear Channel, Entercom -- in fact, most of them -- have structures that make it difficult to survive if debt cannot be restructured. And in case you haven't noticed, money is hard to come by these days......."

"Radio groups are more susceptible because they are leveraged to such a high degree. That's the reason that the stock prices are so low. Shareholder equity is zero as every single penny of cash flow currently goes to servicing debt. Soon, they won't be able to service the debt and/or they will be in violation of covenants with the banks and/or equity lenders who will seek to take the stations back."[119]

If this sounds like cross-Atlantic doom-mongering, I assure you that there are UK banks out there already demanding their pound of flesh from more than one indebted UK radio group. 2009 will not be a pretty year. Particularly when Quarter 4 2008 UK radio revenues were down 15% year-on-year, their lowest quarter since 1999.

In these troubled times, proposing radio sector policies to preserve broadcasters' oligopolies, or to artificially stifle the development of competing delivery platforms, is not what is needed. Sure, you might wish to be the only ship on the ocean but, if your rust bucket has a hole in its hull, you will drown anyway.

---

[113] unpublished report

[114] unpublished report

[115] http://www.culture.gov.uk/images/publications/DRWG_Final_Report.pdf

[116] unpublished report

[117] http://www.fiascoawards.com/continguts/general/fitxa.php?llengua=en&id=6

[118] http://seekingalpha.com/article/121032-tv-the-next-to-fall

[119] http://insidemusicmedia.blogspot.com/2009/01/radio-bankrupt-in-6-to-12-months.html

15.

20 February 2009

Digital radio stations: one step forward, two steps backward

The RAJAR radio audience data for Q4 2008 were published on 29 January 2009. The day's news headlines heralded the success of the digital radio platform.

"Radio surges in popularity thanks to digital", said The Independent.[120]

"Digital Enjoys RAJAR Boost", said Radio Today.[121]

Music Week said: *"the latest Rajars survey revealed that digital broadcasting is growing apace in the UK......"*[122]

The Times said: *"Digital audio broadcasting (DAB) is clearer, truer, purer. Every year its coverage widens. Every year more stations are added to its almost infinite capacity......FM has had its day."*[123]

Bauer Radio's managing director of national brands Mark Story told Music Week: *"The audience love [digital]."*[124]

The audience must have a strange way of showing their appreciation for digital radio. In Q4 2008, listening to digital-only radio stations fell precipitously, both for the BBC and for commercial radio.

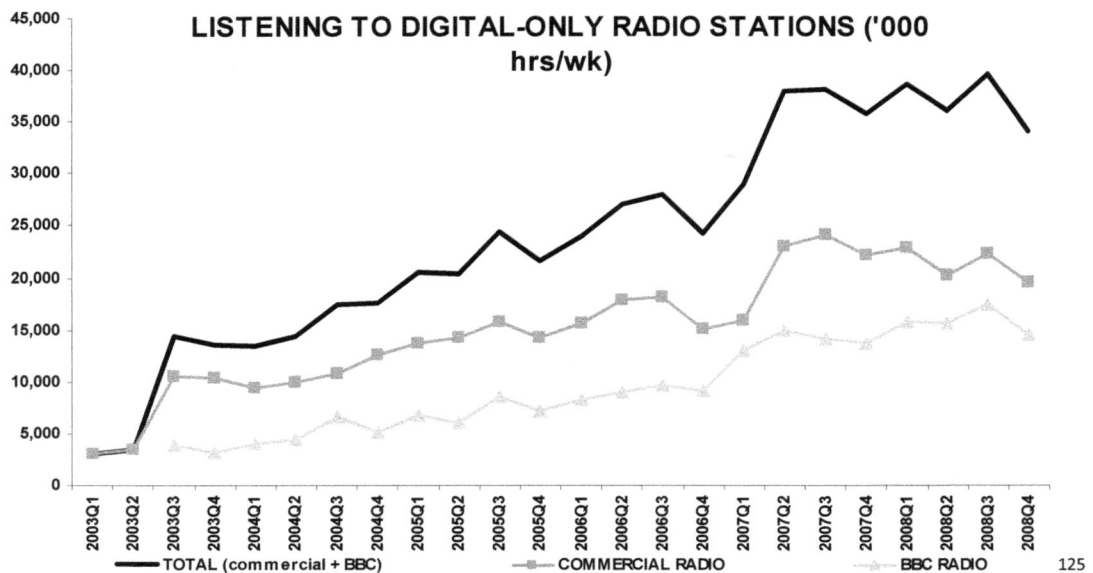

56

This graph illustrates just how sharp was the decline in listening to digital-only radio stations during Q4 2008:

- total digital-only radio station hours listened are down 14% quarter-on-quarter, and down 5% year-on-year to 34m hours/wk, to their lowest level since Q1 2007
- hours listened to commercial digital-only stations are down 12% quarter-on-quarter, and down 11% year-on-year to 20m hours/wk, to their lowest level since Q1 2007
- hours listened to BBC digital-only station are down 17% quarter-on-quarter to 14m hours/wk, their lowest level since Q4 2007.[126]

For the commercial radio sector, 2008 had been the year it finally faced up to the realisation that its digital-only radio stations were not going to break even in the short- or medium-term. This resulted in the closure of digital stations *Mojo Radio, Yarr Radio, TheJazz, Capital Life, Oneword Radio, Virgin Radio Groove* and *Core* during the year. Inevitably, with fewer offerings for consumers, listening to commercial digital-only stations was likely to be impacted.

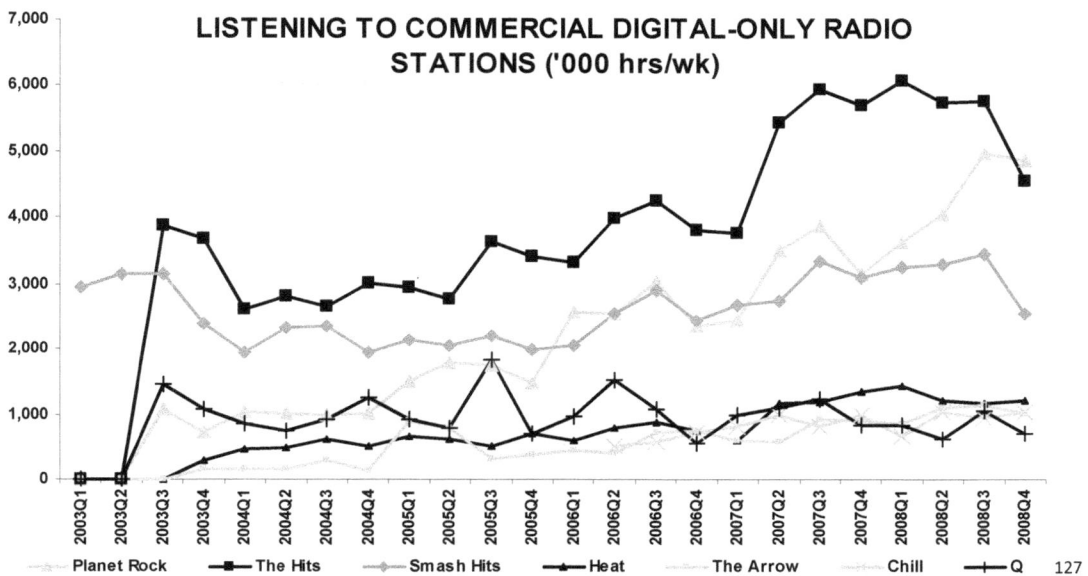

LISTENING TO COMMERCIAL DIGITAL-ONLY RADIO STATIONS ('000 hrs/wk)

Planet Rock · The Hits · Smash Hits · Heat · The Arrow · Chill · Q   127

The surprise result from Q4 2008 RAJAR data is that the sector's remaining digital radio stations have suffered terrible declines. The graph above tracks the largest digital stations, of which only *Planet Rock* achieves a relatively stable performance (and now becomes the sector's most listened to digital station). Otherwise:

- hours listened to *Smash Hits Radio* are down 26% quarter-on-quarter and down 17% year-on-year
- hours listened to *The Hits* are down 21% quarter-on-quarter and down 20% year-on-year
- hours listened to *Q Radio* are down 34% quarter-on-quarter and down 16% year-on-year
- hours listened to *Heat Radio* are down 9% year-on-year.[128]

Although, as the graph shows, the data have always been 'bumpy', the simultaneous decline of listening to all these Bauer-owned stations is a very worrying trend. Bauer is now left carrying the torch for digital commercial radio in the UK, following rival GCap Media/Global Radio's decision last year to close/divest almost all of its digital stations (only *The Arrow* and *Chill* remain).

Planet Rock's owner Malcolm Bluemel said this month that his aim is to make the station profitable by Christmas.[129] The question is: if the UK's most listened to digital commercial radio station is still struggling to break even, what hope is there for the rest of the pack?

It is all very well for Lord Carter's Digital Britain Interim Report to *"expect the radio industry to strengthen its consumer proposition [..] in terms of new and innovative content...."* but, at present, the economics of digital-only radio stations simply do not add up.[130] Not a single digital-only radio station has yet reached break even. How can realistic business plans for new commercial digital services be forged, when nine-year old *Planet Rock* has yet to make an operating profit, let alone recoup its accumulated losses?

If the commercial sector's digital radio audiences offer cause for concern, the BBC's comparable audiences are downright scary. With the exception of *BBC7* (which remains the UK's most listened to digital radio station), audiences for the BBC digital services are down substantially.

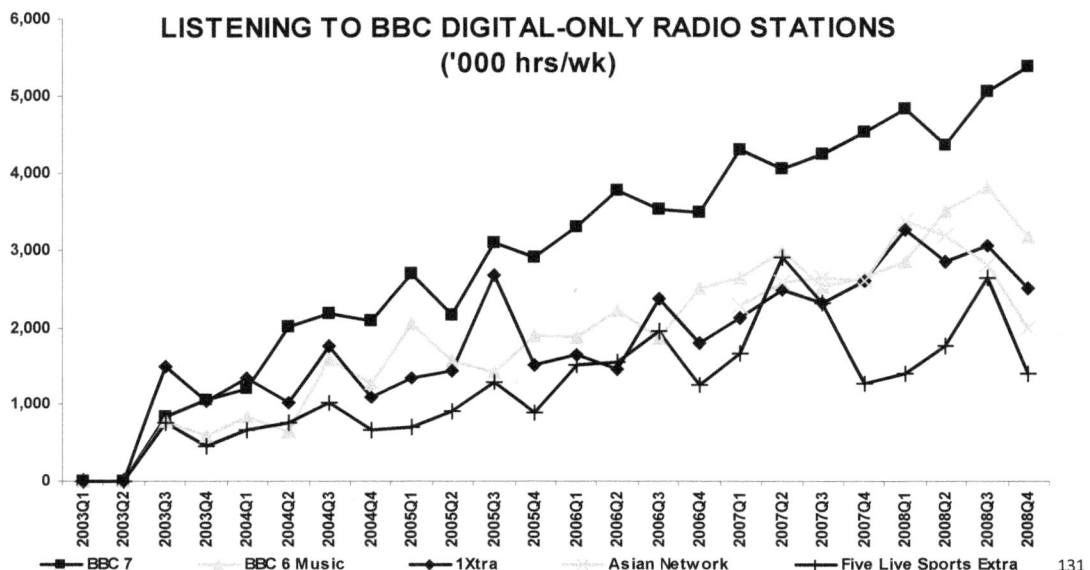

LISTENING TO BBC DIGITAL-ONLY RADIO STATIONS ('000 hrs/wk)

Legend: BBC 7, BBC 6 Music, 1Xtra, Asian Network, Five Live Sports Extra   131

BBC Five Live Sports Extra can be excused because it is a part-time station whose listening fluctuates with the sporting seasons, but elsewhere:

- hours listened to *1Xtra* are down 18% quarter-on-quarter and down 3% year-on-year
- hours listened to *6 Music* are down 17% quarter-on-quarter
- hours listened to *Asian Network* are down 29% quarter-on-quarter and down 23% year-on-year.[132]

Despite the BBC having launched its digital radio stations in 2002 and then having promoted them extensively on TV, radio and online, their growth of listening remains stubbornly linear. One quarter's RAJAR results alone do not a trend make, but the worry must be that the volume of listening to these stations might have already plateau-ed. In other words, if everyone who would be interested in listening to, say *1Xtra*, is already listening to the station after seven years of promotion, then there would be little headroom for further audience growth.

Planet Rock's Malcolm Bluemel pointed out: *"[The BBC] spend £7m a year on 6 Music and another £1m on marketing it. Our annual budget is £1m, plus £20,000 on marketing."*[133]

At some point in time, and sooner rather than later if the audiences of the BBC digital stations show further signs of having plateau-ed, the BBC Trust is likely to want to conduct a cost/benefit analysis to determine if its digital radio stations really offer the Licence Fee payer value for money. In Q4 2008, the peak half-hour audience of *Asian Network* was 29,000 adults, of *1Xtra* 36,000 adults, of *BBC7* 68,000 adults, and of *6 Music* 69,000 adults.[134] The 2008 service budgets for these stations were £8.7m, £7.2m, £5.4m and £6.0m respectively.[135]

Between the BBC and commercial radio, huge sums of money have been spent over the last decade on launching and running digital radio stations that have attracted relatively small audiences. In the meantime, new technologies (on-demand, downloads) have overtaken us. If the radio industry's response to Lord Carter's Digital Britain is simply to launch more new digital stations that will inevitably lose more money, the industry has missed the point.

We now live in an on-demand world where 'content', not 'radio stations', is what consumers increasingly demand. Perhaps we do not need more new radio stations, or even existing local commercial radio brands rolled out nationally as faux new digital brands. What we need is the ability for consumers to access engaging radio content, when, where and how they want it. The days of listener loyalty to one radio station are fading fast.

In these financially hard pressed times, it seems ridiculous to be creating more expensive, new broadcast 'stations', each of which are unlikely to attract significant amounts of listening, but each of which will use a huge amount of scarce radio spectrum. Today, I wanted to listen to the Northern Soul show from *BBC Radio Stoke*, followed by David Rodigan's reggae show on *Kiss*, followed by *WMPR*'s breakfast show. What I need is not a new digital radio brand, but a 'pick'n'mix' menu system where I can easily create my own personal radio station – a bit like a *Pandora* or *Last.fm*, but populated with radio programmes rather than just songs.

This will be the future………. and it will probably arrive as soon as the BBC has finished inventing it. Broadcast radio will continue to be an important medium for mass audience shows like *Today*, Terry Wogan and sports coverage. But, for any content that is remotely specialist, on-demand delivery will have to be the way forward, the result of economic necessity. In 2009, the idea of creating more radio stations, more radio brands, more costly 24-hour broadcast operations has to be wholly redundant. This is an issue that the BBC Trust will have to face up to much earlier than the commercial radio sector. Next quarter's RAJAR could be that touchpaper.

In the meantime, the future of digital-only radio stations hangs in the balance. As Bauer Radio's Mark Story had told Media Week: *"It's going to be a long road for digital radio."*[136]

Then, twelve days after the latest RAJAR results were released, Mark announced he was leaving Bauer after eleven years' service. He said: *"To be brutally honest, it's not the most fantastic time to be in radio….."*[137]

---

[120] http://www.independent.co.uk/news/media/tv-radio/radio-surges-in-popularity-thanks-to-digital-1520405.html
[121] http://radiotoday.co.uk/news.php?extend.4336.7

[122] http://www.drdb.org/newsletter/index.php?id=479

[123] http://entertainment.timesonline.co.uk/tol/arts_and_entertainment/tv_and_radio/article5615143.ece

[124] http://www.drdb.org/newsletter/index.php?id=479

[125] source: RAJAR

[126] source: RAJAR

[127] source: RAJAR

[128] source: RAJAR

[129] http://entertainment.timesonline.co.uk/tol/arts_and_entertainment/tv_and_radio/article5670799.ece

[130] http://www.culture.gov.uk/images/publications/digital_britain_interimreportjan09.pdf

[131] source: RAJAR

[132] source: RAJAR

[133] http://entertainment.timesonline.co.uk/tol/arts_and_entertainment/tv_and_radio/article5670799.ece

[134] source: RAJAR

[135] http://www.bbc.co.uk/bbctrust/assets/files/pdf/regulatory_framework/service_licences/explanatory_note.pdf

[136] http://www.brandrepublic.com/News/877086/Carter-expected-call-BBC-lead-DAB-promotion

[137] http://www.brandrepublic.com/News/879742/Story-leave-Bauer-form-own-radio-consultancy

16.

3 May 2009

DAB: actions speak louder than keynote speeches

Giving the commercial keynote speech at the *Radio Reborn 2009* conference this week in London, Global Radio chief executive officer Stephen Miron banged the drum for the radio medium, banged the drum for Global Radio, and banged the drum for digital radio.

It was the last of these three exhortations that appeared particularly contradictory, given Global Radio's track record with the DAB platform. However, nothing could stop Miron from proclaiming:

- *"At Global, we believe that the government must set a clear and rightfully ambitious programme for digital migration"*
- *"As you would expect from the largest commercial radio broadcaster, we plan to play an active role in helping ensure the successful delivery of that* [digital] *strategy"*
- *"We back digital and we back the* [Digital Britain] *strategy, but we cannot afford to get this wrong"*
- *"Digital Britain has made us focus our minds. Now the government must focus theirs"*
- *"We have embarked on a clear path to digital,* **to DAB**, *and we need to make serious progress and do it quickly"* [emphasis added]
- *"This means naming a date for* [digital] *migration …. A firm date needs to be set"*
- *"The future of our sector is intrinsically linked to the successful implementation of the government's digital strategy and to the successful* **migration to DAB**" [emphasis added]
- *"We need more of this in the coming weeks and months. Not just words, but action"*
- *"We need to get our act together to make the best possible case for consumers to switch to digital"*
- *"Global is up for the challenge and, as the largest commercial player, we are prepared to lead this charge."*[138]

Miron's comments seem particularly difficult to reconcile with Global's 'actions' on DAB, which hardly demonstrate confidence in the platform.

1. Global Radio exits DAB multiplex ownership

On 6 April 2009, it was announced that Global Radio sold its 63% stake in the sole commercial radio national DAB multiplex owner Digital One to transmission provider Arqiva.[139] Global Radio also sold its local DAB multiplex business Now Digital to Arqiva. After almost a decade of operation, these multiplexes were still to generate an operating profit. Global Radio's involvement in DAB multiplexes was thus reduced, at a stroke, from having been the biggest player to much less, writing off most of a decade's worth of massive investment in the process, because the transaction is likely to have happened for a nominal amount.

UK DAB CAPACITY OWNERSHIP

company	% share	
	pre-sale	post-sale
Global Radio	50%	16%
Bauer	16%	16%
Arqiva	14%	47%
UTV	12%	12%
Guardian Media Group	3%	3%
TIML	2%	2%
Carphone Warehouse	2%	2%
UBC	1%	1%
Ford Motor Company	1%	1%
Hopstar Ltd.	1%	1%
Sabras Sound Ltd.	0%	0%

[140]

2. Global Radio/GCap Media closes digital stations

Digital stations *Capital Life* and *TheJazz*, both of which had been carried on the national Digital One DAB multiplex, were closed on 31 March 2008, the day that Global Radio acquired GCap. (GCap had already closed another national digital station *Core* in January 2008).

In a recent interview, Tony Moretta, chief executive of the Digital Radio Development Bureau, tried to explain the closures of these stations: *"Well, the main stations that went away – aside from all the Channel 4 stuff, which never launched and was nothing to do with DAB – where the GCap stations, such as* The Core *and* thejazz *also had nothing to do with digital* [sic]."[141]

3. Global Radio turns digital station *The Arrow* into music jukebox

In December 2007, Global Radio dropped live presenters from the digital radio station *The Arrow* which it had acquired from Chrysalis Radio. *The Arrow* was removed from DAB in London in May 2008, and is now only available over-the-air on the 5 MXR regional DAB multiplexes. However, Global's recent sale of its share in these multiplexes to Arqiva puts a question mark over the station's future. Why would Global Radio pay Arqiva to carry a digital station in which it has demonstrated no interest to develop?

4. Global Radio does nothing with digital station *Chill*

Part of Global Radio's acquisition of GCap Media, *Chill* is also only available over-the-air on the 5 MXR regional DAB multiplexes (and not in London on DAB). Like *The Arrow*, *Chill*'s future looks very precarious. However, it would prove embarrassing to close these two digital stations before Lord Carter's final Digital Britain report is published.

5. Global Radio cancels deal with Sky for digital news radio station

In October 2007, Global Radio cancelled the contract with Sky inherited from its acquisition of Chrysalis Radio that would have created a national *Sky News Radio* station on DAB. A spokesperson said then that *"Global was not prepared to make the necessary investment in this project."*

6. Global Radio scraps digital-only shows on *Galaxy Radio*
In January 2008, Global Radio dropped dedicated shows from the digital version of its *Galaxy Radio* brand, instead simply simulcasting its local FM output on DAB multiplexes that also carry it.

So what is going on here? Miron's speech is a large part of Global Radio's public campaign to cosy up to Lord Carter ahead of the publication of his final Digital Britain report. Global needs a big favour from Carter if it is to retain a shred of intrinsic value on its corporate balance sheet – an automatic renewal of its *Classic FM* national analogue licence. In return for the favour it seeks, Global is responding to Lord Carter's insistence that the radio industry speak with one voice on the issue of the transition from analogue to DAB radio.

The important thing here is to be seen to be saying the right things publicly about DAB – it's great, it's the future, we are committed to it, we love it. Forget the past. Forget our recent 'actions'. Conveniently forget that, less than a month ago, we transformed our company from the leading player in DAB infrastructure into less than an also-ran. DAB is the future – we are part of that future. Our commitment is to say all the right things, and probably to do absolutely nothing. The endgame is to persuade government to amend primary legislation so that Global Radio can hang on to *Classic FM*, as Ashley Tabor explained: "*It is one of those times when common sense has to prevail. Classic FM is a national treasure and to lose it would be tragic.*"[142]

The consumer and trade press willingly obliged by reprinting chunks of Miron's speech without any kind of critique. This ensures that the press cuttings, demonstrating Global Radio's glowing confidence in DAB, will land on Lord Carter's desk and, Global hopes, convince him of the 'common sense' of not bothering to auction the *Classic FM* licence to the highest bidder (which is required by existing legislation). Here is a selection of that press coverage.

Broadcast magazine reported that "*Miron's comments mark the first time that Global Radio – the largest commercial player in the UK radio sector – has come out so strongly in favour of DAB and migration*" under the headline 'Global Radio chief demands DAB deadline.'[143]

Radio Today reported that "*Global Radio has also called on the government this morning to set a switchover date for DAB*" under the headline 'Industry unites for a DAB future.'[144]

Marketing Week reported that Miron wanted the government "*to name a date for a switchover from analogue*" under the headline 'Radio industry needs to be bold, says Miron.'[145]

Media Week reported: "*Global Radio has made one of its biggest interventions in the debate over the future of digital radio, with chief executive Stephen Miron calling on the Government to set a date for digital radio switchover.*" The headline was 'Global boss Miron calls on Government to name digital radio switchover date.'[146]

The Guardian, to its credit, published the only report which acknowledged Global had "*sold its majority stake in national DAB platform Digital One to transmission business Arqiva earlier this month*", though its headline nevertheless read 'Government must be bolder on digital radio, says Global chief Stephen Miron.'[147]

But today's *Sunday Times* developed the theme by including this comment from Global Radio's Ashley Tabor about digital switchover: "*I am really confident now that all the right things are happening that will get us where we need to be. We are in favour of switch-off, so can we do it quickly please?*"[148] Maybe Lord Carter is tiring of Tabor's persistent phone calls, so Ashley is now having to turn to weekend press puff pieces to labour his point.

The *Sunday Times* article's headline, without a hint of irony, is 'Global evangelist for digital radio.' Closing digital stations, selling off DAB infrastructure, baling out of DAB development deals – is this some kind of 'do as I say, not as I do' evangelist?

---

[138] author's recording

[139] http://www.arqiva.com/press-office/press-releases/press-releases-2009/arqiva-to-take-full-ownership-of-digital-one-commercial-dab-mult

[140] source: Grant Goddard

[141] http://www.techradar.com/news/audio/hi-fi-radio/interview-the-future-of-digital-radio-591550

[142] http://business.timesonline.co.uk/tol/business/industry_sectors/media/article6211141.ece

[143] http://www.broadcastnow.co.uk/news/multi-platform/news/global-radio-chief-demands-dab-deadline/2022425.article

[144] http://radiotoday.co.uk/news.php?extend.4659.5

[145] http://www.mad.co.uk/Main/Home/Articles/97459421ef0d493aa2d476a6e8898658/Radio-industry-needs-to-be-bold%2c-says-Miron.html

[146] http://www.brandrepublic.com/News/901143/Global-boss-Miron-calls-Government-name-digital-radio-switchover-date

[147] http://www.guardian.co.uk/media/2009/apr/27/digital-radio-globalradio

[148] http://business.timesonline.co.uk/tol/business/industry_sectors/media/article6211141.ece

17.

13 May 2009

Digital radio: never mind the content, feel the bandwidth?

"It's a simple equation. The BBC has had an unfair share of the analogue spectrum but digital enables the commercial players the space to compete on a much more equal footing."[149]
Steve Orchard, operations director of GCap Media in 2006.

For almost an eternity, the UK commercial radio industry has complained vociferously that it has been discriminated against because the BBC has the use of more analogue spectrum than it does. The argument has been made repeatedly that commercial radio will always 'under-perform' against the BBC as long as the BBC is allocated more space on the FM waveband. To support this argument, its proponents hold up the fact that the BBC has four national channels on FM, whilst commercial radio has only one (they choose to ignore the fact that, additionally, the BBC has 40 local stations on FM, whilst commercial radio has 200+ local stations on FM).

When DAB radio arrived a decade ago, there was a widely held notion within commercial radio that the new technology provided an opportunity to even the score with the BBC. Whereas the government was unlikely ever to re-allocate analogue spectrum to provide equal amounts to the BBC and its commercial competitors, in digital spectrum the commercial sector pushed ahead with DAB (before the BBC did) and a successful 'land grab' rewarded it with much more DAB spectrum than the BBC. The prognosis was that, in the future, DAB would replace analogue usage, and that the commercial sector's dominance of digital spectrum would eventually reward it with the dominance over the BBC it craved.

It is difficult to say precisely how much more DAB digital spectrum the commercial radio sector has than the BBC. With DAB, there is a degree of flexibility because you have the choice to either use a section of spectrum for one station (in high audio quality) or for two or three stations (in lower audio quality). Commercial radio and the BBC each have one national DAB multiplex (though their coverage of the UK is not identical). Additionally, commercial radio has 46 operational local and regional multiplexes that cover the most populous parts of the UK. These multiplexes probably more than double commercial radio's superiority over the BBC in DAB spectrum. But then commercial radio also leases some space on its local multiplexes to the BBC for its local stations. This makes comparisons complicated.

Whatever the detail, it is obvious that commercial radio has control of far more DAB digital spectrum than does the BBC. To compound the situation, commercial radio also has control of far more Freeview digital radio spectrum than does the BBC. So, as had been hotly anticipated a decade ago, surely by now commercial radio must have the upper hand over the BBC in digital radio listening. The answer is 'yes' – commercial radio had almost been winning the digital race – and 'no' – it is no longer. In fact, the latest RAJAR data show that commercial radio's share of digital listening (40.5% in Q1 2009) has fallen below its share of analogue listening (41.6% in Q1 2009) for the first time.[150]

COMMERCIAL RADIO: HOURS LISTENED BY PLATFORM (%)

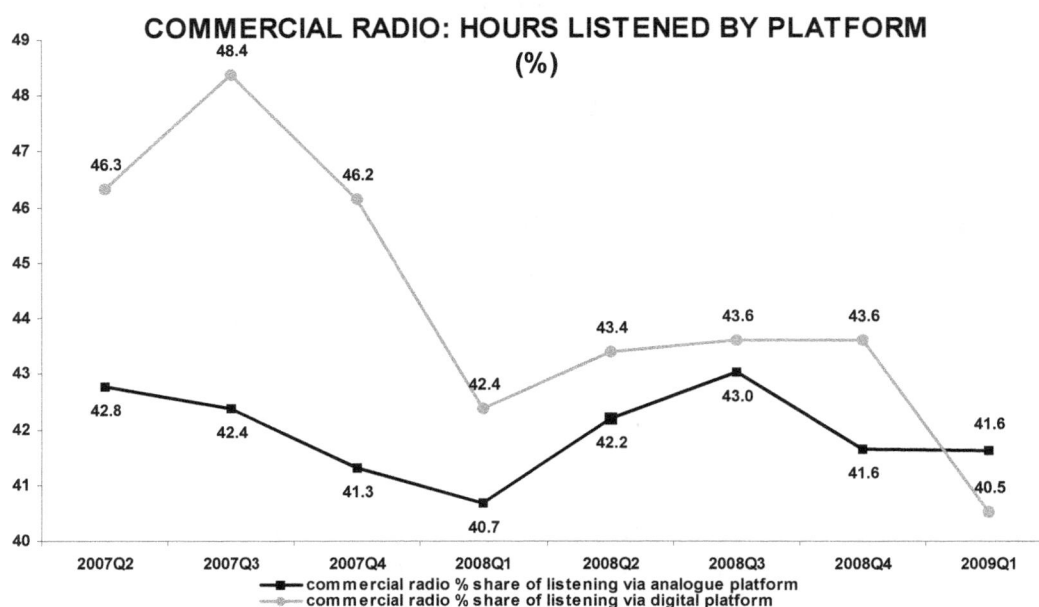

Chart axis values: 49, 48, 47, 46, 45, 44, 43, 42, 41, 40

Digital platform line: 46.3, 48.4, 46.2, 42.4, 43.4, 43.6, 43.6, 40.5

Analogue platform line: 42.8, 42.4, 41.3, 40.7, 42.2, 43.0, 41.6, 41.6

X-axis: 2007Q2, 2007Q3, 2007Q4, 2008Q1, 2008Q2, 2008Q3, 2008Q4, 2009Q1

■ commercial radio % share of listening via analogue platform
◦ commercial radio % share of listening via digital platform

151

These data cover all digital listening to all stations available on digital platforms (including simulcasts of analogue stations). However, because of the RAJAR methodology, the data do not include time-shifted listening to 'listen again' and 'podcast' radio content. These are both areas in which the BBC offers far more content (and markets it much more heavily) than does commercial radio. If it were possible to incorporate this time-shifted listening into the above data (which it is not), it is likely that commercial radio's share of listening would be much lower than its present 40.5% via digital platforms.

The long-held belief that commercial radio would somehow automatically win the war with the BBC on digital spectrum purely because it controlled more spectrum had always been mistaken. This belief assumes, somewhat bizarrely, that each consumer randomly spins their radio dial and then leaves it on whatever frequency the radio has landed on. Only by utilising such a random system of selection would usage ever be proportionate to the amount of spectrum. Unfortunately for the commercial radio sector, consumers are not mindless idiots. Anyone endowed with an Economics GCSE can easily see the gaping holes in this notion. Apparently few in the commercial radio industry could.

Consumers make choices and the radio station they decide to listen to is the one from which they expect to derive the most 'utility'. This is why 'content is king'. This is why *BBC Radio Two* and *Three* both use equal amounts of spectrum, but the former has a 16% share, and the latter 1%.[152] And this is why one fantastic radio station will always attract more listening than any number of mediocre ones (viz *Atlantic 252*, *Laser 558*, *Luxembourg 208*). It is not about how much spectrum you occupy, but about what you do with it. Consumers are motivated to listen by your content, not by your spectrum.

For commercial radio, after a decade of trying to convince itself and others that its abundance of digital spectrum would somehow entitle it to automatically trash the BBC, the dream (and it was a dream) is now over. Belatedly, it is back to the drawing board. As the BBC, *PrimeTime*, Bauer and *Planet Rock* have demonstrated, if you put some content on digital radio that consumers want to listen to, then they will listen (if they are made aware it exists through a marketing campaign). Digital radio would have a lot more listeners today if that simple truism had been understood by more players in the commercial radio sector a decade ago.

---

[149] Music Week, 9 December 2006, p.10
[150] source: RAJAR
[151] source: RAJAR
[152] source: RAJAR

18.

1 June 2009

Exclusive digital radio content: saying it and doing it are two different things

Everyone seems to agree – it is the availability of exclusive radio content on digital platforms that will drive consumer uptake of the hardware and digital listening.

In its Final Report, the Digital Radio Working Group had said in December 2008: *"We must present a compelling [DAB] proposition for consumers not only through new content, but in building a whole new radio experience."*[153]

In its Interim Report, Digital Britain had said in January 2009: *"We will expect the radio industry to strengthen its [DAB] consumer proposition both in terms of new and innovative content and to take advantage of the technological developments that DAB can offer."*[154]

In its report commissioned for RadioCentre, Ingenious Consulting had said in January 2009: *".... there is not as much DAB-only material as hoped, and very little that's truly compelling – there's no 'must have' content as with sports and movies on Sky [TV]."*[155]

In its submission to Digital Britain, Ofcom had recommended in March 2009 *"the creation of new commercial radio stations to create a consumer proposition analogous to Freeview: a wide range of popular and niche services, delivered digitally."*[156]

The Digital Radio Working Group had spent a year meeting throughout 2008 and made its final recommendations in New Year 2009. Five months later, for the consumer turning on their DAB radio, the choices do not seem much different than they were then. While the industry continues to talk and talk and talk and talk endlessly about what should be done, the consumer proposition for digital radio seems to be disappearing down the tubes. The data from the Q1 2009 RAJAR audience survey demonstrate that.

For commercial radio, its digital stations are now capturing a lower proportion of its listening (4.5%) than a year ago (5.5%). Only 23% of listening to commercial radio via digital platforms is to exclusively digital content, compared to 30% a year ago.[157] These results are not surprising, given the closure of many digital stations during 2008 (*Core, Oneword, Life, TheJazz, Virgin Radio Groove, Yarr, Easy, Mojo* and *Islam Radio*). In 2009 so far, Stafford's *Focal Radio* and London's *Zee Radio* have also closed.

For the BBC, the results are almost as disappointing. Its digital stations have recovered from a poor performance last quarter, but it appears that much of this improvement may have been due to heightened public interest in *6Music* following the Ross/Brand affair. BBC digital stations now capture 2.9% of listening to the BBC, compared to 2.7% a year ago. Only 14% of listening to the BBC via digital platforms is to exclusively digital content, compared to 16% a year ago.[158] For the BBC, it is beginning to look as if interest in its digital content is no longer growing as it had been during 2006 and 2007.

COMMERCIAL RADIO: LISTENING TO DIGITAL-ONLY RADIO STATIONS

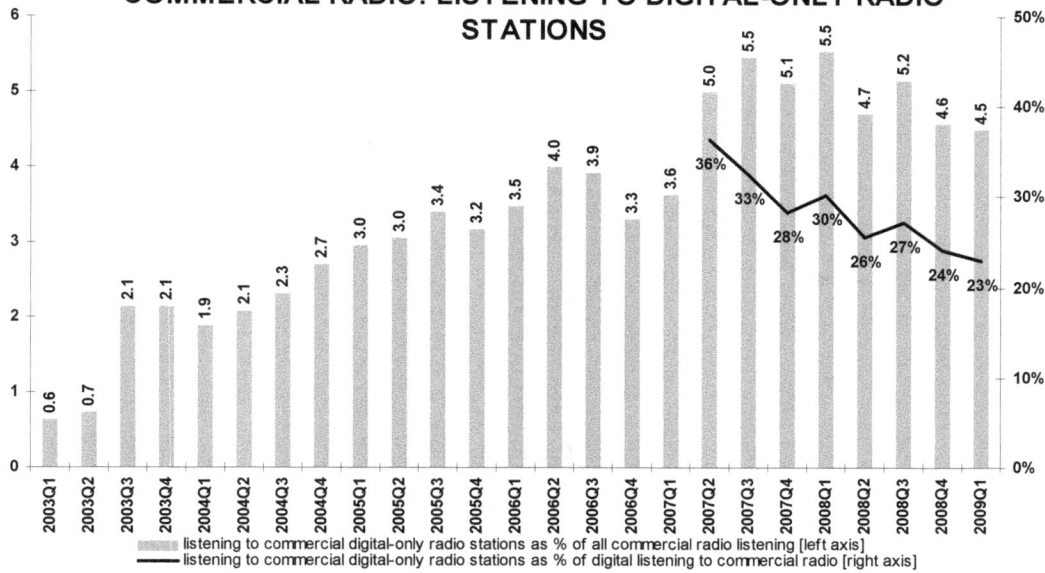

listening to commercial digital-only radio stations as % of all commercial radio listening [left axis]
listening to commercial digital-only radio stations as % of digital listening to commercial radio [right axis]

159

BBC RADIO: LISTENING TO DIGITAL-ONLY RADIO STATIONS

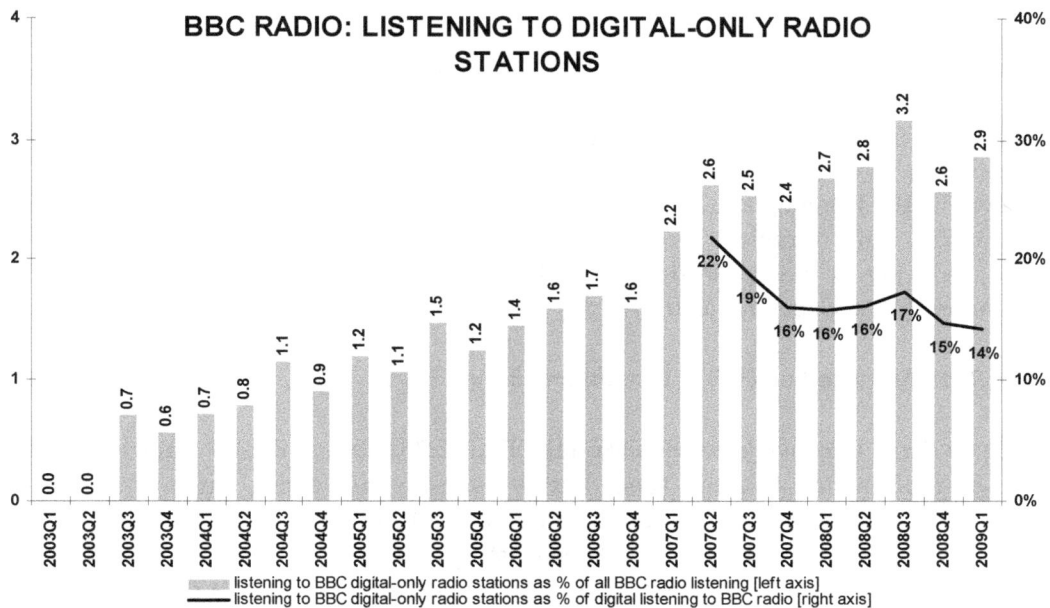

listening to BBC digital-only radio stations as % of all BBC radio listening [left axis]
listening to BBC digital-only radio stations as % of digital listening to BBC radio [right axis]

160

The summary graph (below) of hours listened to exclusively digital radio stations demonstrates the trend's recent tendency to have levelled out, primarily as a result of commercial radio's performance since 2007, but now also as a result of the BBC's performance in recent quarters. Whilst commercial radio experienced significant station closures in 2007/8, the BBC's portfolio has remained constant and is receiving as much cross-promotional marketing exposure as ever.

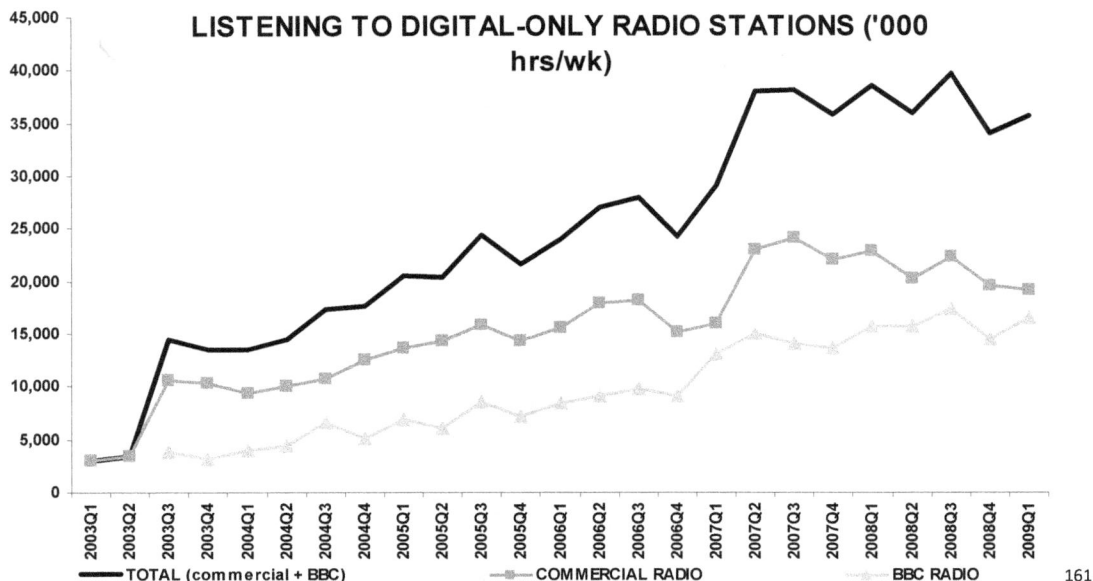

LISTENING TO DIGITAL-ONLY RADIO STATIONS ('000 hrs/wk)

TOTAL (commercial + BBC) ——— COMMERCIAL RADIO ——— BBC RADIO ——— 161

It is true that some new initiatives to provide exclusive digital radio content have happened in recent months:

- *Colourful Radio* launched on DAB in London on 2 March 2009.

- *BFBS Radio* is available nationally on the Digital One DAB multiplex from 20 April 2009. The station is government funded and aimed at British forces and their families. Unfortunately, listening to *BFBS* by the general public is likely to substitute for either commercial radio listening, reducing its ratings and revenues, or substitute for BBC radio, reducing its ratings. In the end, neither result will help commercial radio or the BBC make DAB a successful platform.

- *NME Radio* launched on DAB in London on 13 May 2009.

- *Amazing Radio* is available nationally on the Digital One DAB multiplex from 1 June 2009 on a six-month trial. Amazing Tunes is a UK website showcasing unsigned bands and musicians.[162] This is a great idea for an on-demand internet service but I am not sure this content will prove so appealing as a broadcast station. The problem, as *Xfm* discovered with its own disastrous experiment two years ago, is that listening to a playlist chosen by listeners can be as entertaining as looking through a relative's 300 holiday snaps. Out of several million people's playlists on Last.fm, I find there are no more than a handful of other people's selections that I can sit through. What works well online for Amazing is not necessarily going to work in the broadcast medium.

However, at the same time:

- Bauer Radio has relocated *Q Radio* from London to Birmingham, and *Heat Radio* from London to Manchester, effectively downgrading these digital stations and making redundancies.

- Bauer Radio has removed five stations (*Kerrang!, The Hits, Q, Heat, Smash Hits*) from the Sky platform.

These downgrades are significant because Bauer is easily the biggest player in digital radio, now that Global/GCap/Chrysalis has sold/closed all but two of its digital stations, both of which (*The Arrow* and *Chill*) survive only as music jukeboxes. Commercial radio's commitment to exclusive digital content seems to be hanging by the barest of threads. If Lord Carter decides not to respond positively to the commercial radio industry's demands for some kind of financial support in the Digital Britain report published in a fortnight, that thread is in imminent danger of snapping.

And so the talk about the need for exclusive digital radio content is likely to run and run and run. But, as long as it remains talk rather than significant action, consumers will remain unimpressed and the graphs above will continue their present trajectories. Nobody wants this to be the outcome, but nobody seems to be doing anything concrete to stop it happening.

---

[153] http://www.culture.gov.uk/images/publications/DRWG_Final_Report.pdf
[154] http://www.culture.gov.uk/images/publications/digital_britain_interimreportjan09.pdf
[155] unpublished report
[156] http://www.ofcom.org.uk/radio/ifi/radio_digitalbritain/digitalbrit.pdf
[157] source: RAJAR
[158] source: RAJAR
[159] source: RAJAR
[160] source: RAJAR
[161] source: RAJAR
[162] http://www.amazingtunes.com/

19.

10 June 2009

Digital radio switchover: searching for the 'credible plan'

Ed Richards, Chief Executive, Ofcom [**ER**]
Q&A @ Radio 3.0 conference, London [excerpt]
21 May 2009[163]

Q: *Isn't the big issue with DAB '[FM] switch-off'?....*

ER: *It is one big question but it definitely isn't the only big question. And the difference with TV is very instructive. One of the profound differences with TV, of course, is that in the case of TV you couldn't extend Freeview digital television without turning off the analogue spectrum, and that's a profound difference. One of the other differences, of course, is that the value of the spectrum released by analogue switch-off in television is extremely high. Indeed, people are fighting each other metaphorically to get hold of it and have been ever since we mooted the idea some years ago. So there are some very big differences. The other obvious differences are that people have more radios than they do have TV's, and so on and so forth. It is a very big question but I don't think it's the only one. That is why we put as much emphasis on the inherent sustainability and viability of digital [radio] services. It is always going to be asking people a lot to simply look forward, especially in the context of no switch-off date – and even if there was a switch-off date, it would be some years away – it's always going to be asking them a lot to take losses for a long period. If only we can get to a point where DAB services are essentially at least breakeven, the better because that gives you a base from which to plan the other more challenging things, which include switch-off, and we want to work quite hard at that alongside the debate about Digital Britain.*

Q: *Without a date, it feels like it's almost over the horizon. People I talk to in radio nearly all say 'what we need is a date'. Is Digital Britain going to give us a date, do you think?*

ER: *That is a common theme that you hear, it is true. Before answering the thrust of that, I reiterate that I think you need to address the date and the migration issue, but you need to address the underlying economics first and immediately at the same time. And that means a frequency plan, savings in transmission, and so on and so forth, and it means continuing growth and more listeners on DAB. I hear everybody, a lot of people, say that we have to have a date. Will Digital Britain give us a date? I don't know. There are a number of things we don't know about Digital Britain yet.*

Q: *Would it be helpful if it did give us a date?*

ER: *It depends what the date was. It wouldn't be helpful if the date was next year. I think the most important thing is ... Let me rephrase the question slightly. You can only have a date if you have got a credible plan that delivers that date. So I could give you a date now but it would be meaningless. It would be rather like the television switchover date in one or two* *countries around the world – which I won't name because it would get me into trouble – but they name dates, the governments stand up and puff up their chests and name dates but they are meaningless and, as soon as they have left the room, everybody laughs. So a date is meaningless without a credible plan to get there, so I recognise ….*

Q: *It's a bit 'chicken and egg', isn't it?*

ER: *Well, you have to have one in order to have the other. I think where people really are on this is, when they say we must have a date, that is another way of saying we must have a credible plan which gives us a date, and I would agree with that ….*

Q: *But how close are we to a credible plan?*

ER: *We are getting closer. We are doing a lot of work, as I said in my contribution, around the re-planning and I think the re-planning is very important to it. We need to have a clear set of proposals about quality of service and coverage and all those sorts of things, and those things need to be in place before you can have a credible plan. But there is work actively taking place on that and being driven forward. But better to get that right and to have a sense of urgency and determination, than just to pluck a meaningless date out of the air.*

[edited]

Q: *I'm still struggling slightly with [FM] switch-off only because it strikes me that almost everything hinges upon this and what you say is perfectly sensible – you can't really have switch-off until you have a credible plan – but we know that, in the real world, unless we are forced by one thing or another, we don't actually face this and businesses are very similar to life and everybody is still hedging their bets on FM. I speak to mobile phone manufacturers who say 'well, look, we only have room in our phones for so many transmitters and receivers. We have got Bluetooth, we have got infra-red, we have da-da-da-da-da and all our users tell us they really value FM'. So they are not going to switch it until they have to. People with DAB radios in their cars are still a rarity and the manufacturers are not going to start installing them as standard until someone says 'OK, 2013, 2012, 2011, whatever it is – that's it'.*

ER: *Well, that's the attraction of setting a date and driving everyone to it. But I'm trying to think of something different to say than what I said earlier.*

Q: *Would you favour it as an option?*

ER: *If there is a credible plan, yes. You've got to have a credible plan. And what you can't do is just pluck a date out of the air and say 'we're all going to get there' because I know what will happen under those circumstances. What will happen is that it will be fine for about a month and then, going for coffee outside the conference room, everyone will say 'well, that is not going to happen, is it?'*

Q: *Except in TV, it has and it is.*

ER: *In TV, we wrote the original document which said 'we will push on to digital switchover in this timeframe and here is how you can do it'. We wrote that document and said 'these are the six of seven things you have to do to deliver it' and we knew what you had to do to re-plan, we knew what you had to do to lead people across, we knew about the re-tuning, we* *knew the vast majority of things and there was a plan. That plan was then picked up by the creation of Digital UK, and so on. We've got to get to that next step, so I think it's an exciting prospect but we've got to believe that it's credible and deliverable. So I know that I'm repeating myself and not being particularly helpful but I do genuinely believe that and we need to – senior people in the industry need to sit round, look at this, stare at the steps and say 'will that deliver it, is it consistent with what is in the audience's interest?' There's no point in doing something which audiences then regard as a disaster. We have to do something that audiences, as it took place, will regard as a good thing. That's an acid test and I think that's possible, but there's a lot of work to do and we've got to see if we can get there.*

---

[163] author's recording

20.

17 June 2009

Radio in Digital Britain: sense and sensibleness

In the 13-page radio section of the Digital Britain Final Report published yesterday, there was not one mention of the word 'switchover' in the context of 'digital radio switchover'.[164] Neither was there a single mention of the word 'switch-off', as in 'FM radio switch-off'. Throughout the document's radio section, the new buzz phrase is 'Digital Radio Upgrade', meaning a drive to make DAB radio better and improve its consumer take-up. In Digital Britain, the notion of switching off FM radio broadcasting, notably for local stations, has been buried for good.

Not that you would have realised this fundamental policy shift by reading some of the press reports.

'FM radio switched off by 2015' said the headline in The Telegraph.[165]

'Government sets 2015 as digital radio switchover date' said the headline in Media Week.[166]

'Digital radio switchover set for 2015' said the headline in Broadcast.[167]

'Analogue radio switch-off set for 2015' said the headline in The Guardian.[168]

These bold press assertions are contradicted by the Report's recommendations that "*FM spectrum is to be re-planned to accommodate the current MW services*" (paragraph 43) and that "*a new tier of ultra-local radio [which] will occupy the FM spectrum*" (paragraph 39). The report is perfectly clear that FM is not to be switched off (at least, not in my lifetime).

It was almost as if the lobbyists for FM switch-off – the large commercial radio groups, most notably Global Radio – had written the press headlines the way they had wanted the outcome, regardless of the actuality. This was reinforced by an article that appeared in Media Week yesterday morning – only hours before Digital Britain was published – in which "*a well-placed source*" predicted "*a schedule for the shutdown of FM radio*" under the headline 'Digital Britain to give radio licensees guaranteed protection.'[169] That source proved not to be so well-placed.

The Media Week headline referred to the owners of the three national commercial stations who had been lobbying to have their licences extended by another term in order to avoid the impending auction of their frequencies, as required by existing legislation. I have written previously about Global Radio's determination to seek an automatic renewal of its *Classic FM* licence, which otherwise expires in September 2011. So did Digital Britain give Global, TIML and UTV the renewals that they wanted?

The answer appears to be both 'yes' and 'no'. Digital Britain will:

- extend all commercial radio licences, national and local, *"up to a further seven years"* for stations that simulcast on DAB
- insert a two-year termination clause into all new licences
- review all licences in future and determine whether the Digital Radio Upgrade is likely to be achieved by the end of 2013
- terminate licences if the Digital Radio Upgrade is not achieved
- then re-advertise the national licences under the existing auction scheme.

Not only does this add considerable strings to licence extensions of *"up to"* seven years, not only does it allow those extended licences to be terminated at two years' notice, but it also puts the onus squarely on the licensees to make sure that the DAB platform succeeds (something which has not been achieved in the last decade). If the Digital Radio Upgrade does not hit its targets, the licensees lose their stations. This is a poker game that, whilst offering national stations a potential second life, also threatens to take that life away not so far down the line. For an owner trying desperately to convince its bank lender of the long-term value of its national commercial radio licence, Digital Britain has not offered anything in the way of future guaranteed revenue streams. As a result, indebted radio owners now have two guns pointed at their head – one from their bank manager and the other from Lord Carter.

Worse, even the licence renewals proposed by Digital Britain require new legislation to be enacted. If there is renewed turbulence in government, and with the ever-present threat of a snap general election, it is looking doubtful whether media legislation will be a priority in a Parliamentary timetable that will be rushing to legislate more significant political issues during this government's final days. If new legislation doesn't happen soon, then Ofcom will have to rush to advertise the *Classic FM* licence in an auction by early 2010 at the latest.

Furthermore, even if digital platforms do succeed in accounting for more than 50% of radio listening by the end of 2013, which station owner (either commercial or BBC) is going to be prepared to switch off their analogue signal and lose 50% of their listening at a stroke? In the case of a commercial station, losing 50% of listening would mean losing 50% of revenues, an idea that nobody will entertain. In this way, regardless of the speed with which the 50% criterion is reached, the outcome is the same – stations will have to simulcast on both analogue and digital broadcast spectrum for many years to come, a necessity that is almost doubling transmission costs during a period when sector revenues are falling precipitously.

For smaller local analogue radio stations, the future remains rather unclear. Another Digital Britain proposal (paragraph 26) to amalgamate local DAB multiplexes into bigger geographical units makes sense in order to bring economies of scale to multiplex owners, but unequivocally transforms DAB into a large-scale broadcast platform for national or regional operators. A local analogue station in Bridlington, for example, will find it even more expensive and inefficient to be on a 'Yorkshire' multiplex, thus restricting that local station's future distribution platforms to FM broadcast and online. Neither will such a local station benefit from the automatic analogue licence renewal promised only to stations simulcasting on DAB. If anything, such stations' predicament will ensure that FM continues to be the consumer platform for local radio, which still accounts for 40.7% of all radio listening.[170]

Digital Britain's acceptance of the important citizen benefits of local radio broadcasting is underlined by its (unexpected) proposal to license *"a new tier of ultra-local radio"* on FM and to re-plan the FM waveband if existing stations (ever) migrate from FM to DAB. Although the report is at pains to explain that it does not intend to *"blur the lines between commercial and community stations"*, it makes sense in the long run to consolidate a third tier of radio with the flowering of a whole new set of radio stations that genuinely want to serve local communities. With many small local commercial stations now barely breaking even, it might make sense to turn some of them into companies limited by guarantee and thus let them seek public subsidy from local councils and regeneration schemes.

Such an expansion of radio content in local markets could potentially invigorate the entire radio medium, making 'local radio' more of a 'must have', particularly following cutbacks in local news provision by local newspapers and regional television. It is also a potential antidote to the continuing transformation of many of our former local commercial radio stations into regional or quasi-national services. As Digital Britain commented: *"Today's radio industry has been shaped more by the scarcity of the analogue spectrum than by market demand"* (paragraph 4).

On the issue of public subsidy, the biggest disappointment for commercial DAB radio owners/operators must be Digital Britain's insistence that "the *investment needed to achieve the Digital Radio Upgrade timetable will, on the whole, be made by the existing radio companies*" (paragraph 44). The report acknowledges that *"this will require a significant contribution from the commercial operators"* (paragraph 21) but suggests it should be funded by:

- savings from the negotiated 17% reduction in transmission charges as a result of the Arqiva/National Grid Wireless merger (paragraph 22)
- future savings from the ending of simulcast analogue and DAB transmission (paragraph 22)
- cost savings from the anticipated relaxation of co-location rules and the automatic extension of analogue licences (paragraph 25).

Although there is a brief mention of *"residual access"* to some of the funds left over from the BBC's Digital Switchover Help Scheme being used to support DAB infrastructure build-out, the overwhelming message is 'you guys are on your own to make DAB work'. The worry is that, when times were relatively good in the late 1990s/early 2000s, commercial radio did not manage to develop sufficient traction for the DAB platform. How is it ever going to succeed now in an environment where sector revenues are falling so rapidly?

So the conundrum continues, same as it ever was. Everybody wants DAB to work. Nobody except the BBC wants to pay for it. Commercial radio simply isn't making a profit anymore. We can argue about how/why it got to that desperate situation, but nothing changes the fact that there is no surplus cash slopping around ready to invest in either DAB infrastructure or exclusive digital content. Without an ongoing commitment to both, even the limited migration of national radio services from analogue to digital transmission proposed in Digital Britain is unlikely to ever happen. Consumers follow content, not platforms (or, as Digital Britain says: *"consumers will adopt new technologies when they are affordable and the benefits are clear"* (paragraph 8)).

This is not at all to imply that Digital Britain does not offer a lot of sensible recommendations. Whereas the outcome of the Digital Radio Working Group in December 2008 was a remarkably theoretical report that appeared to bypass the harsh economic realities of the radio sector, the Digital Britain document is realistic and pragmatic, telling the radio sector that much of what it needs to do to make the DAB platform a success is in its own hands. How the radio sector moves forward with these issues in the coming weeks will determine how much further we continue to plod along the long DAB road. There is an increasingly stark choice for commercial radio – to give up now and accede the DAB platform to the BBC and Arqiva, or to press on and further endanger the viability of the entire commercial radio sector.

Lord Carter proffered a lot of home truths in Digital Britain and he threw down this gauntlet: *"Any good business will invest in its future if it understands that future and the potential returns from its investment"* (paragraph 8). What he did not do was throw commercial radio a map to get it to the buried treasure.

On a purely personal level, I was pleased to see Digital Britain embracing several policies I had advocated for the radio sector:

- the two-year pilot scheme for an output focused radio regulatory regime takes up the idea of the Local Impact Test I proposed in November 2007[171]
- the proposal to use the surplus from the Digital Switchover Help Scheme and the savings from the Arqiva/NGW merger for DAB infrastructure build-out was a strategy I suggested in October 2008[172]
- the notion that 'localness' will prove a commercial radio station's Unique Selling Point in the future global media village is a scenario I have included in client briefings and conference presentations for several years.

For the purpose of transparency, I contributed radio sector analysis to two documents that were part of the Digital Britain process – a pre-consultation overview[173] and the regulation of local radio.[174]

[164] http://www.culture.gov.uk/images/publications/digitalbritain-finalreport-jun09.pdf
[165] http://www.telegraph.co.uk/finance/newsbysector/mediatechnologyandtelecoms/digital-media/5552937/FM-radio-switched-off-by-2015.html
[166] http://www.brandrepublic.com/News/913660/DIGITAL-BRITAIN-Government-sets-2015-digital-radio-switchover-date
[167] http://www.broadcastnow.co.uk/news/radio/digital-radio-switchover-set-for-2015/5002536.article
[168] http://www.guardian.co.uk/technology/2009/jun/16/digital-britain-analogue-radio-switchoff
[169] http://www.mediaweek.co.uk/news/913266/Digital-Britain-give-radio-licencees-guaranteed-protection
[170] source: RAJAR Q1 2009
[171] Grant Goddard, "UK Commercial Radio: A New Way To Regulate 'Localness'", unpublished report, November 2007.
[172] http://www.endersanalysis.com/publications/publication.aspx?id=620
[173] http://www.culture.gov.uk/images/publications/digital_britain_interimreportjan09_annex1.pdf
[174]

http://www.culture.gov.uk/images/publications/An_Independent_Review_of_the_Rules_Governing_Local_Content_on_Commercial_Radio.pdf

21.

23 June 2009

Digital Britain: is the 50% criterion for digital listening achievable?

Both Digital Britain[175] and the Final Report of the Digital Radio Working Group[176] which preceded it have placed considerable emphasis on one performance metric – the date when the proportion of listening to all radio via digital platforms surpasses 50%. This date will be a 'trigger point' for policy changes that will impact the entire radio broadcast industry and it is therefore important to ensure that the data used to determine it are entirely correct.

Figure 6 on page 93 of the Digital Britain Final Report is a graph that shows three things:

- historical data for the share of all radio listening listened to on digital platforms from 2005 to date
- a trendline demonstrating how this share would continue to increase through *"organic growth"*
- a forecast demonstrating how this share would grow faster if a *"drive to digital"* were to be pursued.

In the Digital Britain graph, the historical data for digital platforms' share of radio listening is shown as 7% at year-end 2005, 12% at year-end 2006, 17% at year-end 2007 and 20% at year-end 2008.[177]

However, the historical data I have from RAJAR (the radio industry's official radio ratings body) show these figures as 11.0% for year-end 2005, 12.5% for year-end 2006, 16.6% for year-end 2007 and 18.3% for year-end 2008.[178]

So there may be minor differences in the precise numbers for each year, but is that really such a big deal in the overall scheme of things? Well, in this case, yes it is. Whilst the numbers look relatively close on paper, it is only when you draw them onto a graph that you can see the significant differences.

In the Digital Britain report, the trendline for continuing 'organic growth' demonstrates that the important 50% criterion would be reached in 2015. The government is proposing that, through a concerted *"drive to digital"*, that date could be brought forward to 2013. Left to its own devices, the 50% criterion would be reached six years from now, but concerted action could reduce that time to four years.

However, instead of using the data in the Digital Britain report, if a graph is constructed of the official quarterly data from RAJAR, the resultant trendline displays a noticeably less steep gradient. Using this industry data, the 50% criterion is unlikely to be reached through 'organic growth' until 2018. In this scenario, the government's concerted *"drive to digital"* would pose the challenge of reducing the interim period from nine years to four years.

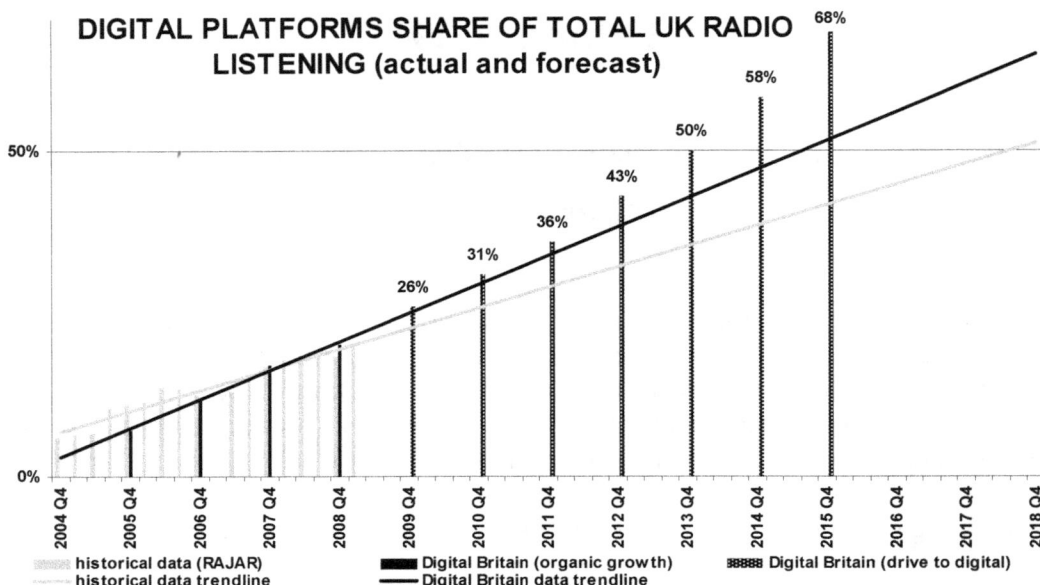

DIGITAL PLATFORMS SHARE OF TOTAL UK RADIO LISTENING (actual and forecast)

	historical data (RAJAR)	▬ Digital Britain (organic growth)	▦ Digital Britain (drive to digital)
	historical data trendline	— Digital Britain data trendline	

179

Whilst it might seem realistic in Digital Britain to propose reducing a six-year period to four years through concerted action to push digital radio, the reduction of a nine-year period to four years represents a considerably more substantial challenge for the industry to achieve. Some might say it could prove impossible.

The issue becomes even more critical for the commercial radio sector when you realise that the Digital Britain report threatens to revoke stations' licence renewals if the radio industry as a whole does not succeed in achieving this 50% criterion by 2013. In other words, the government is holding a gun to the radio industry's head – either achieve this specific goal by 2013, or you may lose your livelihoods.

This is what makes this single dataset so critical to the future of the radio industry. Is the 2013 goal a reasonable target that can be realistically attained, as Digital Britain argues, or is it unrealistic if the interim period has to be somehow slashed from nine years to four years?

When you agree to join a game of poker, you should always check first that the odds are not overwhelmingly stacked against you winning.

[NB: The trendlines in the above graph are straight-line trendlines generated automatically by Microsoft Excel from the datasets, not my subjective judgement.]

---

[175] http://www.culture.gov.uk/images/publications/chpt3b_digitalbritain-finalreport-jun09.pdf
[176] http://www.culture.gov.uk/images/publications/DRWG_Final_Report.pdf
[177] http://www.culture.gov.uk/images/publications/chpt3b_digitalbritain-finalreport-jun09.pdf
[178] source: RAJAR
[179] source: RAJAR and Digital Britain Final Report

22.

30 June 2009

Commercial radio in Germany and Switzerland reject DAB

The commercial radio industries in Germany and Switzerland have both rejected proposals that they should invest in developing the DAB digital radio system in their countries to replace existing FM/AM transmissions. The German argument against DAB was that the significant investment required simply did not justify the lengthy wait for a financial return, based on evidence from other European countries that have already introduced DAB radio.

This news is a blow to UK broadcasters and technological companies who have long hoped that the DAB system would become the pan-European digital radio broadcast standard. In June 2009, the Digital Britain Final Report had proposed the government would *"work with our European partners, including the European Commission, to develop a common European approach to digital radio."*[180]

This proposal drew on the work of its predecessor, the Digital Radio Working Group, whose Final Report had noted in December 2008 that *"Germany has plans to launch DAB+ across the country in 2009, while France will launch DMB audio services at around the same time."*[181]

Not only do the German and Swiss announcements impact the prospects of UK consumers benefiting from economies of scale that could have reduced the retail prices of DAB receivers. They also throw doubt over the willingness of European car manufacturers to install DAB radios in new cars, if the broadcast technology is still only implemented in a handful of countries.

A week ago, UK technology company Imagination Technologies, whose processors are used in over 80% of DAB radio receivers, had said that *"recent announcements from France, Germany, Denmark and Eastern Europe …. mean that the global market for digital radios and digital radio technology is due to take off."*[182]

Frontier Silicon, the UK's leading supplier of DAB radio chips, had announced a US$10m investment in production of a new advanced DAB chip at the beginning of 2009 and had noted that *"penetration of DAB radios in the year continues to rise, with ageing analogue broadcasting systems [due to be] switched off in Switzerland …."*[183]

The profitability of both companies is very dependent upon the uptake of DAB technology more widely than only their home market.

In Germany, the association of private broadcasters (VPRT) issued a statement on Thursday which said:

"The conditions required for a successful introduction, always a prerequisite, have not been met. … For VPRT's private radio companies, the significant initial and operating costs are too great. Against the backdrop of the economic crisis, such investments are a certain risk. … The VPRT member radio companies have, therefore, concluded that DAB+ has no economically viable future. Even with significant promotion of the system by public funds for at least the next five to ten years and under regulatory pressure, there is only a slim chance of partially recovering (the costs) within the market. Against this background, the VPRT speaks against the planned introduction of DAB+ in the autumn of 2009." [184]

The World DMB Forum, the international agency promoting the adoption of DAB technology, describes Germany as *"among the leading European proponents of DAB Digital Radio"* with 546,000 DAB radios sold to date and 116 different radio services available on the platform. [185] Its June 2009 update said that *"it is planned that by 2012 most of the German population will have access to the [DAB] services."* Without the co-operation of commercial radio operators, it now looks unlikely that this target will be met.

In Switzerland, the Association of Private Radios (VSP) issued a statement the same day as the Germans, which said: *"Today's ruling by the VPRT makes even more difficult the launch of DAB+ in the whole German-speaking world and VSP recommends that all members use realistic calculations before beginning."* [186] VSP said that the pursuit of DAB radio could create an additional cost of 5 to 8 million Swiss francs *"until break even is reached."* Whilst it acknowledged that such an investment could *"make sense for strategic market reasons"* for one or two players, for the rest of the commercial sector it felt that the financial requirements *"exceed the entrepreneurial risk."*

Switzerland presently has around 20 million FM radio receivers, but only 300,000 DAB receivers and an unknown quantity of newer DAB+ receivers. The commercial radio industry there noted that it anticipates greater competition for radio listening will derive from internet-delivered services. Both German and Swiss commercial radio have warned that a phasing out of FM technology would lead to lower revenues, reduced investment and fewer jobs in their companies, and would thus reduce diversity of media voices in their markets.

At the same time, elsewhere in Europe, the decision by the French government that every new car in France will have to include a digital radio from 2012 is looking increasingly challenging. At the recent EBU Digital Radio conference, it was revealed that the decision had been made by the Ministry of Industry without the benefit of prior consultations with technology companies. The French media regulator, the CSA, is only now meeting industrialists this month to discuss the urgent requirement to manufacture car radios by 2012 that include the T-DMB digital standard (a variant of DAB) adopted in France.

Although both the DAB+ and the T-DMB technologies are part of the DAB family of standards, the overwhelming majority of the 9m DAB radios purchased to date in the UK are unable to process either DAB+ or T-DMB signals and would therefore be of no use in Germany or France. Swiss commercial radio, meanwhile, has expressed more interest in using another technology, 'HD Radio', which is not part of this DAB family of standards but is the digital radio broadcast system already used in the US and which requires altogether different radio receivers.

---

[180] http://www.culture.gov.uk/images/publications/chpt3b_digitalbritain-finalreport-jun09.pdf
[181] http://www.culture.gov.uk/images/publications/DRWG_Final_Report.pdf
[182] http://www.imgtec.com/News/Release/index.asp?NewsID=467
[183] http://www.frontier-silicon.com/media/releases/09/0202_chorus3.htm

http://www.frontier-silicon.com/media/releases/09/0126_touchscreen.htm

[184] http://www.vprt.de/index.html/de/press/article/id/197/or/2

[185] http://www.worlddab.org/country_information/germany

[186] http://www.kleinreport.ch/meld.phtml?id=52281

23.

1 July 2009

DAB radio in cars by 2013? – "extremely challenging" say car makers

The UK association of car makers, the Society of Motor Manufacturers and Traders [SMMT], has cautiously welcomed the Digital Britain report but has expressed *"reservations about the ambitious timetable"* to ensure that DAB radios are available in all new cars by 2013.[187] It has also expressed concern about the 32 million vehicles already on the road, of which it says *"only a small percentage"* already have DAB radios fitted, noting that the timetable to fit them with *"aftermarket devices"* is *"extremely challenging."*

SMMT has emphasised that the government's ambition to accelerate the take-up of DAB radio will be *"contingent on all national and commercial broadcasters investing in content."* Its Chief Executive Paul Everitt said: *"The long-term challenge will be for the broadcasters to invest in content and coverage to create demand for these [DAB radio] products to be provided as standard."*

The commercial radio industry has yet to make explicit statements, in the wake of the Digital Britain report, as to how it plans to enhance its exclusive digital radio content to accelerate consumer interest in the platform, or how it plans to finance the build-out of necessary DAB infrastructure upgrades to improve UK coverage.

The Digital Britain report had set out a five-point plan to encourage take-up of DAB radio receivers in cars:

- to work with car manufacturers so that vehicles sold with a DAB radio are available by the end of 2013
- to support a common logo for DAB car radios
- to encourage the development of portable analogue-to-digital radio converters
- to promote the introduction of more sophisticated traffic information within DAB broadcasts
- to work with European partners to develop a common European approach to DAB radio in cars.[188]

The last of these points has already received a setback, following the decision last week of commercial radio in Germany and Switzerland not to commit investment to the development of DAB as a replacement platform for their existing FM/AM services. An announcement from Austria is anticipated soon.

Asked about the DAB situation with cars, Tony Moretta, Chief Executive of the Digital Radio Development Bureau [DRDB], the agency charged with marketing DAB in the UK, had said on BBC Radio 4's 'You & Yours' show last week:

"One of the things that has held back the car industry slightly with DAB in the UK is that the UK has been ahead of the rest of the world in going to digital radio. Now if you're a mainstream car manufacturer, you want to be able to manufacture a car with a radio that will work all around Europe. It's only been relatively recently that you've seen France and Germany and other countries commit fully to digital radio. And so the car manufacturers now have a common standard they can build a radio into their car and it can work across the whole of Europe. So you're starting to see a big change now. Most car manufacturers now offer DAB as standard in a car or as a factory-fitted option starting for as little as £55. So that's for new cars, and we saw the other day Ford and Vauxhall announce their support for Digital Britain's recommendations. What we are going to have to do is look at adapting those cars that haven't been changed by that point."[189]

The DRDB has cited the more enthusiastic Ford and Vauxhall responses to Lord Carter's Digital Britain report, but has not yet mentioned the considerably more *"cautious"* SMMT response. It should be noted that Ford has been a long time minority shareholder in the MXR regional DAB multiplexes, and thus would benefit financially from improved uptake of the DAB platform in the UK, whether in-car or otherwise.

In-car DAB radios are still a rarity in the UK:

- out of 2.4m new vehicles registered in the UK in 2007, only 20,000 buyers chose to install a DAB radio
- out of 34m cars on the road in the UK in 2007, it is estimated that between 170,000 to 200,000 had DAB radios fitted.

[187] http://www.smmt.co.uk/articles/article.cfm?articleid=19795
[188] http://www.culture.gov.uk/images/publications/chpt3b_digitalbritain-finalreport-jun09.pdf
[189] http://www.bbc.co.uk/iplayer/episode/b00l36gn/You_and_Yours_Call_You_and_Yours/

24.

2 July 2009

Funding DAB radio improvements: who pays?

At the Radio Festival[190] in Nottingham, the final session on Wednesday 1 July 2009 at 1215 was a discussion about the future of UK radio that was broadcast live on BBC Radio 4's Media Show[191] and hosted by Steve Hewlett. Part of the discussion was about DAB in the UK following the publication last month of the Digital Britain report.[192]

Amongst its range of proposals, Digital Britain had recommended:

- *"at a national level, we will look to the BBC to begin an aggressive roll-out of its [DAB] national multiplex to ensure its national digital radio services achieve coverage comparable to FM by the end of 2014"*
- *"where possible, the BBC and national commercial multiplex operator should work together to ensure that any new transmitters benefit both BBC and commercial multiplexes"*
- *"further investment is required if local DAB is ever to compare with existing local FM coverage."*

How will this improved DAB infrastructure be paid for? Digital Britain had suggested:

- in some geographical areas, *"the BBC will need to bear a significant portion of the costs"*
- *"however, the full cost cannot be left to the BBC alone"*
- *"some [commercial radio] cost-savings must support future [DAB] transmitter investment by the local multiplex providers"*
- *"the investment needed to achieve the Digital Radio Upgrade timetable will on the whole be made by the existing radio companies"*

Interviewed about these issues for the Media Show were:
- **Tim Davie**, Director of Audio & Music, BBC [**TD**]
- **Phil Riley**, former Chief Executive, Chrysalis Radio [**PR**]

Q: *Tim, where is this money coming from?*

TD: *The truth is that I can't say I can find it. What I have been saying very clearly is that I can make a case for it. And, where the money comes, or could come, from I think is pretty well articulated in public debate, which is… We have been spending money against broader digital distribution projects – the digital television switchover – and where we spend the Licence Fee beyond content, it's this thing called the ring fenced fund where we've been investing in digital television switchover. Now, as the radio guy, it's saying 'we have a case for this medium'. We love radio. We think there's a really good case for it being there as an investment ….*

86

Q: *This investment has got to happen pretty quickly to stand any chance of getting us to the 2015 date which the government have set us as their target for switching from analogue to DAB. That means quite a lot of things have got to happen by 2013. That's into the next Licence Fee settlement. So you need to find £100m for your 600 extra [DAB] transmitters, or whatever it is, in this settlement. Have you got it?*

TD: *Well, we have said that we don't think – and we're yet to see what that looks like because we haven't done TV switchover that ….*

Q: *Have you got the money? You have to start spending now, you can't leave it because [otherwise] you're never going to get there, are you?*

TD: *We've said that as part of Digital Britain – it's all in the report – it says that in the course of the next 12 months, even if we wanted to spend money at this point, we don't quite know what we are spending it on. Without getting too technical, if you look at the ….*

Q: *On 'Feedback', you've said 600 transmitters are needed to get to an equivalent coverage of FM and you said the BBC wouldn't go there unless coverage was roughly equivalent to FM.*

TD: *Specifically, the minority of money is those 600 transmitters that gets you on the national multiplex, which is what the big stations like Radio 4 go on, that gets you to 98% cover. The bigger money is in sorting out the regional and local stations which are a bit of a patchwork and that investment – the numbers are loose because we are going to be doing some detailed planning with the commercial sector on ….*

Q: *Very briefly, a one-word answer. Do you have any money set aside now to spend on this purpose?*

TD: *No.*

Q: *Splendid.*

………………………..

Q: *Does commercial radio have any money to spend on this proposal?*

PR: *If you read the Digital Britain report in its totality, there are a number of proposals for changing the way commercial radio operates, in terms of co-location and regional licences becoming national networks. Now, bringing all of that together as a piece, will that free up sufficient additional funds for the commercial sector to be able to roll out more digital? I don't know. You'll have to ask the other commercial players.*

Q: *What's your guess?*

PR: *'No' is the answer at the moment.*

Q: *Because one of the issues with DAB surely is that the commercial side of the equation has already, in commercial terms, failed. Increased costs, but no increased revenues. Not even Channel 4 was able to galvanise it to make it change. Is there a commercial model in DAB at all, do you think?*

PR: *I think DAB is a terrific platform. The die has been cast. 9 million sets, 20% of all listening. DAB is here to stay. So, we can't go back to not having DAB so actually we've got to go forward and we've got to go forward with as sufficient a pace as we can. My concern would be trying to go forward too fast and falling over ourselves.*

Q: *The government has said they want to do this in 2015. They have said that, by 2013, they won't press the button to switch off [FM] until …. They will give 2 years' notice. So, by 2013, I think they want 50% of listening to be on DAB, and 90%+ coverage of DAB across the country. Is that timetable in any way realistic? The BBC say they have no money set aside just now. You say that you don't. How's it going to happen?*

PR: *I think Tim [Davie] famously used the euphemism 'ambitious' yesterday and I think 'ambitious' is the right word for it. Personally, I can't see us getting to 2013 although, to be fair to the Digital Britain report, it says it will test it every year from 2013 and when we get there, then we will move on to Phase Two.*

Q: *Tim, lots and lots of listeners have contacted this programme and other programmes whenever they have been asked and are very very worried about this. They think they might have 3, 4, 5, 10 – 15 in one case – analogue radio sets and they have been asked to go through all the rigmarole of changing them and 'what for?' is the question they ask. 'Why are you asking me to do this? It's not broke, don't fix it'.*

TD: *If you look at the industry as a whole, you could argue that we are not ready for the future. Actually, although we have some fantastic services on-air now, we have just talked about commercial radio – their financial model looks pretty broken at this point.*

Q: *Isn't the key question 'content'?*

TD: *I think the case to the listener is really clear, which is – digital radio can present a much wider range of national stations, it can offer functional benefits. We've seen what that can bring in something like television. There is a real challenge for the industry to step up to the plate and deliver that content, and that has to happen. And, to be very clear, I am very worried, like the listeners, that if you have all these old [analogue radio] sets and there is no benefit, we should not be moving. What's happened in the last few weeks though, and months, is that the radio industry as a whole has said 'we're going to go for DAB and we're going to try the transition to digital'. We haven't said that it is actually happening until we've earnt that, which will be at a threshold level.*

---

[190] http://www.radioacademy.org/events/radio-festival-2009/radio-festival-09-running-order/
[191] http://www.bbc.co.uk/programmes/b00dv9hq
[192] http://www.culture.gov.uk/images/publications/chpt3b_digitalbritain-finalreport-jun09.pdf

25.

3 July 2009

DAB radio switchover: BBC listener opinions offer exit strategy

The BBC is in a tight corner over DAB. It played a significant role in developing the technology in the 1980s, in experimenting with the earliest DAB transmissions in the UK in the 1990s, and in launching a portfolio of exclusively digital radio stations in the 2000s. During that long period, management teams within the Corporation have come and gone, yet the commitment to DAB as a future technology to replace FM/AM analogue radio has remained resolute. Until now.

Realism eventually rears its ugly head, even in the BBC. And a changing of the guard at the top of the BBC radio division offers a timely opportunity to re-evaluate a strategy for DAB that must have been first decided almost two decades earlier. Across the meeting room conference table, the question is eventually asked by the newcomer – exactly why did we decide to commit so much time and so much money to DAB in the first place? The answers are many and various and have inevitably become muddled over time. The one thing that is certain is that nobody in the BBC could have believed back in the 1980s that we would still be, arguing **in 2009** as to whether implementation of DAB radio technology is worth the effort. Back then, the bright digital radio future looked attainable within a matter of years, rather than decades. How wrong they were.

The longer you have peddled away, the harder it is to stop and get off the bicycle. Having thrown decades of resources at DAB technology, it would be almost impossible for the BBC to say 'whoops, it didn't quite work out so we'll stop now'. The ire from DAB radio receiver purchasers, the backlash from Licence Fee payers, and the possibility of an incoming Tory government potentially using it as a stick with which to beat the Corporation for wasting money are all too horrible to consider.

So it was interesting to hear Tim Davie, Director of BBC Audio & Music since September 2008, on BBC Radio 4's 'Feedback' programme, ingeniously beavering away at building a potential DAB exit strategy by invoking the will of the listener.[193] As everyone working in BBC radio understands, its listeners are extremely resistant to change – almost however minor it is – and are not afraid to voice their opinions in the media at the slightest inconvenience. It was therefore appropriate that the 'Feedback' programme itself should be used to suggest that, if BBC listeners did not want to change over completely to DAB radio, then the BBC might decide it should not happen. Tim Davie said:

TD: *We support the idea of switchover to digital. In terms of the switchover date, our position has always been that 2015 is ambitious. We think that the listeners need to be reassured that coverage levels, quality levels are at a point where switchover is realistic. So we are totally focused on delivering a position where we have hit certain thresholds, we know that we are in a place where switchover can happen without widespread disruption.*

Q: *Are you going to make that judgement yourself or are you going to consult your listeners, many of whom dispute claims that are made by BBC spokesmen about the quality of reception and other things. Have you any plans to consult the audience about whether the time has come when switchover is possible?*

TD: *Absolutely. We are talking to government now about how consultation should take place. From a BBC perspective, whether it be 'Feedback' or our constant audience research, the idea that we would move to formally engaging switchover without talking to listeners, getting listener satisfaction numbers, all the various things we do, would be not our plan in any way. We would be – we are – in dialogue now for the next six years.*

Q: *But consultation implies the possibility of changing policy, and a lot of our listeners are sceptical ….*

TD: *I think we are pretty committed to digital. Having said that, **since I have arrived at the BBC, I certainly haven't seen it as inevitable that we move to DAB.** We do believe that, if radio doesn't have a digital broadcast platform, it will be disadvantaged. I'm pretty convinced of that logic. What I'm not saying is that we have to move at 2015 if we haven't delivered the thresholds – the right levels of listening to digital radio and to DAB. I don't think we are on a course that is unstoppable to 2015 although we are pretty committed to a DAB switchover over time.* [emphasis added]

Q: *Do you accept, at the moment, that DAB is often inferior to the existing [FM] sound?*

TD: *DAB doesn't have the coverage of FM at this point, and it's really straightforward that the quality of your audio is related to how close you are to a transmitter. So, DAB currently has less transmitters. So those people who are further away from a transmitter aren't getting as good sound. One of the things I've been very clear on in my position is – we will not even entertain a switchover unless the level of quality coverage is at 98%, which is in line with FM. So we, as the BBC have said, without the extra 600 transmitters that we would need to put in place, DAB switchover will not be a reality.*

In terms of BBC public pronouncements, these viewpoints on DAB are revolutionary. Under Tim's predecessor, Jenny Abramsky, public dissent about the DAB future was simply not permitted. Last year, after I had been interviewed for an item on Radio 4's 'Today' programme about the problems facing switchover to DAB, I never again heard a similar item about DAB on the show. Asked about the BBC's commitment to DAB at conferences, BBC staffers would look sheepish and admit they had been told to make no comments.

What a difference a year makes. The last ten days have witnessed a blizzard of managed dissent on BBC radio. The 'Today' programme yesterday morning ran a substantial piece in the important pre-0830 slot that was very critical about the pitfalls of DAB reception in cars. This week's 'Media Show' on Radio 4 devoted considerable time to the DAB issue. Last week's 'You & Yours' on Radio 4 discussed listeners' issues with DAB in gory detail. And the weekend's 'Feedback' has opened up the possibility of BBC listener revolt on DAB translating into a policy change.

It feels almost as if a subtle marketing campaign is now going on from within the BBC as a response to the radio proposals in the Digital Britain report, softening up the outside world for the BBC to be able to downgrade/dump DAB at some future time. Of course, Tim is a clever marketer from the real world (Pepsi, P&G), whereas his predecessor was a (very successful) career BBC apparatchik. What we might be seeing is the opening salvo of an action folder marked 'Possible DAB Downgrade/Exit Strategy'. The nuclear button might never have to be pressed, but it's always useful to know where the exit doors are and how you are going to reach them, however little you might want to think about the DAB plane going down in flames.

---

[193] http://www.bbc.co.uk/iplayer/episode/b00l63vl/Feedback_26_06_2009/

26.

5 July 2009

DAB radio switchover: the Conservative Party viewpoint

Jeremy Hunt, Shadow Culture Secretary, speaking on The Guardian's 'Media Talk' podcast:[194]

JH: *We support the idea of [digital radio] switchover. We have more concerns about [FM] switch-off. There are 120 million analogue radio sets and, if we were to tell consumers that, after 2015, those are going to be useless and you have got to chuck them away, people would I think be very angry. And so there's a lot of work that needs to be done before we can even think about switch-off.*

Q: *What sort of incentives do you think you could give to the public to be attracted to digital radio?*

JH: *I think the most important thing is not something the government can do, but something the industry can do, which is to develop new services on digital platforms that actually mean there is a real consumer benefit to DAB. At the moment, the benefits are marginal. I mean, there are some benefits in terms of quality, but your batteries get used up a lot more quickly, the reception is a lot more flaky, and a lot of the things that make digital switchover attractive on TV don't apply to radio in the same way. So I think the industry needs to do a lot more to make it in consumers' interests to have that switchover. That's one thing. I think what the government can do, though, is work much more closely with car manufacturers. The French government has bitten the bullet on this. I think we should do a lot more.*

Q: *The French government has mandated car manufacturers to put digital radios in cars. Should the British government follow suit?*

JH: *Given the French government has done that, there may be no marginal cost to car manufactures were the British government to say the same thing. But, at the very least, we should be looking at incentives to encourage car manufacturers to standardise on DAB because, until you do that, we are not going to get the network to the 97% or 98% coverage that we really need.*

Q: *What about all those old [radio] sets? You raised it in your speech here at the Radio Festival – old analogue sets that could become obsolete.*

JH: *Well, exactly, and there is an environmental consideration with that as well, because I think people would be very, very concerned at the environmental cost of having to get rid of 120 million sets. So I think we have got to think about that. We have also got to think about consumer anger. Consumers are people that the radio sector needs. It's going through a very tough patch. We don't want to switch off listeners by suddenly saying that we are not going to – that we are going to force you to have a new radio, and there's a real danger, if we do that, that they might start listening to their iPods and their CD players instead.*

Q: *You mentioned a possible swap scheme. How would that work? You take your old analogue radio into Currys and Dixons and get a shiny new digital one?*

JH: *Yes, I think this is something that I don't think is really for the government to do. But I'm just really putting it on the table. I think it's the kind of thing the industry might think about. If you could swap your analogue radio for a digital one, people might think 'wow, there's a benefit to switchover'. At the moment, we seem to be getting into this mindset where we want to force it on the public, even though the public can't really see what the benefits are.*

---

[194] http://www.guardian.co.uk/media/organgrinder/audio/2009/jul/02/podcast-digital-radio-festival-tim-davie

27.

9 July 2009

Paying for Digital Britain's 'Digital Radio Upgrade': who, me?

The Digital Britain Final Report[195] published in June 2009 proposed that the UK radio industry embark on a 'Digital Radio Upgrade' which would seem to involve (take a deep breath):

- providing greater choice and functionality for listeners (para.15)
- listeners who can currently access radio can still do so after Upgrade (para.15)
- building a DAB infrastructure which meets the needs of broadcasters, multiplex owners and listeners (para.21)
- redrawing the regional DAB multiplex map (para.21)
- the BBC beginning *"an aggressive rollout"* of its national DAB multiplex to ensure its coverage achieves that of existing FM by 2014 (para.23)
- commercial radio to extend the coverage of its national DAB multiplex and to improve indoor reception (para.21)
- investment to ensure that local DAB multiplexes compare with existing FM coverage (para.24)
- the extension and improvement of local DAB coverage (para.25)
- measures to address the existing failings of the existing DAB multiplex framework (para.26)
- the merger of adjoining local DAB multiplexes and the extension of existing multiplexes into currently unserved areas (para.26)
- the existing regional multiplexes to consolidate and extend to form a second national commercial radio multiplex (para.26)
- convincing listeners that DAB offers significant benefits over analogue radio (para.28)
- DAB to deliver *"new niche [radio] services"* and to gain better value from existing content (para.29)
- DAB to offer more services other than new stations (para.30)
- DAB to offer greater functionality and interactivity (para.31)
- implementation of digitally delivered in-car traffic and travel information (para.31)
- DAB radio receivers to be priced at below £20 within two years (para.32)
- introduction of add-on hardware (similar to Freeview boxes) to enable consumers to upgrade their analogue receivers (para.32)
- energy consumption of DAB radio receivers to be reduced (para.33)
- new cars to be sold with digital radios by 2013 (p.99 box)
- a common logo to identify and label DAB radios (p.99 box)
- development of portable digital radio converters (p.99 box)
- integration of DAB radio into other vehicle devices such as 'SatNav' (p.99 box)
- work with European partners to develop a common approach to digital radio (p.99 box).

A lengthy list. And who is going to pay for all this? Digital Britain stated that *"the investment needed to achieve the Digital Radio Upgrade timetable will on the whole be made by the existing radio companies"* (para.44).

This means the BBC and the commercial radio sector. And what exactly do these radio broadcasters think about having to pay for all these proposals without the aid of specific government funding? A seminar organised by the Westminster Media Forum this morning gave us an opportunity to find out.[196] Here's what was said about the Digital Radio Upgrade issue (speech excerpts):[197]

Caroline Thomson, Chief Operating Officer, BBC [CT]: *The [Digital Britain] report is clear that there is an ambitious target for analogue switch-off in 2015. It is an ambitious target. Radio switch-off is a very different issue from television switchover, but we are supportive of this ambition and we will work with partners in the industry towards delivering it. And we have already made a lot of progress working with commercial radio to develop the policies on this. But, at the heart of it, we must remember that we must put listeners first and be careful not to damage the ability of listeners to tune in to the content they love. Working with commercial radio to secure the digital future in a way that will work for all our listeners is a crucial part of this. As my colleague Tim Davie, Director of [BBC] Audio & Music, said recently: 'unless we huddle together for scale, we are going to be in trouble'. The BBC is drawing up our digital rollout plans in radio to see where and when it is possible to extend DAB coverage, and how much it would cost. We are willing partners, and DAB is a good example of an area of the Digital Britain report where we are helping to meet the charge.*

Andrew Harrison, Chief Executive, RadioCentre [AH]: *The real choice, which Digital Britain identifies, is which broadcast platform do we want – FM or DAB. And here, the genie is out of the bottle. DAB now exists on 10 million sets, the BBC will not withdraw 6Music and BBC7 or the Asian Network or Five Live Extra – it never withdraws services – and commercial services will not fold DAB-only stations like Planet Rock or Jazz FM. Digital Britain has been clear in its aspiration – national, regional and larger local stations will have a clear pathway to upgrade to DAB and switch off FM. Smaller players will have a clear opportunity to remain on FM without an obligation to move across to DAB. Strategically, that's a simple resolution – both will co-exist. So, next we need a plan to work out how we might achieve the migration criteria – on transmitter coverage, set sales and in-car penetration. The devil inevitably will be in the detail. But we need two strong interventions from government – on coverage and on cars – before any migration plan will be taken seriously. On cars, Digital Britain falls short of mandating manufacturers, unlike in France, to put digital radio in all cars from 2013. Encouragingly, Ford and Vauxhall have both confirmed their intent to upgrade in line with the timeline for 2013, but we need government to force the pace. On coverage, Lord Carter has ducked the funding issue. The commercial sector has already built out its national and local multiplexes as far as is commercially viable. So I'm delighted to hear Caroline emphasise that the BBC is supportive of the direction and ambition for digital radio and are willing partners helping to fund the change. It's now time for the BBC and government to stop their wider dance around the BBC's future role and theoretical possible future uses of the Licence Fee which have never been paid for before, and [to] instead consider how to broker a coverage plan for digital radio that will make it happen.*

Carolyn McCall, Chief Executive, Guardian Media Group [CM]: *It's hard to escape the feeling that what the Digital Britain report has done is just gone: 'we recognise the issue, big issue DAB'. They said something like that, which is pretty important, but they have just gone: 'Ofcom, deal with it'. That's how it strikes me. It just seems that so much of this on radio is being left to Ofcom to deal with. And, if what I read is true, David Cameron doesn't want an*

Ofcom anyway. So that is quite a serious issue for us as an industry. The most worrying aspect of the report in relation to radio is the assertion that investment needed to achieve the Digital Radio Upgrade will be made by existing radio companies. Effectively, the promise of deregulation is being made conditional on commercial radio funding digital [upgrade], stumping up more money that the commercial industry simply cannot afford. We've always had too much regulation for a small industry struggling in an unregulated digital world. While we back DAB, I don't think any commercial broadcaster is going to feel comfortable about paying for those developments. The final point on radio is that, at a time when that industry in particular needed some clarity, the report does not give us any clarity. What new powers will Ofcom have, what role will they be expected to play, what is the position on the vital issue of Format change, what is meant by greater flexibility in relation to co-location, and mini-regions? The list goes on. I would say to Stephen [Carter], or Ben [Bradshaw], or indeed Jeremy Hunt, we need urgent clarifications on these issues and quickly.

Q&A Session [excerpts]:

Q: *Is analogue radio switch-off going to include the [BBC] Radio 4 Long Wave signal?*

CT: *That is the government policy. The policy is to switch off all analogue radios.*

Q: *Existing DAB coverage is not good enough?*

AH: *Right now, self-evidently, DAB coverage is not good enough for anyone to consider switchover. There is a bill to be paid to deliver that public policy imperative. As long as that bill is met and covered, I think the BBC and the commercial sector would confidently switch over knowing the coverage is better ….*

Q: *Unless you start spending money now, and if you are, where is it going to come from, it's not going to happen, is it?*

CT: *First of all, we will not do the analogue switch-off unless it is the case that there are very big thresholds that have already been passed, particularly about car radios. And the challenges of getting to those thresholds by 2013, which is what we've said, are enormous, even if we build out the transmission. So let me just be clear. It is not the BBC's policy to switch off FM or Long Wave until we are secure and clear – that is why I made the reference to listeners in my speech – that that is the policy which will work for listeners. On the money, for now we don't have the money to build out beyond 90% – that is our current build-out – and the final 10% costs much more per percentage than the previous 90%, but we will look forward to a discussion with the government about it. We would like to be able to do it because, in the long term, as for commercial radio, running dual illumination [FM/DAB simulcasting] costs a lot of money so a switchover in 2020 costs us more than a switchover in 2015. But we won't do the switchover in 2015 unless we believe particularly that car radios are up…..*

CM: *This point about digital radio [switchover]. There are no funds. I am not really convinced […noise…] and margins are slim because everyone has been hit by the recession quite badly. I don't know where the money is going to come from for digital switchover of radio.*

AH: *I remain confident that where we are now with Digital Britain from the radio perspective is into the negotiation now – who pays for this? Frankly that is a negotiation that is far more* *likely to be concluded positively in the next few months between the BBC and a Labour government than under a Conservative government, so I remain optimistic that both sides will be brought to the table. In terms of who pays and who can afford this, the reality is that the BBC Licence Fee is £3.5bn, that's seven times the total income of commercial radio. The cost of DAB coverage build-out is about £5m a year – that's less than Jonathon Ross' salary or Michael Lyons' pension fund – so it's purely a question of priorities for the BBC. I would have thought that it is quite within the limit of the BBC's talented management to come up with a solution that can meet the public purposes set out for DAB and still deliver all the wonderful content that we enjoy.*

---

[195] http://www.culture.gov.uk/images/publications/chpt3b_digitalbritain-finalreport-jun09.pdf

[196] http://www.westminsterforumprojects.co.uk/eforum/digitalbritain2agenda.pdf

[197] author's recording

28.

11 July 2009

Digital Radio Upgrade: everyone's a winner?

For every winner, there is inevitably a loser (or three). The 'Digital Radio Upgrade' proposals contained in the Digital Britain Final Report are no exception. It is relatively easy to see who the winners will be from its proposals, as some of these are made explicit in the accompanying Impact Assessment:

- *"the beneficiaries of these proposals are primarily [DAB] multiplex operators"* (p.12)
- *"benefits of £38.9m per annum [to broadcasters] for each year after dual transmission on analogue and DAB ceases"* (p.12)
- *"cost savings to [commercial radio] national broadcasters of licence extensions approximately £10m"* (p.12)
- *"cost savings [to local commercial radio stations] of co-location and increased networking £23m"* (p.12).[198]

However, the losers are made far less explicit in the fine print of the Impact Assessment:

- *"merging [DAB] multiplexes will reduce the overall capacity available for DAB services, therefore reducing the potential for new services"* (p.117)
- *"reduced capacity on local multiplexes might result in some services losing their current carriage on DAB"* (p.117)
- *"extending the licence period of existing analogue services would reduce the opportunities for new entrants"* (p.119).

There would appear to be a degree of contradiction here. Digital Britain also insisted that:

- *"DAB should deliver new niche services, such as a dedicated jazz station …. The radio industry has already begun to agree a pan-industry approach to new digital content …"* (p.98 main report).[199]

However, the Impact Assessment admits that amalgamation of existing local DAB multiplexes will reduce their capacity, *"therefore reducing the potential for new services."* Worse, it states that some existing stations broadcasting on DAB will have to be bumped off as a result of local multiplex amalgamation.

So the potential losers from Digital Radio Upgrade would seem to be:

- commercial stations presently carried on local DAB multiplexes who might have to be bumped because there is no longer the capacity after amalgamation
- local commercial stations presently carried on their local DAB multiplex who will have to quit DAB because they do not wish to serve the enlarged geographical area after amalgamation of multiplexes (for example, the cost of DAB carriage for Kent/Sussex/Surrey is likely to be considerably higher than Kent alone)

- new entrants.

The local commercial radio stations bumped from DAB will fall into two types:

- digital-only stations (such as Yorkshire Radio) whose current regional multiplex will be transformed into a national (or quasi-national) multiplex under Digital Britain proposals – such stations have no analogue broadcast licence and could lose their radio broadcast platform altogether
- analogue local stations who were simulcasting on DAB, but whose multiplex has either bumped them post-amalgamation, or who are not in the market to pay more for increased coverage across a much larger area – many of these stations have had their Ofcom analogue licences renewed on condition that they simulcast on DAB. If they are now forced off DAB, will Ofcom take their licences away?

In the rush to frame proposals in Digital Britain that respond to the circumstances of the large radio players with substantial investments in DAB infrastructure, it might appear that the voices of the smaller local commercial radio stations have got lost in the stampede of lobbying. These stations might be small in number but many of them remain standalone, so they will not benefit financially from the relaxation of co-location rules. Digital Britain is condemning many of them to remain on FM (or AM), leaving the large radio groups to dominate the DAB platform.

Although the proposals in Digital Britain have been framed to 'help' local commercial radio, overwhelmingly they will reduce the financial burden of group radio owners with local station operations in adjacent areas, and of group owners who have invested in DAB infrastructure. There is little in the way of financial benefits for independent local commercial stations, or for potential new entrants, both of whom face being crowded out of the DAB platform.

---

[198] http://www.culture.gov.uk/images/publications/digitalbritain_impactassessment.pdf
[199] http://www.culture.gov.uk/images/publications/chpt3b_digitalbritain-finalreport-jun09.pdf

29.

19 July 2009

DAB radio in Germany: further public funding rejected

The organisation that funds public radio in Germany has rejected a request for €30m from state broadcasters to develop DAB broadcasting between 2009 and 2012, and has rejected an additional request for €12m to fund digital switchover. Following its meeting on 15 July, KEF announced that the funds for DAB development *"will not be released because substantial elements of the criteria agreed previously with broadcasters had not been met and the viability of the projects could not be demonstrated."*[200]

According to Follow The Media, which broke the story online today, more than €200m of public money has already been spent developing DAB broadcasting in Germany.[201]

In April 2008, twelve criteria had been agreed between KEF and the broadcasters that would need to be met for funds to be released for digital radio projects:

- concrete agreements from public and private broadcasters to launch digital radio services, with a rollout plan
- statements regarding the content of these digital radio services and their value to listeners as a nationwide offering, compared to existing FM stations
- plans for added value services, such as Visual Radio, TPEG traffic data and podcasts
- evidence of the extent of DAB usage, both in Germany and abroad
- statements from manufacturers regarding their DAB radio receivers, delivery dates and retail prices
- statements on the future of FM broadcasting
- statements on the marketing strategy and necessary budgets for DAB
- plans for the development of DAB broadcast infrastructure in metropolitan areas and their service quality
- total costs of the proposed projects
- implementation time of the proposed projects
- milestones to be met in the implementation of the project, with KEF auditing their achievement
- compliance with the KEF checklist and responses to additional KEF questions.

At its meeting last week, KEF decided that *"the criteria had still largely not been met."* A forecast of the total cost of implementing DAB in Germany was not offered to KEF, although transmission costs for the period 2009 to 2020 were estimated by state radio to be €163.6m. However, KEF was told that FM radio broadcasts could not be ended until digital platforms accounted for 90% of radio listening, which was anticipated by 2020. The public radio companies expected to make a further application to KEF for funds of approximately €300m to complete the switchover from FM to DAB beyond 2012.

The earlier decision by Germany's private radio sector not to invest further funds in DAB development weighed heavily on the KEF decision, as it concluded that FM switch-off would be *"unthinkable"* without the participation of commercial radio in the DAB platform. KEF also made it clear that the financial savings anticipated from the ending of FM/DAB dual transmission were a pre-requisite for further investment in DAB, as was *"a minimum diversity of programme offerings significantly above those currently offered on FM."*

Follow The Media reported:

"There must be no more time wasted with this project now," said media spokesperson Thomas Jarzombek of the CDU party in North Rhine-Westphalia to Wolbeck-Münster (July 17). *"Instead, all the resources are now directed to the internet. …. After the exit of private radio stations and the rejection by the KEF, digital radio on DAB+ died."*[202]

---

[200] http://www.kef-online.de/

[201] http://followthemedia.com/payonlypage.php?referer=mediarules%2Fkefgermany19072009.htm

[202] http://followthemedia.com/payonlypage.php?referer=mediarules%2Fkefgermany19072009.htm

30.

22 July 2009

Digital Radio Switchover: Parliamentary Question

20 July 2009 : Column 561[203]
House of Commons
Monday 20 July 2009

The House met at half-past Two o'clock
Prayers
[Mr. Speaker in the Chair]
Oral Answers to Questions
Culture, Media and Sport

The Secretary of State was asked —

Digital Radio Switchover

1. Sir Nicholas Winterton (Macclesfield) (Con): *What his most recent assessment is of progress on digital radio switchover; and if he will make a statement.*

The Parliamentary Under-Secretary of State for Culture, Media and Sport (Mr. Siôn Simon): *The "Digital Britain" White Paper set out the Government's vision for the delivery of the digital radio upgrade by the end of 2015. We have committed to a review of the progress towards that timetable in spring 2010, and we have also asked Ofcom to review and publish progress against the upgrade criteria at least once a year, starting next year.*

Sir Nicholas Winterton: *Is the Minister not aware that "Digital Britain" has in fact failed to address the inadequacies of digital radio broadcasting coverage? I am sure that he will agree with that comment. Representations made to me so far suggest that the idea of a switchover is currently very unpopular. Instead of rushing ahead with the switchover, will he take positive action to allow people to see some tangible benefits?*

Mr. Simon: *I am disappointed that the hon. Gentleman thinks that we are rushing ahead. We have said that we will move Britain to digital by 2015. That gives consumers and the industry six years to make the upgrade, which we are doing because we are committed to radio, we believe in radio and we love radio, and radio will not have a future unless it goes digital. We are not switching off FM, and we are putting new services on the FM spectrum that is vacated by the services which move to digital audio broadcasting, because we want to see radio prosper and grow in the digital age.*

Mr. Barry Sheerman (Huddersfield) (Lab/Co-op): *Is my hon. Friend aware that switchover is affecting valued services on both radio and television? I have been lobbied by Teachers TV, which fears that it will lose an enormous part of its audience because the Department for Children, Schools and Families is stipulating that it must switch over totally to digital.*

Mr. Simon: *We are ensuring with radio switchover that community organisations and small community radio stations, which might currently be able to broadcast for only two weeks a year, will inherit the FM spectrum currently taken up by big regional and national FM broadcasters. Precisely such small, commercial, local community organisations will be able to flourish in the digital future in a way that they are technologically constrained from doing now.*

Adam Price (Carmarthen, East and Dinefwr) (PC): *The Minister is a Welsh speaker, so is he aware of the fears for the future of Radio Cymru, the BBC's Welsh language national service? It is not currently available on digital and will not be available in large swathes of western Wales for reasons of topography.*

Mr. Simon: *I have, with personal regret, to tell the hon. Gentleman that I am not really a Welsh speaker. [Hon. Members: "Ah!"] Dwi'n dysgu, 'de? I should have been a Welsh speaker. We are alive to the particular problems of Wales. There are serious problems with coverage, not just with respect to Radio Cymru but with digital coverage throughout Wales. We have made it clear that the nations and regions that are furthest behind in digital coverage will be the first priority for the most serious intervention, to ensure that they are not left behind when we move to digital. We have made it clear also that we will not move to digital unless 90 per cent coverage at the very least is achieved.*

Mr. Jeremy Hunt (South-West Surrey) (Con): *I start by welcoming you to your post, Mr. Speaker — an elevation that was only marginally more likely than man walking on the moon, which happened 40 years ago today. I offer you my congratulations. I am sure that you will want to join me in offering the congratulations of the whole House to the England cricket team, which won an historic victory today — their first victory over the Australians at Lord's for 75 years. We would also like to congratulate the Minister on taking up his post in the DCMS team. The Government's own figures state that there are 65 million analogue radios in circulation, and they hope that the cost of digital radios will fall to £20 a set. That means that the cost of upgrading the nation's analogue radio stock will surpass £1 billion. Who will pay that £1 billion? Will it be the Government, or will it be consumers?*

Mr. Simon: *Mr. Speaker, I should apologise for having forgotten to congratulate you; I thought that we were taking your position for granted by now, but it is my first time speaking under your chairmanship. I offer my very sincere congratulations. I never thought that your elevation was unlikely.*

Mr. Edward Vaizey (Wantage) (Con): *What about cricket?*

Mr. Simon: *The hon. Gentleman shouts "cricket" from a sedentary position. I can tell him that the Under-Secretary of State for Culture, Media and Sport, my hon. Friend the Member for Bradford, South (Mr. Sutcliffe), was at the cricket, which almost certainly accounts for the first English victory at Lord's since, I believe, 1934. In response to what we might call the "Tory sums" of the hon. Member for South-West Surrey (Mr. Hunt) — [Interruption.] No, Tory sums. We do not know how many analogue radios are in circulation; it may be 65 million. The first point to make is that those sets will not become redundant. The FM spectrum will be well used for new services that are currently squeezed out. We are working with industry to come up with sets that are consistently priced at £20 or less. That will enable consumers to add to the 9 million digital sets —*

Mr. Speaker: *Order. May I gently say to the hon. Gentleman, who has been extremely generous in his remarks that I do not want to have to press the switch-off button, but I am a bit alarmed that he has a second point in mind? It might be better if he kept it for the long winter evenings.*

Mr. Hunt: *The point is that, if people use their analogue sets, they will be able to listen to new radio stations, but not the radio stations that they have been listening to for a very long time. Was it not the height of irresponsibility to announce the phasing out of analogue spectrum without announcing any details or any funding for a help scheme, similar to the one that was in place for TV switchover? Will that not cause widespread concern among millions of radio listeners, who will feel that they are faced with the unenviable choice of either paying up or switching off?*

Mr. Simon: *I shall try to squeeze in my answer at the end of that extraordinarily long question. We will do exactly the same with radio as we did with television: we will carry out a full cost-benefit analysis of exactly what kind of help scheme might or might not be required, and we will proceed accordingly. There are 9 million digital sets in use already. Consumers have six years to decide how much they want to pay, for what equipment, to receive which services.*

---

[203] http://www.publications.parliament.uk/pa/cm200809/cmhansrd/cm090720/debtext/90720-0001.htm

31.

22 July 2009

Digital One: an end to wishing and hoping

Today, transmission company Arqiva announced that it had finally acquired the remaining 63% stake that it did not own of Digital One, the national commercial radio DAB multiplex, from Global Radio. Tom Bennie, Arqiva CEO said: "*Arqiva now plans to invigorate DAB with new channels and services and, as an independent operator, we're in a good position to realise the full potential of the Digital One multiplex.*"[204]

Let's go back in time.

In March 2007, National Grid Wireless had applied to Ofcom for a new licence to operate a second national commercial DAB radio multiplex and it noted in its application that:

- "few of the digital-only services on Digital One have been marketed aggressively"
- "aqwareness and reach conversion [of digital-only stations] is not keeping pace with the rise in DAB digital radio penetration"
- "over the past three years, there is no discernable positive [listening] trend for any of the [digital-only] services on Digital One, except for Planet Rock"
- "despite increasing DAB penetration, the proportion of listening generated by DAB homes to these [Digital One digital-only] services has not altered significantly"
- "DAB digital radio listeners are primarily using their DAB radios to tune in to established [analogue] services"
- "newcomers to DAB digital radio are primarily replacement set purchasers who have not been motivated by the prospect of new channels or improved functionality"
- "the lack of development of DAB digital radio in cars is also a possible threat to its development"
- "there is [advertising agency] dissatisfaction not only with the current digital radio offering as an advertising medium …. [but also] that too many of the existing stations sound alike and are trying to appeal to the same people."[205]

National Grid Wireless did not win the licence, as Ofcom awarded it to Channel 4 in July 2007. Then, Arqiva acquired National Grid Wireless. Then, in 2008, Channel 4 returned its licence to Ofcom unused. Ofcom has not re-advertised this second DAB multiplex licence, so there remains only one multiplex, owned by Digital One.

Now it has been two years since National Grid Wireless identified the problems with Digital One, and its successor – Arqiva – is suddenly in a position where it owns Digital One and it is in the driving seat to do something to fix it. The question is whether that two-year gap has now made it too late in the day for Arqiva/National Grid Wireless to fix things. Two years is a long time in technology, and time has not been kind to DAB. There are significantly fewer digital radio stations on-air now, there is less appetite for investment in new ventures, and commercial radio is suffering badly from the recession.

One wonders what might have happened subsequently if:

- Ofcom had not advertised a second national DAB multiplex?
- Ofcom had not awarded that licence to Channel 4?
- Channel 4 had not burnt through up to £9m of funding before deciding to scrap radio?[206]
- commercial radio had got on with the task of fixing DAB itself, instead of hoping that Channel 4 would kick-start the platform?
- Fru Hazlitt had stayed at GCap Media long enough to offload Digital One to Arqiva a year ago for £1?

With hindsight, it is already beginning to look as if that two-year period (March 2007 to July 2009) offered a critical opportunity for DAB. Critical in the sense that a lot needed to be achieved, that there was a lot of wishing and hoping for things that never materialised, and much seemed to eventually go backwards, instead of forwards, during that time. If you re-read the bullet points listed above from National Grid Wireless' application, you realise that these issues have still not been resolved during the last two years. In many ways, regrettably little of significance has yet changed. We are still waiting.

It's like a DAB Groundhog Day. Every day you wake up wishing and hoping things will be different, but every day the same issues still need solving, exactly as they were the day before, and everyone ends up talking again about finding solutions, but the day eventually comes to an end. And then tomorrow it starts all over again.

---

[204] http://www.arqiva.com/press-office/press-releases/press-releases-2009/global-radio-and-arqiva-conclude-landmark-dab-deal
[205] http://www.ofcom.org.uk/radio/ifi/rbl/dcr/awards0708/applications/app_national/ngw.pdf
[206] http://www.guardian.co.uk/media/2008/oct/21/channel4-digital-radio

32.

24 July 2009

Digital radio: a European update

This month's decision by Germany not to invest further public funds in developing the DAB radio platform has inevitably caused reverberations around Europe during the last fortnight. In an article headlined 'There will always be FM', Geneva-based Follow The Media notes that Germany is *"Europe's richest ad market for radio"*, ensuring that what happened there would inevitably influence other territories.[207]

In Austria, it is understood that the private and public stakeholders in DAB held an emergency meeting on 17 July to discuss the fall-out from the German decision. Nothing has yet been announced publicly.

In Spain, the Association of Spanish Commercial Radio [AERC] held a General Assembly this week which, amongst other things, considered the progress of DAB in Spain. AERC general secretary Alfonso Ruiz de Assin concluded: *"The DAB system is obsolete in Spain and we have conveyed to the authorities that it is a road to nowhere."* He added that *"traditional and digital [radio] will co-exist for a long time."*[208]

In France, the timetable for implementation of its T-DMB digital radio system still looks challenging. The average French household has six radios and it is estimated that the replacement cycle for these will be ten years. From 1 September 2010, radios with display screens will incorporate a digital tuner. From 1 September 2012, all media players, mobile phones and GPS hardware will include digital radio. From 2013, all new cars will be sold with digital radios.

Although digital TV switchover in France is happening in autumn 2011, there has been no date set yet for digital radio switchover. Radio station owners have applied to the government for a €16.5m grant to contribute to the costs of simulcasting on T-DMB over the next eight years (estimated at €30k per annum per station per market). The headline of a recent French article asked 'Is digital radio success guaranteed?' and commented that *"given the financial constraints required by this new method of distribution, the answer is not so obvious."* It noted that *"FM radio will not disappear in the near future and that radio via the internet is increasingly popular."*[209]

Also in France, the National Union of Free Radios has expressed concern that the T-DMB standard (like DAB) will require small stations to broadcast over a large coverage area as part of a cluster of broadcasters from each multiplex. It notes that such an arrangement will prove too expensive for small stations which are seeking an opportunity to go digital at low cost. The Union is advocating the DRM+ standard be used in France alongside T-DMB, and conducted a test broadcast in Paris this week. As one article noted, *"DRM+ has the advantage of being more flexible – it is an opportunity for radio to be broadcast independently outside the big [T-DMB] multiplexes."*[210]

Meanwhile, back in Germany, the Financial Times ran a story today headlined 'Digital radio fails in Germany.' Asked about the prospects there for DAB radio, Hans-Dieter Hillmoth, deputy head of the German private broadcasters association [VPRT] said bluntly: *"Currently there is no viable business model."*[211]

The article noted that, after ten years of DAB in Germany, only 600,000 DAB radios have been sold. In neighbouring Switzerland, it is anticipated that 300,000 DAB radios will have been sold by year-end. DAB radio receiver manufacturers, including the UK's Pure, had expected to sell 300 million units in Germany. Asked what importance it attached to the German DAB market, global audio manufacturer Pioneer commented *"absolutely none"*, and it added that the death of traditional analogue radio receivers is *"absolutely not in sight."*[212]

[207] http://www.followthemedia.com/payonlypage.php?referer=mediarules%2Ffm22072009.htm
[208] http://www.xornal.com/artigo/2009/07/22/sociedad/comunicacion/aerc-rechaza-publicidad-radio-publica/2009072219291088534.html
[209] http://telecom.sia-conseil.com/index.php/uncategorized/090710-01-best-of-radio-numerique-succes-garanti
[210] http://www.satmag.fr/affichage_module.php?no_theme=1&no_news=10934&id_mod=50
[211] http://www.ftd.de/it-medien/medien-internet/:kein-markterfolg-digitalradio-scheitert-in-deutschland/544181.html
[212] http://www.ftd.de/it-medien/medien-internet/:kein-markterfolg-digitalradio-scheitert-in-deutschland/544181.html

33.

1 August 2009

Digital radio switchover: "you can't move faster than the British public want you to move"

Feedback, BBC Radio 4, 31 July 2009 @ 1330[213]

Sir Michael Lyons [ML], chairman of the BBC Trust, interviewed by Roger Bolton and listeners:

Q: *Do you think the principle of moving across to DAB is a good one?*

ML: *The BBC has been a strong supporter of digital radio, believing that it will actually offer an improved service, and …*

Q: *Improved in what way? The quality of the existing services will be made better? Or it allows you to provide a range of other services as well?*

ML: *I think both. But, of course, you only satisfy the first of those two tests when you've actually got the same sort of coverage [on DAB] that you've got on FM. And indeed, it's important to say that the BBC has already picked up what commercial radio was going to do in terms of more investment to get to 90% of the population, and that will be achieved by 2011. But I think we're going to go on to the question of '[FM] switch-off' because actually that's a different issue altogether ….*

Q: *Well, one of the key things of public service is universal access and, clearly, a lot of people are saying [that] until 2015 there won't be one because, unlike a television set, perhaps we've got five or six radios around the house and a different radio in the car. And are you telling us we are going to have to buy five or six new radios and a new radio for the car in order to listen to something we might not want in the first place? That's the argument.*

ML: *Well, let me underline that I'm not saying that. That's actually in the government's Green Paper – they propose a date of 2015. The Trust is very clear actually. Who comes first in this? Audiences and the people you pay the Licence Fee. It is an extraordinarily ambitious suggestion, as colleagues have referred to, that by 2015 we will all be ready for this. So you can't move faster than the British public want you to move on any issue. So there's no doubt that 2015 looks challenging.*

Q: *Chairman, are you prepared to say, on behalf of the listeners, to the government, whichever government is in power, if they are insistent in pushing this through and you believe that listeners will be significantly disadvantaged, are you prepared to say "no, the BBC can't go along with this"?*

ML: *Well, as things stand at the moment, [in] the Digital Britain report, it seems that the BBC will find the money for this final stage, so there are serious discussions to be had about how it's going to be funded, as well as whether actually 2015 is in any way a realistic timescale. Now, what I can say now, is that those have already formed part of our discussion with Ministers and will continue to form part of our discussions with Ministers.*

Q: *But, to repeat my question, are you prepared to say at some point, or countenance saying, to a Minister "no, we can't go along with this because, in doing so, we will provide a disservice to our listeners"?*

ML: *Well, I think I've said as much I need to say today …..*

Q: *…. as a diplomatic chairman …..*

ML: *…. and also, you know, it's very important that I don't try and conduct any discussion I'm having with Ministers over the air.*

---

[213] http://www.bbc.co.uk/programmes/b00lt16f

34.

14 August 2009

Digital Britain: the implementation plan

The government has published the Implementation Plan for Digital Britain, setting out its action plans for the proposals made in June 2009's Final Report.[214] These are the sections that directly concern the radio sector:

PROJECT 1: DIGITAL ECONOMY BILL
LEAD: Colin Perry

GOVERNANCE
- Bill Project Board oversees the delivery of the Bill. Members are David Hendon (BIS)/Jon Zeff (DCMS) – joint SROs, Carola Geist-Divver (DCMS legal), Eve Race and Jose Martinez-Soto (BIS legal), Colin Perry (Bill Team Leader), Laura Williams (secretariat)

- Bill Management Group tracks progress and drives delivery of the Bill. Members are Colin Perry (Bill Team Leader) chair, Deputy Directors BIS/DCMS, Carola Geist-Divver (DCMS legal), Eve Race and Jose Martinez-Soto (BIS legal), Laura Williams (secretariat). Other policy leads attend as appropriate.

ACTIONS COVERED FROM THE FINAL REPORT [excerpts]:
- Amending the Communications Act 2003 to make the promotion of investment in communications infrastructure and content one of Ofcom's principal duties.

- Ensure the Board of Ofcom has a statutory obligation to write to the Government alerting Secretaries of State to any matters of high concern regarding developments affecting the communications infrastructure and in any event to write every two years giving an assessment of the UK's communications infrastructure.

- Encouraging, where appropriate, adjoining radio multiplexes to merge and extending existing multiplexes into currently un-served areas rather than awarding new licences. Grant Ofcom powers to alter multiplex licences which agree to merge.

- We will make an amendment to the existing legislation to support a change in the localness regulatory regime to allow location in mini regions defined by Ofcom.

- Grant a further renewal for up to seven years of analogue radio licences for broadcasters which are also providing a service on Digital Audio Broadcasting (DAB).

- Grant Ofcom new powers to insert a two year termination clause into all radio licences awarded or further renewed before the Digital Radio Upgrade date.

PROJECT 6: DIGITAL RADIO UPGRADE
LEAD: John Mottram

ACTIONS COVERED FROM FINAL REPORT [in full]:
- Develop Action Plan for Digital Radio Upgrade, including a Cost/Benefit Analysis.

- Invite Consumer Expert Group to extend its current scope to inform the development of the Digital Radio Upgrade.

- Facilitate the roll-out of the BBC's national multiplex to ensure it achieves coverage comparable to FM by the end of 2014.

- Encourage, where appropriate, adjoining local multiplexes to merge and extend coverage into currently un-served areas. Grant Ofcom powers to alter multiplex licences which agree to merge.

- Allow for the extension of multiplex operators' licences until 2030, if part of an agreed plan towards Digital Radio Upgrade.

- Consider with Ofcom the case for delaying the implementation of AIP on DAB multiplexes until after the Digital Radio Upgrade is completed.

- Grant Ofcom new powers to extend the licence period of all national and local licences, broadcasting on DAB, for up to a further seven years, although this decision will be kept under review. In addition, amend the rules under which Ofcom grants analogue licence renewals to ensure that regional stations which do become national DAB stations do not lose their current or future renewal.

- Grant Ofcom new powers to insert a two year termination clause into all licences awarded or further renewed before the Digital Radio Upgrade date.

- Work with broadcasters and vehicle manufacturers to implement the 'Digital Radio in vehicles: a five point programme'.

- Agree with Ofcom a two-year pilot of a new output regulatory regime.

- Reduction in number of locally-produced hours in exchange for enhanced commitment to local news.

- Ofcom to consult on a new map of mini-regions which balances the potential economic benefits but also the needs and expectations of listeners. We will make an amendment to the existing legislation to support this change.

- Consultation seeking views on proposals for a new licence renewal regime for community radio. This consultation will include proposals to remove the 50% funding limit from anyone source and the restriction preventing a station being licensed in an area overlapping with a small commercial service and extending our commitment to promoting best practice within the community sector and encouraging self-sustainability by allocating a small portion of the Community Radio Fund to support the work of the industry body, the Community Media Association.

- Insert two year termination clause into all new licences.

- Grant Ofcom new powers to extend the licence period of all national and local licences, broadcasting on DAB, for up to a further seven years (keep this decision under review). If by the end of 2013 it is clear the Digital Radio Upgrade timetable will not be achieved we will use the powers, set out above, to terminate licences and the existing licensing regimes will apply.

- Amend the rules under which Ofcom grants analogue licence renewals to ensure that regional stations which do become national DAB stations do not lose their current or future renewal.

---

[214] http://www.culture.gov.uk/images/publications/DB_ImplementationPlanv6_Aug09.pdf

35.

22 August 2009

Paying for DAB radio carriage: god only knows

Premier Christian Radio, the London AM station, is planning to broadcast on the national DAB platform from 21 September 2009. In an e-mail to listeners, its chief executive Peter Kerridge explained:

"Beginning in September, we will start to incur the cost to transmit on this digital platform – £650,000 per annum – which is an expense that is over and above our current operating costs. The only way the £650,000 in transmission costs will be covered is through the generosity of friends like you. It is fantastic that God has moved in such an amazing way to provide Premier this national digital licence! Now may you and I be found faithful as we steward this new resource for His glory and for the advancement of His Kingdom!"[215]

DAB carriage remains a costly business. Digital One, the owner of the sole national commercial DAB multiplex, fixes the carriage costs for content providers such as Premier Christian Radio. If £650,000 seems like a lot of money for broadcast on a platform that reaches 33% of adults in the UK and accounts for only 13.1% of radio listening, understand that this is a bargain compared to the expensive contracts some content providers had signed previously.[216] In January 2009, Digital One responded to the government's Digital Britain initiative by cutting its prices. Acting chief executive Glyn Jones had said:

"We're turning the ideas set out in the Digital Radio Working Group's report into actions. That includes looking hard at how Digital One can offer lower carriage costs. In turn we're expecting that stakeholders involved in the Working Group, and other companies with the ambition to launch new national radio stations in 2009, will step up and engage with a view to adding compelling new choice for consumers. We're expecting that prices will initially be set below Digital One's 2008 rate card. One reason for that is to help provide an incentive for people to invest in high quality services. But, over time, companies providing new services will be expected to contribute to the costs of a transmitter roll-out plan which was something also identified by the DRWG as important."[217]

Digital One's January 2009 press release was ambitiously headlined 'New National Radio Stations To Launch In 2009'. Seven months later, what stations have stepped forward to take advantage of the Digital One offer? Government-funded BFBS Radio started DAB simulcasting on 20 April 2009, following a three-month trial in 2008. Amazing Radio launched on DAB in June 2009 for a six-month trial period, playing unsigned artists from its music web site. Also in June 2009, Fun Kids, which is normally on DAB only in London, launched a fourteen-week trial simulcast on national DAB. Neither BFBS nor Amazing Radio are participating in RAJAR radio audience research, so it is impossible to know how much listening these services are attracting on the DAB platform.

Have we seen any major media players step forward and put a new mass market radio service on the national DAB platform? Not yet. Why? Because, even at the knockdown rate of £650,000 per annum, it still proves impossible to make a profit from offering radio content on DAB. The table below offers very rough estimates of what digital stations measured in RAJAR (and carried on a mix of broadcast platforms including DAB and digital TV) should and might be earning in revenues. The second column lists the total hours presently listened to each digital station. The third column uses the average commercial radio sector yield (how much revenue was generated from how much radio listening in 2008) to estimate, in theory, what these stations' revenues should be.

DIGITAL RADIO STATION REVENUES			
digital station	listening hours per week ('000) [RAJAR Q2 2009]	hypothetical annual revenues (£ '000) from 2008 sector average yield	guesstimate annual revenues (£ '000) from Ingenious sector data
Planet Rock	5,009	6,548	402
The Hits	4,833	6,318	388
Smash Hits Radio	4,288	5,606	344
Heat	2,073	2,710	166
Jazz FM	1,695	2,216	136
Chill	1,019	1,332	82
Absolute Radio Classic Rock	967	1,264	78
Q	640	837	51
NME Radio	493	644	40
Panjab Radio	487	637	39
Yorkshire Radio	465	608	37
Absolute Radio Xtreme	372	486	30
Fun Radio	178	233	14
Punjabi Radio	171	224	14
TOTAL	22,690	29,662	1,820[218]

However, the 'Commercial Radio: The Drive To Digital' report commissioned from Ingenious Consulting by RadioCentre in January 2009 told us that:

"Incremental revenue from DAB-only stations is negligible at ~£130k per 'bespoke' station ...[219]

The list above comprises the 14 digital radio stations that subscribe to RAJAR. Not all of these stations broadcast on DAB (Smash Hits Radio is only on digital TV), not all of them are national (Yorkshire Radio is only on the Yorkshire DAB multiplex, for example), but let us be generous and assume that each station earns revenues of £130,000 per annum. In total, these stations combined would generate £1.82m per annum of revenue. This is substantially less than the £29.7m revenues that would be expected to be generated from them attracting 22.7m hours per week of listening.

The final column in the table estimates how much revenue each station might be earning from the £1.82m total, if revenues were proportionate to hours listened. I must stress again that this only a rough estimate – none of these stations, nor Ofcom, publishes the actual revenues of digital radio stations. What these estimates demonstrate is that, if Planet Rock were (like Premier Christian Radio) paying £650,000 per annum for its carriage on the national DAB multiplex (the financial details of its "long-term" deal with Digital One were not made public), the station is still nowhere near breaking even, not even after ten years on-air.[220]

The Ingenious Consulting report found that DAB-only stations are spending £25m per annum on operating expenses.[221] The above table shows that, if these stations were attracting revenues proportionate to the listening they presently enjoy, collectively they would then be profitable (£29m revenues minus £25m operating expenses). But, in fact, their revenues are presently less than £2m. The Ingenious Consulting report concluded that, as a result, the *"annual negative cash flow impact of DAB"* on the commercial radio sector is around £27m per annum.[222]

This £27m annual loss attributable to digital radio stations represents around 5% of commercial radio's revenues, a significant impact on an industry which is only marginally profitable overall at present. The nub of the problem is this: digital radio stations presently account for 5.3% of listening to commercial radio, but digital radio stations attract only 0.3% of commercial radio revenues. Here is a massive economic disconnect that requires much more than a mere increase in productivity or some kind of performance improvement. Doubling or even tripling these stations' revenues would barely dent the problem.

Maybe DAB is simply not a platform where the traditional commercial radio model can be made to work – the old model of 'give away free content, pay for it by attracting advertisers to buy on-air spots'. Maybe DAB is not a medium from which traditional UK commercial broadcasters can generate profits from offering content, as they had anticipated in the 1990s. Commercial broadcasters are pushing no commercial product other than their on-air brand (and some music downloads, concert tickets and click-through purchases). Instead, perhaps DAB can only be made to work as a marketing tool to assist companies selling (non-radio) products. So, for example, it would make sense for Universal Music to have a DAB radio station to expose directly to the public the CDs/videos/movies they are currently selling. It would make sense for Amazon to have a DAB radio station to promote all the consumer products it is selling. Then, the £650,000 carriage cost could be considered an additional 'marketing expense' for these companies' core business, rather than a direct operating expense that had to be recouped **on-air**.

The other possibility is for DAB to be used predominantly by organisations whose objective is something other than breaking even financially. In January 2008, I had written:

"Worryingly, this sudden flowering of ethnic, religious and publicly-funded radio stations on the DAB platform echoes the fate of the 'AM' waveband in the 1990s, at a time when the radio industry and the regulator had become convinced that audiences were deserting that platform for the improved audio quality offered by the 'FM' waveband. By 2002, declining audiences of 'AM' stations had persuaded the regulator to suggest that the platform be used in future 'for better serving minority, disadvantaged or currently excluded audience groups, whether defined by their interests, demographics or ethnicity.' The 'DAB' platform of 2008, particularly in London, is already starting to resemble the 'AM' platform of 1998, suggesting that 'DAB' might have already been written off by the sector as a means to reach the 'mass market' audiences that national advertisers desire from the medium."[223]

This trend towards non-commercial content has developed further since then. The national DAB platform has added BFBS Radio (government-funded) and now Premier Christian Radio (religious), but no new permanent digital radio stations operating on a commercial model. Local DAB multiplexes have added Traffic Radio (government-funded), Colourful Radio (ethnic) and UCB (religious). Interestingly, UCB has taken two channels on each of the regional MXR DAB multiplexes, giving it a substantial amount of DAB spectrum. But there have also been ethnic DAB radio casualties since my earlier report – Islam Radio in Bradford closed its DAB service in December 2008, and India's Zee Radio closed its London DAB service in April 2009. Even for ethnic broadcasters locked out of analogue radio, DAB can prove a struggle.

Premier Christian Radio's Peter Kerridge hit the DAB nail on the head when Media Week reported:

"Kerridge said Premier Media's funding meant it was in a better position than other media organisations, as the 'ad-funded model is smashed' ....."[224]

The available financial data confirm that, certainly for the DAB platform, an ad-funded model simply is not viable at present. To make DAB work for your content, you need government funding, direct listener financial support, a sugar daddy, or some kind of god smiling benevolently down upon you.

[215] http://www.guardian.co.uk/media/2009/aug/19/premier-christian-radio-national-dab
[216] source: RAJAR Q1 2009
[217] http://www.ukdigitalradio.com/news/display.asp?searchnews=&year=&id=310
[218] source: Grant Goddard
[219] unpublished report
[220] http://www.ukdigitalradio.com/news/display.asp?searchnews=&year=2008&id=304
[221] unpublished report
[222] unpublished report
[223] http://www.endersanalysis.com/publications/publication.aspx?id=514
[224] http://www.mediaweek.co.uk/News/MostEmailed/928163/Premier-Radio-launch-nationwide-Christian-DAB-station

36.

2 September 2009

DAB European update

NORWAY

The newspaper Aftenposten reported that *"sales of DAB receivers are still at a snail's pace"*, with only 61,000 sold in Norway in 2008, compared to eight times that number of analogue receivers sold.[225] Culture Minister Trond Giske said that, if his party wins the election this autumn:

"We will present a white paper on DAB in 2010 which, amongst other issues, will discuss whether the government can contribute more actively to promote the digital migration of the radio medium. We now have good experience from the digital migration of television, though the radio medium will take longer and require more preparation. Among other things, there are many more radio receivers to be replaced than there were TV sets, so it is extremely important that this transition occurs at a socially acceptable pace."[226]

The following day, in an article headlined 'Poor Sales of DAB Radios', Norway's Kampanje magazine reported that sales of DAB radios are only 40,000 to 60,000 per annum out of a total 700,000 to 800,000 radios sold annually.[227] Cumulatively, over the last decade, 300,000 to 400,000 DAB radios have been sold out of a total 8,000,000 radio receivers. Synnove Bjoke, managing director of electronics trade organisation Elektronikkbransjen, said:

"We believe sales will increase in the years ahead. The day we are given a [FM] switch-off date, we will sell many more DAB radios, but we need a date. There has been uncertainty amongst people, and also in our industry, as to whether we're ever going to switch off the FM band, and that uncertainty makes people buy regular FM radios."[228]

SWITZERLAND

Speaking at Swiss Radio Day 2009 held in Zurich last week, Swiss Radio German-language station DRS director Walter Ruegg announced the introduction of DAB broadcasts from 15 October and said that the platform would also be made available to local commercial stations in Switzerland.[229] English-language public station World Radio Switzerland will also be broadcast nationally on DAB from the same date.[230]

IRELAND

RTE Radio boss Clare Duigan told the Irish Independent newspaper that the absence of commercial stations on the DAB platform was a *"big issue."*[231] She said: "*We've begun to talk to the Independent Broadcasters of Ireland [IBI] and we're very much hopeful that over the next couple of months we'll be able to work something out. DAB is one of those areas where we really need to work together as an industry.*"[232]

But IBI boss Willie O'Reilly responded that commercial stations are not interested in rejoining the DAB platform *"at the moment"* because *"the return on investment looks poor."* UTV head of Irish radio Ronan McManamy said that DAB is *"not a priority"* for UTV in the *"current marketplace."*

[225] http://www.aftenposten.no/kul_und/article3244129.ece
[226] http://www.aftenposten.no/kul_und/article3244129.ece
[227] http://www.kampanje.com/medier/article496321.ece
[228] ://www.kampanje.com/medier/article496321.ece
[229]

http://www.bluewin.ch/it/index.php/1137,174863/Swiss_Radio_Day_2009__annunciata_lera_digitale/it/digitale/hitech/articol igguerra
[230] http://genevalunch.com/blog/2009/08/24/radio-cite-bottom-of-geneva-listeners-wrs-gains-ground/
[231] http://www.independent.ie/business/media/players-polls-apart-on-digital-radio-1870840.html
[232] http://www.independent.ie/business/media/players-polls-apart-on-digital-radio-1870840.html

37.

10 September 2009

Digital radio: Parliamentary Question

House of Commons
Written Ministerial Statements
9 September 2009[233]

Digital Broadcasting: Radio

Tim Farron: *To ask the Secretary of State for Culture, Media and Sport whether his Department's proposals for the analogue radio switch-off in 2015 have been submitted for rural proofing to the (a) Commission for Rural Communities and (b) Rural Advocate.*

Mr. Simon: *The Digital Britain White Paper set out our commitment to a full impact assessment of the Digital Radio Upgrade; including consideration of the rural impact. To inform these assessments we will work closely with the relevant stakeholders, such as the Commission for Rural Communities and the Rural Advocate.*

Tim Farron: *To ask the Secretary of State for Culture, Media and Sport what assessment he has made of the merits of providing financial assistance to (a) low-income households and (b) households in hilly rural areas in respect of the analogue radio switch-off in 2015.*

Mr. Simon: *The Digital Britain White Paper set out our commitment to conduct a full impact assessment, including a cost benefit analysis of Digital Radio Upgrade. The results of this impact assessment will help determine whether there is a case for a Digital Radio Help Scheme, and if so, what its scope would be. In addition, the Consumer Expert Group, which brought together key consumer representatives to inform the Digital TV switchover process, has been invited to extend its scope to cover radio and will ensure that the Digital Radio Upgrade programme takes account of the wide range of listener needs.*

---

[233] http://www.publications.parliament.uk/pa/cm200809/cmhansrd/chan117.pdf

38.

10 September 2009

Funding DAB radio infrastructure: still "no"

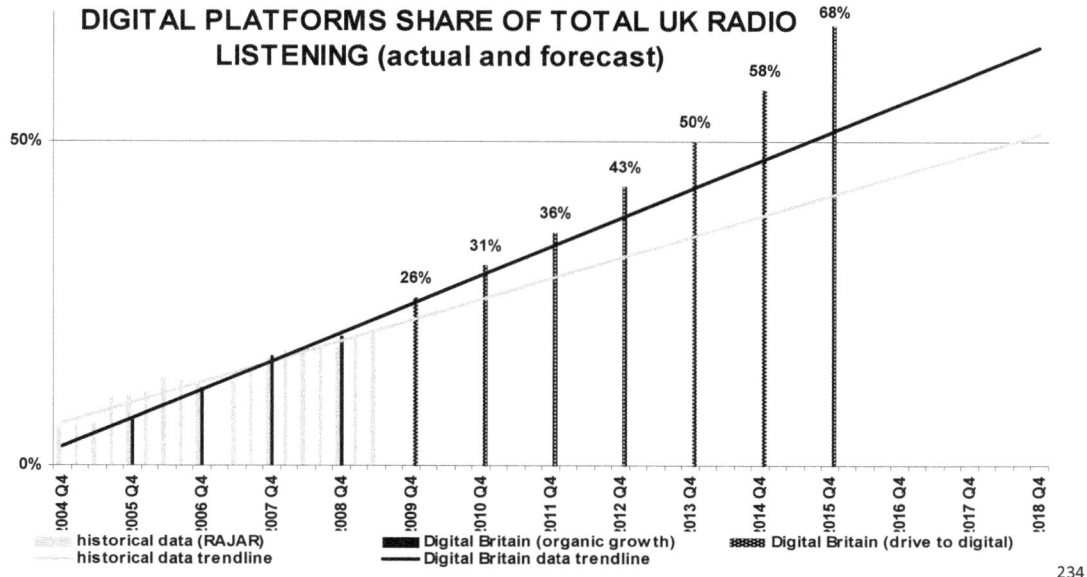

DIGITAL PLATFORMS SHARE OF TOTAL UK RADIO LISTENING (actual and forecast)

234

The Media Show, BBC Radio 4, 2 September 2009 @ 1330[235]
Steve Hewlett interviewed Tim Davie [TD], Director of BBC Audio & Music

Q: *We talked at the Radio Festival a few months ago and you talked a lot about DAB. The criteria have been stated now for moving forward to switchover, or before anyone contemplates switching off the analogue FM signal, of 50% of listening and 90%+ of coverage. Do you think that's realistic by 2015?*

TD: *I use the word 'ambitious' and I mean it. I think it's tough. It is possible. I think the radio industry to date has shown an incremental path towards digital and, unless you get a big step change, you'll never get there. And, to be fair, the BBC has driven this harder than anyone.*

Q: *When we last spoke about it, there was a discussion of £100m or so being needed to pay for the rollout of not the BBC stuff but whatever is necessary for the commercial sector to go digital. At that time, I asked you specifically whether there was any money in your budget identified for that purpose and you said 'no'. Has anything changed since we last spoke?*

TD: *It's another 'no'. No, nothing has changed and until the plan ….*

Q: *This is not going to happen, is it?*

TD: *I think that radio will move to digital, and I think that ….*

Q: *Will it be DAB?*

TD: *I think at this point, it will be …. I believe in DAB. I say 'at this point' because I think we have hurdles to jump over.*

---

[234] source: RAJAR and Digital Britain
[235] http://www.bbc.co.uk/programmes/b00mbz1f

39.

17 September 2009

The demand for DAB radio: where is it?

Most of the current debate on the challenges facing DAB radio seems to be focused on 'supply side' issues, such as upgrading existing DAB transmitters, making DAB radio receivers available in cars and the creation of another national DAB multiplex. Surprisingly little of the talk is about the 'demand side' issues facing the DAB platform. What are consumers demanding from DAB radio? And how great is that demand?

There are two types of consumer demand for DAB: the demand for content broadcast on the DAB platform, and the demand for DAB radio receiver hardware. The two are inextricably linked. Consumer demand for DAB hardware is largely a function of demand for DAB content. You will only want to buy a DAB radio if you believe there is something interesting enough to listen to on it. Let's examine some of the available data on these two issues.

CONSUMER DEMAND FOR DAB CONTENT

Nobody is going to be motivated to spend money on a DAB receiver for listening to the radio if the platform only offers the same content already available to them on analogue receivers. Therefore, it must be the exclusive digital-only content available on DAB (and other digital platforms) that will persuade consumers to both use the DAB platform and to purchase a DAB radio receiver.

So how dissatisfied are consumers by the radio content choices (the range of radio stations) available to them on existing analogue radio receivers? Ofcom research shows that 91% of

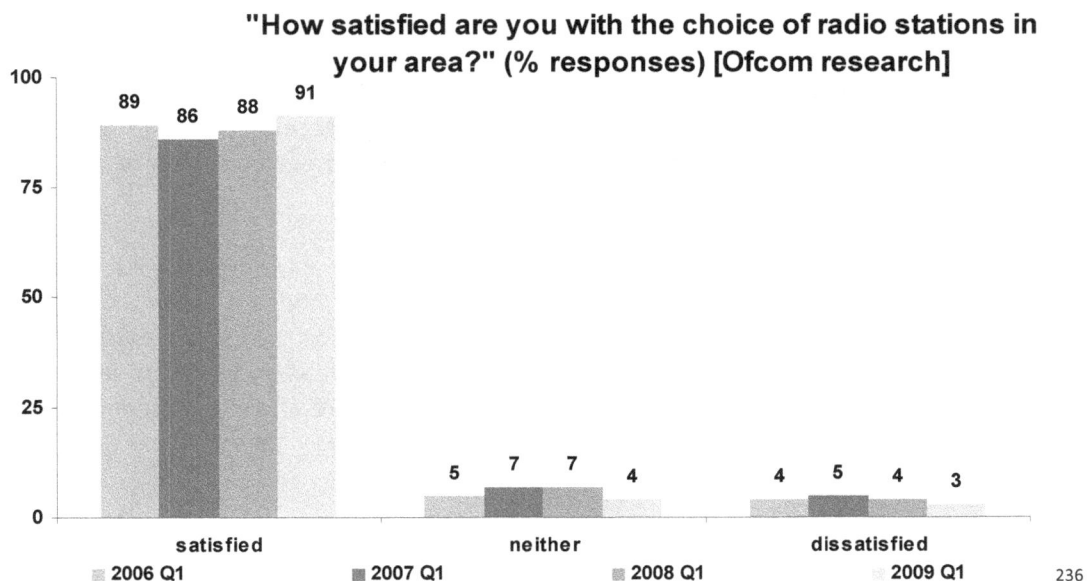

"How satisfied are you with the choice of radio stations in your area?" (% responses) [Ofcom research]

	satisfied	neither	dissatisfied
2006 Q1	89	5	4
2007 Q1	86	7	5
2008 Q1	88	7	4
2009 Q1	91	4	3

236

123

adults are satisfied with the existing choice of radio stations offered to them (see chart above), a proportion that has risen in recent years. This demonstrates that dissatisfaction with existing radio provision is extremely low, making it very difficult for any new platform to attract a substantial audience by offering content that will gratify consumers' few unsatisfied demands.

[In case you are wondering if the increasing satisfaction with radio stations might be a direct result of the exclusive digital-only stations already offered on the DAB platform, it is worth noting that only 3.9% of hours listened to radio are attributed to digital-only stations.[237]]

Ofcom data show that the average consumer listens to very few radio stations. Two thirds of the population listen to only one or two different radio stations in an average week, and the majority of these two-thirds listen to only one station.[238] So, not only are the overwhelming majority of consumers satisfied with their existing choice of radio stations, but most people listen to a very narrow menu of stations.

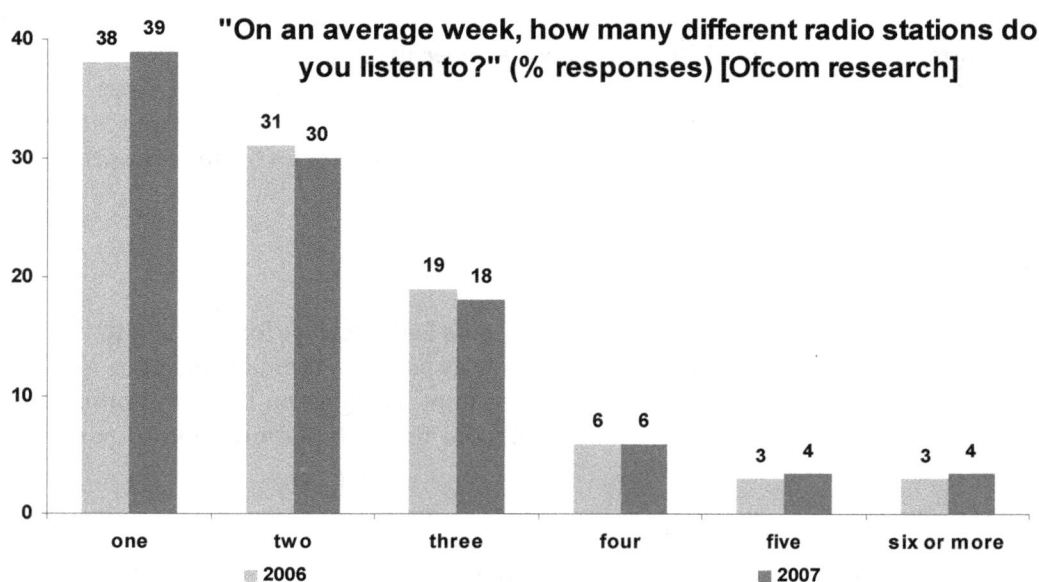

"On an average week, how many different radio stations do you listen to?" (% responses) [Ofcom research]

These phenomena are not the outcome of consumers only being offered a limited choice of radio stations on the analogue platform. Ofcom data demonstrate that, in addition to the 5 BBC radio stations and 3 commercial radio stations available nationally on analogue radio (with near universal coverage) in the UK, there are a significant number of local radio stations available to consumers in most areas of the UK. The average consumer in the UK has a choice of 8 national radio stations and 6 local radio stations.[240]

This existing wide choice of radio stations makes the plan for migration to digital platforms very different for the radio medium than it is for the television medium. In the UK, only four (five in some areas) TV stations are available via analogue, making the wider choice available on digital platforms seem very attractive to consumers. Whereas, in radio, an average 14 stations are available to consumers on analogue, and these are already satisfying the vast majority of consumer demands. As a result, there is only a very tiny untapped consumer market for radio content not already available via analogue.

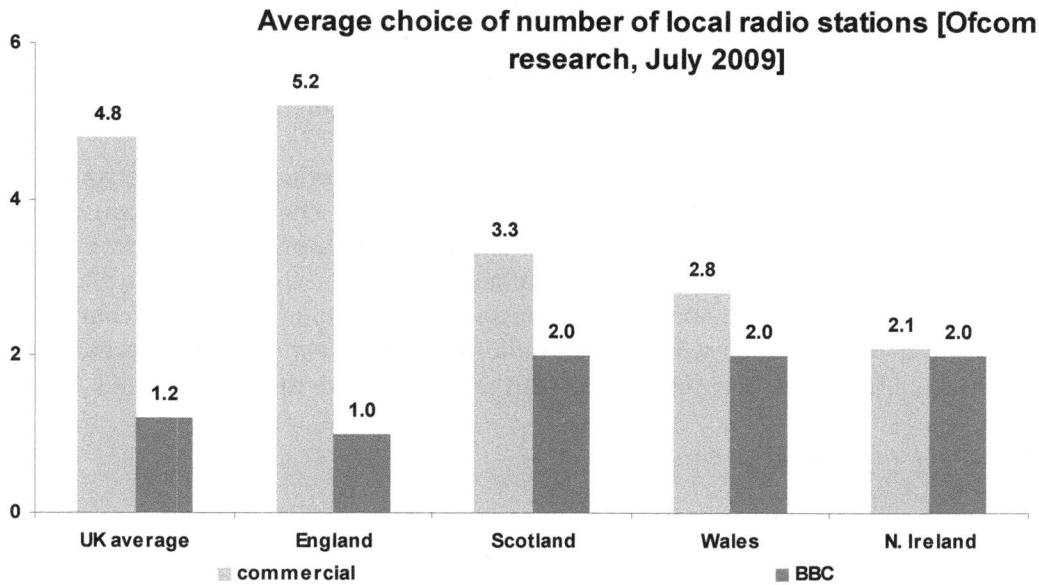

Average choice of number of local radio stations [Ofcom research, July 2009]

	UK average	England	Scotland	Wales	N. Ireland
commercial	4.8	5.2	3.3	2.8	2.1
BBC	1.2	1.0	2.0	2.0	2.0

241

This is demonstrated by analysis of the largest UK radio market, London, in which consumer choice is at its greatest. There are 29 licensed radio stations available on the analogue platform in London (excluding community radio and out-of-area stations), but the top 3 stations account for just under a third of all radio listening in London, and the top 6 stations account for almost half of all radio listening.[242] The radio market in London, as in most of the UK, is dominated by a tiny number of mainstream stations, whilst the remaining radio offerings comprise a 'long tail' that fulfils more specialist consumer needs.

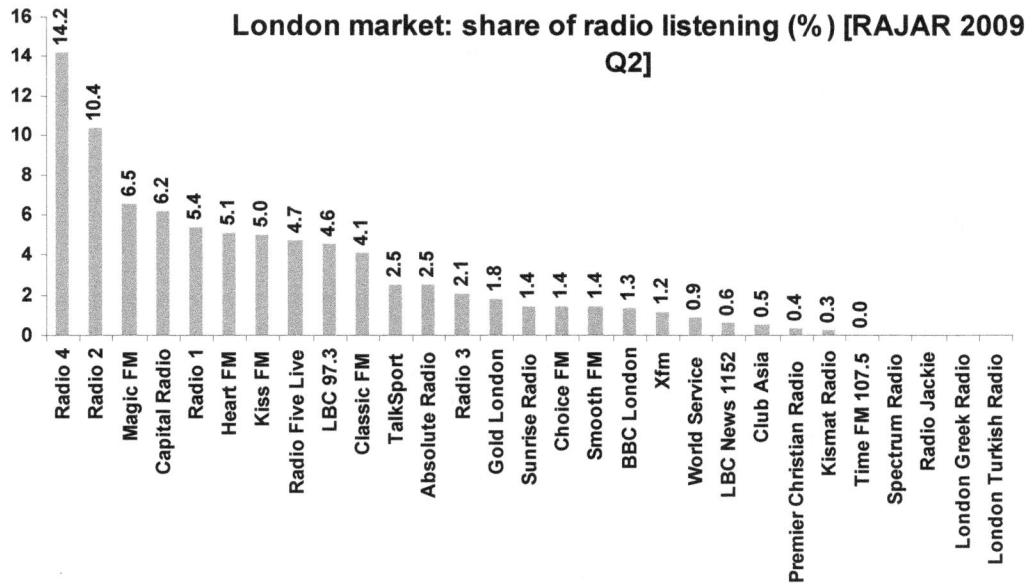

London market: share of radio listening (%) [RAJAR 2009 Q2]

Station	Share
Radio 4	14.2
Radio 2	10.4
Magic FM	6.5
Capital Radio	6.2
Radio 1	5.4
Heart FM	5.1
Kiss FM	5.0
Radio Five Live	4.7
LBC 97.3	4.6
Classic FM	4.1
TalkSport	2.5
Absolute Radio	2.5
Radio 3	2.1
Gold London	1.8
Sunrise Radio	1.4
Choice FM	1.4
Smooth FM	1.4
BBC London	1.3
Xfm	1.2
World Service	0.9
LBC News 1152	0.6
Club Asia	0.5
Premier Christian Radio	0.4
Kismat Radio	0.3
Time FM 107.5	0.0
Spectrum Radio	
Radio Jackie	
London Greek Radio	
London Turkish Radio	

243

The dramatic consumer skew towards mainstream radio means that, even in a radio market as developed as London, it proves difficult for incremental, digital-only stations to draw significant amounts of listening. The most listened to exclusively digital radio station in London is BBC 1Xtra, which ranks 22nd and attracts only a 0.5% share of listening in the market.[244] 1Xtra's content (UK black music) is barely duplicated by any other legal radio station available in London, and yet its 'success' remains slight in a very multicultural market that is already crowded with myriad radio options for consumers. The recent decision by London station Club Asia to enter administration, combined with the closures earlier this year of South London Radio and Time 106.8, demonstrate the challenge for stations to find a 'monetisable' audience in London, even on the analogue platform.

It might be easy to assume that Londoners, offered the widest selection of radio stations on the analogue platform, would be more satisfied with their choice in comparison with consumers in other, less well served parts of the UK. The surprising result from Ofcom research is that Londoners are, in fact, less satisfied with their choice of radio than most other parts of the UK. The chart below (extracted directly from a recent Ofcom report) demonstrates that satisfaction with existing radio provision is almost evenly spread across the whole UK, but consumers in London and Northern Ireland are the least satisfied.[245]

Satisfaction with choice of radio stations

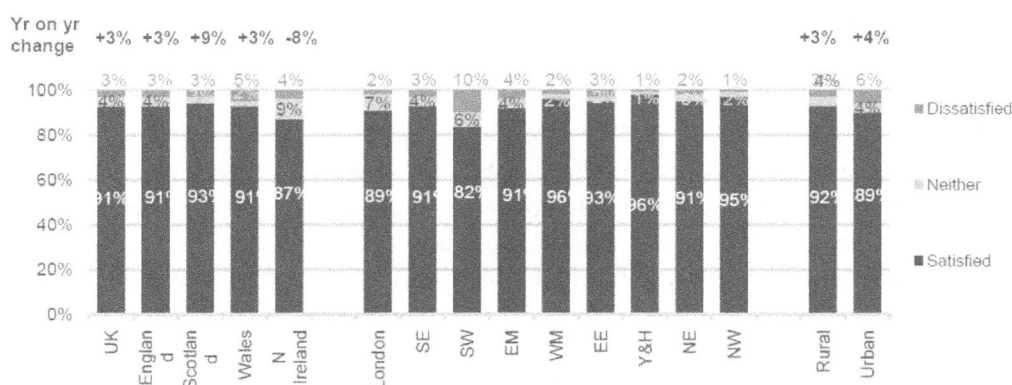

246

In summary, radio in the UK has been a victim of its own success. The universal availability of a range of both BBC and commercial 'national' stations, combined with the extensive choice of local stations available in most markets, mean that consumers are already relatively spoilt for choice on the analogue radio platform. There is very little unsatisfied demand for radio content because the UK already has such a comprehensive choice of radio content on offer. As a result, any new radio platform (DAB, satellite, online, etc) is going to find it hard to compete with the high quality and diverse choice of what is already on offer.

This was always going to make it tough for the DAB platform to entice consumers to purchase DAB receivers as anything other than a 'replacement' for their existing analogue radios. Unfortunately, the natural replacement cycle for radio receivers is so slow (maybe ten years or more) that it will never prove sufficient for a complete UK digital switchover to be co-ordinated for radio, as is happening in the television market. The UK has some of the best radio in the world – ironically, this has been our digital downfall.

CONSUMER DEMAND FOR DAB RADIO RECEIVERS

As noted earlier, consumer demand for DAB hardware is largely a function of demand for DAB content. You will only want to buy a DAB radio if there is something interesting enough you want to listen to on the DAB platform.

Ofcom research demonstrates clearly the lack of interest amongst consumers in purchasing DAB radio receivers. In this year's survey, only 16% of consumers (without a DAB radio) say they are likely to purchase a DAB radio within the next **12** months. Two years ago, 19% said they would be likely to purchase a DAB radio within the next **6** months. This is very bad news for manufacturers and retailers of DAB radios. Worse, this year not only do 64% of consumers say they are unlikely to purchase a DAB radio, but 20% say they don't know – a demonstration that a DAB radio is far from being a 'must have' gadget on consumers' wants lists.

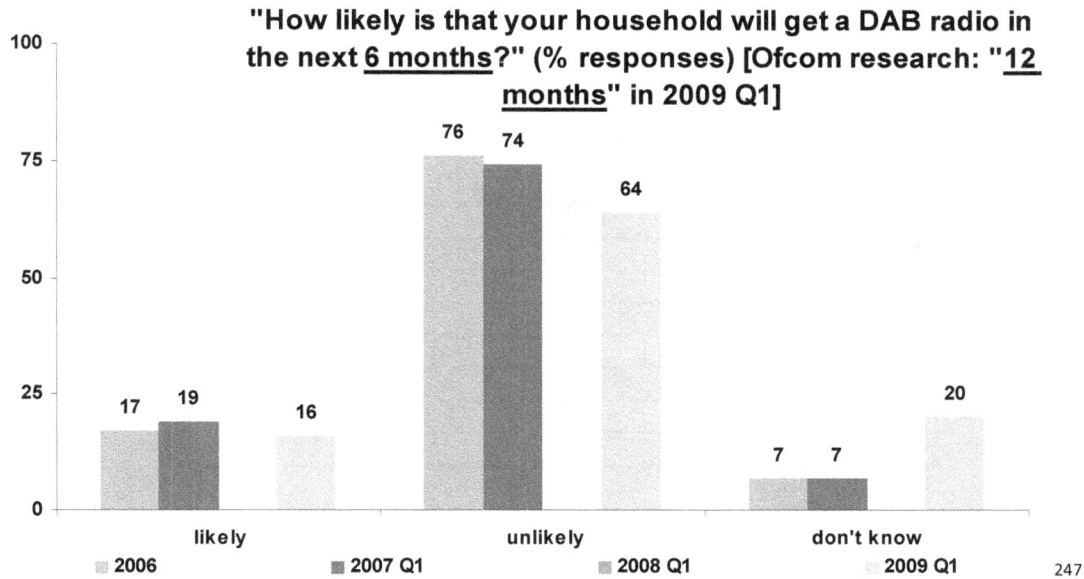

"How likely is that your household will get a DAB radio in the next 6 months?" (% responses) [Ofcom research: "12 months" in 2009 Q1]

likely: 17 (2006), 19 (2007 Q1), 16 (2008 Q1)
unlikely: 76 (2006), 74 (2007 Q1), 64 (2008 Q1)
don't know: 7 (2006), 7 (2007 Q1), 20 (2009 Q1)

▪ 2006   ▪ 2007 Q1   ▪ 2008 Q1   ▪ 2009 Q1   247

The data for current levels of DAB radio receiver ownership are not very helpful in determining the demand for DAB radio receivers. The quarterly survey by RAJAR found in 2009 Q1 that 32.1% of adult respondents claimed to own a DAB radio. However, the annual Ofcom survey found in the same quarter that 41% of adult respondents claimed to have a DAB radio in their household. This disparity between the results from RAJAR and Ofcom would appear to be widening over time.

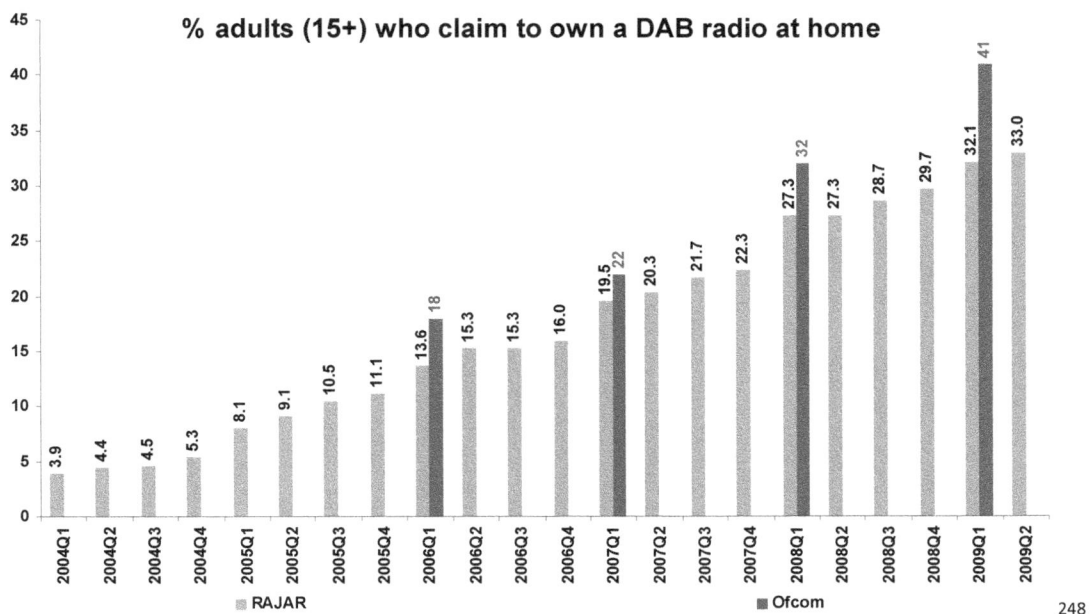

% adults (15+) who claim to own a DAB radio at home

	2004Q1	2004Q2	2004Q3	2004Q4	2005Q1	2005Q2	2005Q3	2005Q4	2006Q1	2006Q2	2006Q3	2006Q4	2007Q1	2007Q2	2007Q3	2007Q4	2008Q1	2008Q2	2008Q3	2008Q4	2009Q1	2009Q2
RAJAR	3.9	4.4	4.5	5.3	8.1	9.1	10.5	11.1	13.6	15.3	15.3	16.0	19.5	20.3	21.7	22.3	27.3	27.3	28.7	29.7	32.1	33.0
Ofcom									18				22				32				41	

▪ RAJAR   ▪ Ofcom   248

127

The uncertainty in the data regarding ownership levels of DAB receivers is not surprising, given the evident level of consumer confusion. Firstly, many radios on the market have the words 'digital' or 'digital radio' written on them, meaning that they either incorporate a digital clock (for radio alarm clocks) or that they offer 'digital' tuning of analogue wavebands, despite them not offering DAB reception. Secondly, the majority of 'DAB radios' presently on sale in the UK offer DAB reception in combination with analogue radio and/or internet radio. When DAB radio receivers were first introduced a decade ago, all the models offered were DAB-only. Nowadays, it is harder to find a DAB-only model in shops. Earlier this year, I surveyed the radio hardware on sale from UK retailers (see chart below) and found that the most common DAB consumer proposition is now an 'FM + DAB' radio.

NO. OF MODELS OF ELECTRONICS HARDWARE INCORPORATING RADIO PLATFORMS			
radio platforms	Argos	Currys	Comet
FM and/or AM	113	48	59
FM + DAB	56	55	43
DAB	2	9	10
DAB + FM + internet	4	4	2
DAB + internet	1	0	0
internet	3	2	0
internet + FM	1	2	2

excludes in-car and mobile phone hardware
web sites @ 10Feb2009   249

In its latest consumer research on take-up of digital radio, Ofcom said that the result of its survey (see below) "highlights the continued lack of awareness among consumers of ways of accessing digital radio."[250] Consumers have low awareness of their ability to already access digital radio, and It appears that the words 'digital radio', 'digital audio broadcasting' and 'DAB' are not yet precisely understood. This uncertainty makes the results of market research about ownership levels of DAB radio hardware somewhat unreliable.

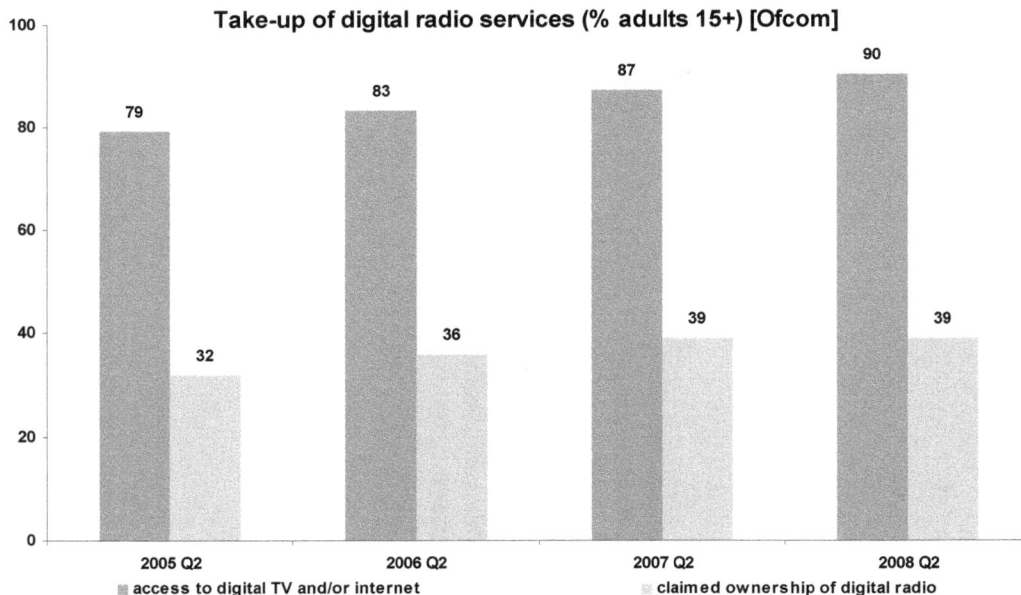

Take-up of digital radio services (% adults 15+) [Ofcom]

	2005 Q2	2006 Q2	2007 Q2	2008 Q2
access to digital TV and/or internet	79	83	87	90
claimed ownership of digital radio	32	36	39	39

251

One of the targets set by the Digital Radio Working Group at the end of 2008 for the implementation of digital radio was that DAB radios should reach 50% of radio receiver sales by volume by the end of 2010. However, if the current rate of growth continues, this target is unlikely to be reached until 2016 (see chart below).

DAB RADIO RECEIVER SALES (% total receiver sales)

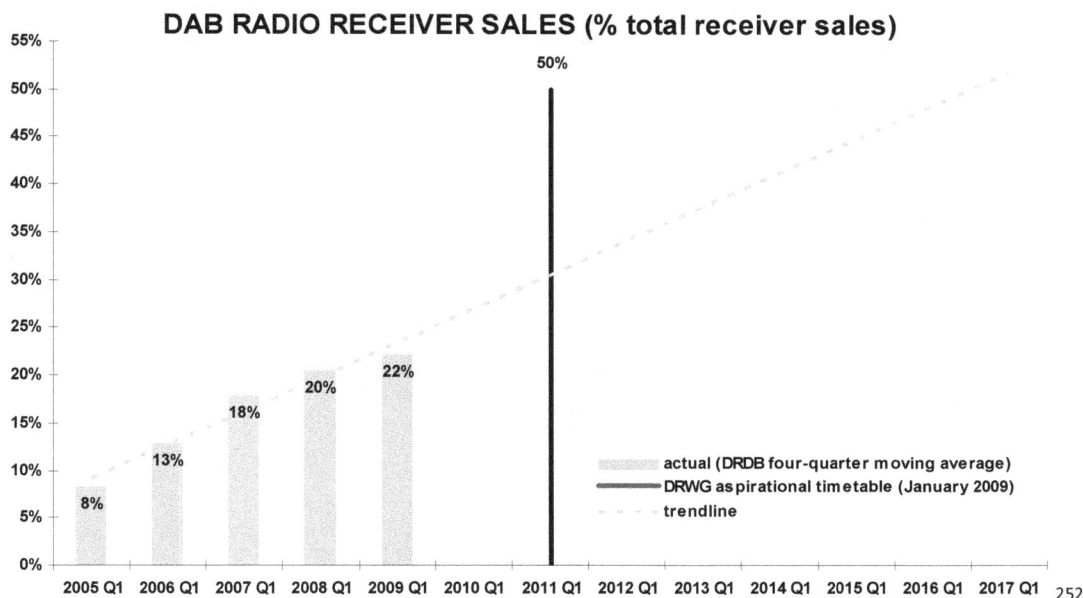

Besides, this target is largely irrelevant to digital switchover because it seems to assume that consumers are making a definitive choice between the purchase of a DAB radio **or** an analogue radio. In fact, as the earlier chart shows, the majority of DAB radios presently offered by retailers also include FM radio. Although the DRDB data state that 22% of radios sold in the UK incorporate DAB, the vast majority of those include FM too. So, for every 100 new radios sold, you are probably adding to the UK's inventory of receivers between 95 and 98 new FM radios, at the same time as adding 22 new DAB radios. In other words, the household penetration level of analogue radio receivers is barely diminishing at all, a fact that will ensure that FM broadcasting remains as vital to our radio system as it has always been.

In summary, the DAB platform seems to be developing slowly as a supplementary platform to existing analogue radio reception. Far from DAB radios 'replacing' analogue radios, the overwhelming majority of new radios purchased in the UK are still analogue-only. The remainder are mostly DAB/analogue combination receivers. In this way, DAB has much in common with 'Long Wave' radio, where consumers for a long time were offered a choice of 'FM+AM' or 'FM+AM+Long Wave' receivers in retail stores. Like Long Wave, for a minority of consumers DAB may be a 'must have' when purchasing a new radio, but for the majority it is merely an optional extra whose purchase is likely to be very dependent on the comparative prices of available options.

Conclusion

The publicly available data on the demand for DAB are not particularly encouraging for the platform's future. Much of the implementation of DAB to date in the UK has focused on 'supply-side' issues, without seeming to determine whether there is sufficient demand from consumers for new content, and without determining whether that new content would prove sufficiently attractive to lure consumers into shops to purchase DAB radios. Ironically, it appears that if our existing system of analogue radio broadcasting had been less well developed in terms of both the range of available content and its near universal delivery, DAB might have been better able to address any pent-up demand from consumers. As it is, the majority of consumers seem very content with their existing radio options. Our pursuit of excellence in radio over the last 80 years has created something we can be proud of – but it has also made it hard for it to be bettered by a 'new' system such as DAB.

---

[236] source: Ofcom

[237] source: RAJAR 2009 Q2

[238] source: Ofcom

[239] source: Ofcom

[240] source: Ofcom

[241] source: Ofcom

[242] source: RAJAR, Q2 2009

[243] source: RA|AR, Q2 2009

[244] source: RAJAR, Q2 2009

[245] http://www.ofcom.org.uk/research/cm/cmrnr09/charts/radio.pdf

[246] http://www.ofcom.org.uk/research/cm/cmrnr09/charts/radio.pdf

[247] source: Ofcom

[248] source: RAJAR and Ofcom

[249] source: Grant Goddard

[250] source: Ofcom

[251] source: Ofcom

[252] source: Digital Radio Development Bureau + Digital Radio Working Group

40.

19 September 2009

Digital radio in France: cold feet, no funding, sue the regulator

On 10 September, the French secretary of state responsible for the digital economy, Nathalie Kosciusko-Morzet, organised a seminar 'Digital: investing now for tomorrow's growth.'[253] The objective was to lay out to the 1000 attendees the costs and opportunities necessary to create an integrated digital economy.

On 16 September, Jean-Luc Hees, head of state broadcaster Radio France, addressed the National Assembly's Committee on Finance & Cultural Affairs.[254] He told them that the broadcaster's advertising revenues were forecast to decline in 2009 by 20 to 30% year-on-year (advertising comprises 8% of revenue, the remainder from the state). He said that the rollout of digital radio in 2010 would require adding 2m to 3m Euros to the budget.

Hees told the National Assembly:

"We now know fairly well the timing of the introduction of digital terrestrial radio, with launches in the coming months in three areas – Paris, Marseille and Nice. …. Our goal is to achieve 95% coverage of France by the end of 2013, according to the CSA's [France's media regulator] schedule. …. We must understand that everything has a cost, and the impact on Radio France's finances means that this house will have to fund dual transmission [analogue and digital] for some time. Analogue transmission presently costs Radio France 80m Euros per annum. The rollout of digital radio will entail additional costs and this is one of the things that require funding in our next budget. I want to emphasise this."[255]

Amongst commercial radio operators, opinions on digital radio appear increasingly ambivalent. Franck Lanoux, deputy director of NextRadioTV said that digital radio *"will not affect 95% of radio listening. It's hard to identify how digital radio will develop – the receivers do not exist, yet the broadcasters are being asked to make significant investments. Consumers have nothing to listen with."[256]*

The publication 'mediasactu' commented this week that digital radio in France has become ensnared in a quagmire and that reservations amongst commercial broadcasters are becoming stronger:

"After putting all their weight behind persuading the government to adopt the T-DMB standard for digital terrestrial radio in December 2007, which is now nearly two years ago, radio broadcasters, particularly those that are members of GRN [France's Digital Radio Group] and the Bureau de la Radio [France's newly created radio trade body comprising the four largest commercial owners], are much more dubious about the real chance of succeeding with the transition to digital radio. Although they were unwavering only a few months ago, now major national radio groups, along with SIRTI [the French broadcasting trade union] and the regional stations, seem determined to thwart digital radio, or at least

seriously slow down its development. Angered by the CSA's decision to only select a handful of new markets to launch digital radio, as well as the costs that will be inherent with dual transmission for several years, not to mention the CSA's cancellation of applications for 16 of the first 19 areas designated for digital radio, the broadcasters are now showing real reluctance."

"The [radio] sector, already suffering from its current lack of revenues, is still waiting for the financial aid promised by the government to fund the migration to digital radio. Now, it seems clear that the total cost of the implementation of digital radio will be made greater by the choice of the T-DMB standard by the Ministry of Culture, at the request of GRN, and that the rollout will be much more expensive than it would be for the DAB+ standard. Moreover, according to our sources, the major radio groups are now putting all their weight behind challenging the decision of the CSA about the channel composition of the first multiplexes in order to delay their launch and buy extra time. According to our sources, some have already initiated legal action against the CSA." [257]

At the government seminar on 10 September, CSA president Michel Boyon reportedly took the opportunity to try to 'save' digital radio, suggesting that part of a new significant loan raised by the French government should be allocated to the rollout of digital radio. But, as mediasactu commented,

"The problem is whether public funds can be used to finance the construction of radio networks intended to broadcast commercial stations. Will the public agree to fund not only a digital transmitter network for commercial radio, but also the purchase, at great expense, of receiver hardware that offers a range of stations almost identical to what is already offered on FM?" [258]

Interviewed in Le Figaro, RTL president Christopher Baldelli said:

"The timetable [for digital radio] will be respected if the CSA believes it is right, but migration to digital terrestrial radio is not a matter of principle. It involves a different economic issue than digital television. One impact is higher transmission costs, at approximately 3m Euros per channel, costing us 12m Euros for the whole [RTL] network (RTL, RTL2, Fun Radio, RTL-L'Équipe). The economic difficulty must be taken into account. It is a matter for the radio groups, the CSA and the government. It will take a lot of consultation." [259]

As mediasactu concludes in its article:

"Overtaken by the internet, mobile phones and MP3 players, does digital terrestrial radio still stand any chance of seducing the general public?" [260]

[253] http://www.tech.youvox.fr/Seminaire-Numerique-Grand-Emprunt,1207.html
[254] http://www.latribune.fr/entreprises/communication/publicite-medias/20090916trib000422780/baisse-des-recettes-publicitaires-de-radio-france.html
[255] http://www.satmag.fr/affichage_module.php?no_theme=1&no_news=11262&id_mod=50
[256] http://www.radioactu.com/actualites-radio/114103/csa-le-grand-emprunt-pour-financer-la-radio-numerique/
[257] http://www.radioactu.com/actualites-radio/114103/csa-le-grand-emprunt-pour-financer-la-radio-numerique/
[258] http://www.radioactu.com/actualites-radio/114103/csa-le-grand-emprunt-pour-financer-la-radio-numerique/
[259] http://www.lefigaro.fr/medias/2009/09/03/04002-20090903ARTFIG00180-baldelli-nous-allons-reduire-les-couts-de-rtl-.php
[260] http://www.radioactu.com/actualites-radio/114103/csa-le-grand-emprunt-pour-financer-la-radio-numerique/

41.

18 October 2009

Sweden: digital radio to be distributed by digital TV network, not DAB+

In December 2005, the Swedish government had announced that it would not expand the existing DAB radio transmission system that already covered 85% of the country and would not propose a consumer migration from FM to DAB radio.[261] Instead, it suggested that the radio industry should focus on a mix of digital platforms including podcasts, mobile phones and distribution via TV.

In June 2008, Sweden's broadcasting authority suggested to the government that the DAB+ codec should replace the country's existing DAB system.[262] However, the government's IT consultant Patrick Fallstrom said that both DAB and DAB+ were an old-fashioned solution and that today's consumers were more likely to listen to audio via the internet, mp3's or their mobile phone.

Now, after Swedish culture minister Lena Adelsohn Liljeroth spoke at the recent 'Radio Day of European Cultures' event, it was suggested that there is no need for a DAB+ radio network in Sweden.[263] Instead, it was proposed that digital radio be carried on the existing DVB-T digital television network which is about to be upgraded to DVB-T2, creating 80% more space for high definition TV and radio channels. Test transmissions of DVB-T2 are scheduled to start in Stockholm and Uppsala before Christmas.

The perceived advantages are: no dual TV/radio transmission system, reduced transmission costs, the digital TV transmission network is already built, DVB-T already provides better coverage (99.8% of the population) and sound quality than the DAB network, increased energy efficiency, plenty of available spectrum, and mobile reception is supported (cars and phones). The perceived disadvantages are: the DVB-T2 system is not yet established, no radio receivers have yet been developed, and there is no lobby group for the system (as there is for DAB+).

One Swedish media commentator noted:

"We do not need more transmission networks in Sweden. We must share the resources that we have. This is what I hope the government has understood. It seems as if it has. ... It is clear to me that Swedish state radio should not be trying to build its own distribution mechanisms, and especially not its own transmission networks. Its money should be used as efficiently as possible to ensure that Swedish state radio programmes can be heard in as many places as possible. This would include DVB-T2, mobile phone networks, the internet, and FM for the foreseeable future."[264]

Per Gulbrandsen of Swedish state radio commented:

"The government is providing no more money for digital broadcasting and wants the market to decide upon the digital migration of radio. But it is no secret in the industry that the Ministry of Culture dislikes the ageing DAB digital radio system, even in its newer DAB+ form."[265]

---

[261] http://www.sr.se/sida/artikel.aspx?ProgramId=2205&Artikel=555676

[262] http://www.sr.se/cgi-bin/International/nyhetssidor/arkiv.asp?ProgramID=2054&formatID=1&Max=2008-07-01&Min=2003-09-14&PeriodStart=2008-06-29&Period=1&Artikel=2160222

[263] http://www.sr.se/sida/artikel.aspx?programid=1012&artikel=3132830

[264] http://www.medievarlden.se/component/rssfactory_pro/viewrss/mindpark-026-nandaumlr-dab-fandoumlrandaumlndrade-sveriges-radio_109672

[265] http://www.sr.se/sida/artikel.aspx?programid=1012&artikel=3132830a

42.

20 October 2009

Norway: government commissions cost analysis of FM switch-off

In April 2009, NRK [state radio] chief Hans-Rore Bjerkaas had written to the Culture Minister asking if the government could persuade consumers to purchase DAB radios rather than analogue radios. He wrote:

"In order to reduce consumer disappointment in the final stages of digital migration, a clear political signal must be given that, when choosing a radio, an FM/DAB radio should be selected rather than a pure FM radio."[266]

Culture Minister Trond Giske responded:

"My responsibility lies with radio listeners. There are two reasons why I think we should progress more slowly than with the introduction of digital TV. It will be more costly for consumers to switch to DAB because most people have multiple radio receivers. In addition, there is no significant improvement in audio quality. When half the population has purchased a digital radio, we would be willing to discuss a specific date for switching off the FM network."[267]

At present, 17% of the population in Norway has a DAB radio.

This month, the Culture Minister reiterated that the government will not consider switching off FM until half the population has purchased a DAB radio.[268] Both NRK and P4 [commercial network owned by MTG] have told the minister that their existing FM networks are expensive to maintain and that they want to move to DAB as quickly as possible. NRK spends NRK120m per annum on FM radio transmission. Transmission provider Norkring [owned by Telenor] says it will need to make a major upgrade to the FM network in 2014 to keep it in service. So the government has tendered for an *"independent and expert"* assessment of the costs associated with switching off FM broadcasts in either 2014 or 2020. The analysis will be part of the White Paper on digital radio to be published by the government next year.

---

[266] http://www.tu.no/it/article206802.ece
[267] http://www.tu.no/it/article206802.ece
[268] http://www.tu.no/it/article225033.ece

43.

21 October 2009

Germany: national public radio to end DAB broadcasts year-end

At the end of September 2009, the Radio Council of Germany issued a public statement in which it objected to the withdrawal of further funding for DAB radio by the KEF [the organisation that allocates funds for public radio] in July 2009 and it made a direct appeal to the Prime Minister to create an independent digital platform for radio broadcasting.[269] It said the rollout of DAB+ to replace DAB would now be cancelled. It argued that the phased closure of analogue radio broadcasting between 2015 and 2020 was realistic and that the government should adopt a legal framework for digital radio migration.

Now, in an interview this month with 'Digitalmagazin', the director of Deutschlandradio (German national public radio) Willi Steul confirmed that its two national stations will end broadcasting on the DAB platform at the end of 2009.[270]

Q: *'Deutschlandradio Kultur' and 'Deutschlandfunk' will not be available on DAB radio after the end of the year. What led to this bitter decision?*

Steul: *We need to face up to the fact that the KEF removed funding for further DAB broadcasting, including the DAB broadcasts of our two channels. This is all the more regrettable because DAB will no longer be available for the digital distribution of our new, knowledge-based, educational station.*

Q: *To what extent does this sound the final death knell for DAB?*

Steul: *So far, not yet. However, DAB – although vital – is in intensive care and living a sad, hospitalised existence. A paradoxical situation!*

Q: *We are meant to be re-launching with DAB+. Is this country making the transition to the age of digital radio, or will Germany remain an 'analogue island'?*

Steul: *At the moment, we are on the low road to becoming a glorious analogue island. In the medium- and long-term, there is no alternative to the age of digital radio. Let's see when the powers will be prepared to take responsibility for media policy and offer some certainty.*

Q: *The Radio Council has criticised the KEF decision as "unacceptable interference in the broadcasting policy of the German states." Why?*

Steul: *The role of the KEF is to identify the financial needs of the public service broadcasters. Its job is not to forge media policy, even when there are fiscal issues, as with DAB. This is unacceptable.*

Q: *But is it not the responsibility of the KEF, in the absence of sound arguments – and this seems to be the case with DAB – to pull the plug?*

Steul: *You mean 'economically'? If you invest properly for the future, you cannot expect immediate returns on your investment. The essence of investing is that money needs to be spent on innovation and a process of transformation that will only bear fruit at a later date. The digitisation of radio is, in many ways, a lengthy process and so it is premature to pull the plug now and jeopardise the investments made to date. This is definitely not economical!*

Q: *The [KEF] budget for digital radio has not been scrapped, but assigned to new initiatives. What options are there now for digital radio?*

Steul: *Digital radio is already available via cable and satellite. Also, internet streaming has a great future. There will be further developments, particularly in mobile internet access. At the moment, internet 'on the move' – in spite of flat tariffs – is still a very expensive pastime.*

[edited]

Q: *The Radio Council considers it is realistic to propose a switch-off of analogue radio between 2015 and 2020. How will this be achieved, given the estimated 300 million FM receivers in Germany?*

Steul: *It will not happen without appropriate incentives for the transition from analogue to digital. Such issues are not the major responsibility of radio people. They are an issue for media policymakers and receiver manufacturers. When people are offered something of interest to them, they will give up their old FM radios. Examples from other sectors should encourage us.*

---

[269] http://www.digitalfernsehen.de/news/news_832252.html
[270] http://www.infosat.de/Meldungen/?msgID=55341

44.

23 October 2009

France: digital radio launch postponed to mid-2010

The launch of digital terrestrial radio in France has been postponed from December 2009 to mid-2010.

"It will take us, I think, until the middle of next year," said Rachid Arhab, president of the CSA [France's broadcast regulator] digital radio working group. *"I have learnt not to trust dates."* He continued: *"What we had not anticipated was the impact of the credit crunch on advertising revenues, particularly in the radio sector, so the particular speeds of the different stakeholders are unknown."*[271]

Speaking at the Siel-Satis-Radio event in Paris, Arhab switched on France's first digital terrestrial radio transmitter and said:

"The greatest difficulty is knowing if all the radio groups want to migrate to digital radio at the same speed. … Today, I feel and I know that some of you are telling us 'we are ready'. We are delighted. A few months ago, this was not the case. … One must not be scared of analogue radio switch-off. Digital radio will not be a success if it has to co-exist with analogue radio for fifteen to twenty years."[272]

According to SatMag, at the beginning of November, there will be 'round table' meetings at the CSA with all the licensed digital radio operators, the set manufacturers, the transmission providers, and representatives from the Ministry of Culture & Communication and the Ministry of Finance & Industry.[273] The CSA is awaiting two reports: one from Marc Tessier on the economic conditions for the rollout of digital radio and on competition issues; the other by Emmanuel Hamelin on the funding of community radio.

It is reported that licences have been signed, but it will take two months for the multiplex operating companies to be formalised for the launch of digital radio in the first three areas. The composition of the multiplexes has not yet been determined. The time period between which the multiplex contracts are signed and the content providers launch digital stations still needs to be fixed, probably around six months.

Elsewhere at the Paris event, SatMag reported that the issue of the T-DMB digital radio standard adopted in France was back on the table. Its report said:

"Is it right that, in France, we are using a standard that is different from the rest of Europe? Should we not be offering radio receivers that are compatible with DRM+? Alan Mear [of the CSA] says that this is not a taboo subject and will be revisited, but he also agreed with Mathhieu Quetel of SIRTI [the trade body for independent regional and local stations] that it was essential to launch digital radio and not to revisit the question of the adopted standard. Besides, Rachid Arhab agreed yesterday that the DRM+ standard was expensive and of no interest."

"There was a big surprise from Michel Cacouault of the Bureau de la Radio which represents the main French commercial radio groups. Remember that it was they who said France had to adopt a particular digital radio standard as it was essential to transmit additional data. Today there was a complete turnaround. Michel Cacoualt reminded us that commercial radio had lost around 18% of its revenues in the credit crunch. Now, the owners want to cut their costs and are willing to choose a different standard that is less expensive. Those in the conference room familiar with this issue were amazed. Well yes! The credit crunch does makes you think. The cost of dual transmission [analogue and digital] for a single national network is estimated to be 2 to 4 million Euros [per annum], though what it will actually be we will only know when it happens." [274]

271

http://tempsreel.nouvelobs.com/depeches/medias/20091020.REU6626/le_lancement_de_la_radio_numerique_terrestre_attendra_m.html

[272] http://www.radioactu.com/actualites-radio/116615/

[273] http://www.satmag.fr/affichage_module.php?no_theme=1&no_news=11429&id_mod=50

[274] http://www.satmag.fr/affichage_module.php?no_theme=1&no_news=11433&id_mod=50

45.

25 October 2009

Culture Secretary speaks about digital radio

The House of Commons Culture, Media & Sport Committee [excerpts]
20 October 2009 @ 1100 in the Thatcher Room, Portcullis House

John Whittingdale MP, Chairman [JW]
Ben Bradshaw MP, Secretary of State, Department for Culture, Media & Sport [BB]

JW: *You have announced very ambitious plans to deliver the Digital Radio Upgrade programme by 2015 and have most of the national stations to move off analogue to digital by then. That will require extensive investment in the digital transmission network. What estimate do you have of what it is going to cost to do that?*

BB: *The current estimate that we are working on is about, I think I'm right in saying, is it £10m per year to build out the DAB multiplexes? Is that the figure that you were interested in?*

JW: *Actually, the one I've heard is rather more than that. Where is that money going to come from?*

BB: *It will come from a mixture of sources. We expect the BBC to play a significant role in this, commercial radio, and there may be public funds as well.*

JW: *I think the current state of commercial radio means that their ability to invest any more is almost zero. Do you foresee, therefore, further government investment, maybe from the Licence Fee?*

BB: *We are not currently intending to spend …. [laughs] That's one of the things we are not intending to spend a share of the Licence Fee on, but if there is an even bigger underspend in the Digital Switchover Programme than we are currently expecting, who knows, Mr Chairman?*

JW: *The Digital Switchover Programme appears to be earmarked for quite a large number of purposes.*

BB: *[laughs] Well, there is quite a significant underspend.*

JW: *But you are confident that it can be delivered. And what are you going to say to all the people that haven't bought a new car in the last two years by 2015?*

BB: *We are working with the motor manufacturers, both to ensure that future new cars do [have DAB radio], but also to ensure that there is this – I can't remember what it is called –*

but it is some sort of gadget that you will be able to use in your existing car to make sure that you can pick up digital radio. One of the things we say quite clearly is that we won't go ahead with this unless, by 2013, certain conditions are reached ie: we have more than 50% digital radio ownership and that [DAB] reception on all of our main roads is not going to be a problem. So we have put conditions down but, at the same time, we felt that it was important to provide market certainty that we specified an end-date by which time this should happen.

46.

29 October 2009

Digital platforms: commercial radio losing share to BBC

Today's RAJAR data demonstrate that a gulf is opening up between BBC radio and commercial radio in their ability to attract listening to digital platforms. Over the last year, the BBC is accelerating away from commercial radio in its audience's usage of DAB, digital television and the internet to listen to live radio programmes. The significance of this growing gulf is reinforced when one remembers that the main RAJAR survey, from which the data below is taken, only measures 'live' radio listening and does not incorporate listening to either time-shifted, on-demand radio ('listen again') or to downloaded podcasts, both forms in which the BBC offers a much greater volume of content than UK commercial radio.

The danger here is that the BBC is poised to dominate listening on digital radio platforms in the long term, exactly as it already dominates listening on analogue radio platforms. One of the main reasons that the commercial radio sector invested so heavily in digital platforms during the last decade was the opportunity it offered to compete more effectively with the BBC for audiences. In the analogue world, the commercial sector has always argued that the BBC (having been there first) was allocated more and better spectrum for its radio stations. 'Digital', particularly DAB, seemed to offer the commercial sector a chance to 'even the score' with the BBC. The RAJAR data show that this ambition is not succeeding.

Across all digital platforms aggregated, commercial radio is losing ground, with the latest quarter reducing its share of listening to 41%, versus the BBC's 56% share.[275]

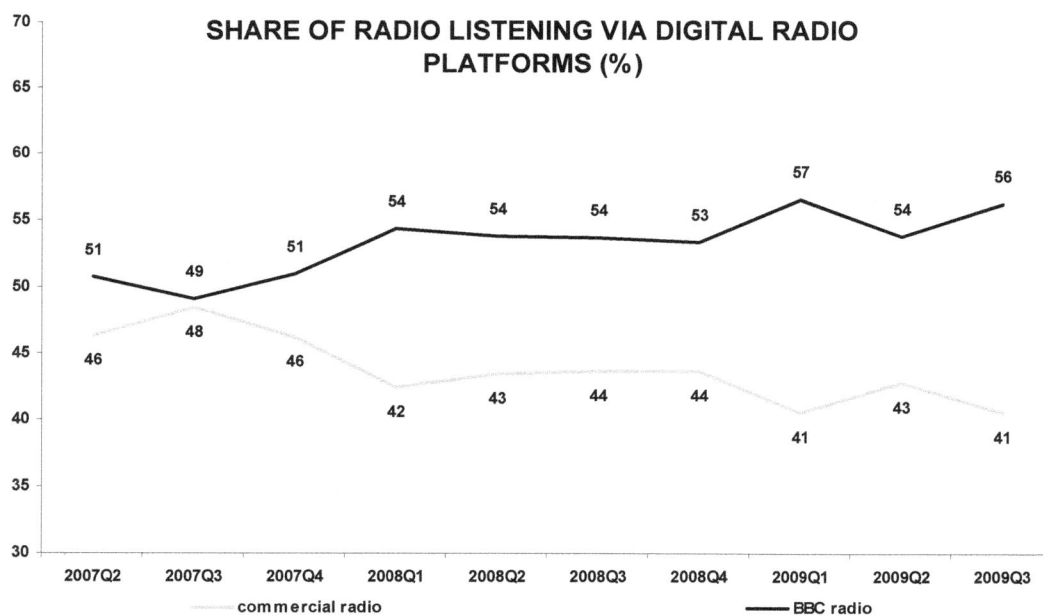

SHARE OF RADIO LISTENING VIA DIGITAL RADIO PLATFORMS (%)

276

142

Taking each digital platform in turn, commercial radio's share of listening on the DAB platform fell to 33% in Q3 2009, compared to the BBC's 65%.[277] This is not surprising because the age profile of DAB purchasers tends to be older listeners who are statistically more likely to listen to BBC stations. However, it does pose a grave question as to the return that commercial radio can expect from its substantial investment to date in DAB infrastructure, if listening on that platform is dominated so much by the BBC.

SHARE OF RADIO LISTENING VIA DAB PLATFORM (%)

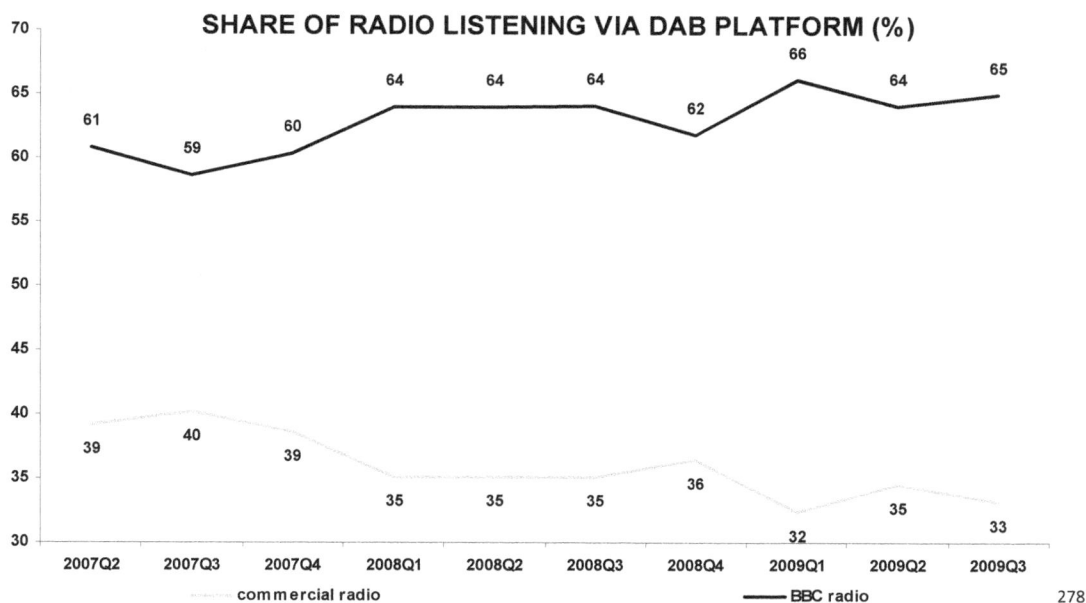

commercial radio   BBC radio   278

The digital TV platform is one that commercial radio has long dominated because of the large amount of spectrum it leased in the early days of Freeview. However, the increasing popularity of digital terrestrial television has already substantially increased the cost of spectrum on Freeview for the radio industry when its contracts come up for renewal. Furthermore, the forthcoming re-ordering of the multiplexes to accommodate HD television and new compression codecs is likely to squeeze

SHARE OF RADIO LISTENING VIA DIGITAL TV PLATFORM (%)

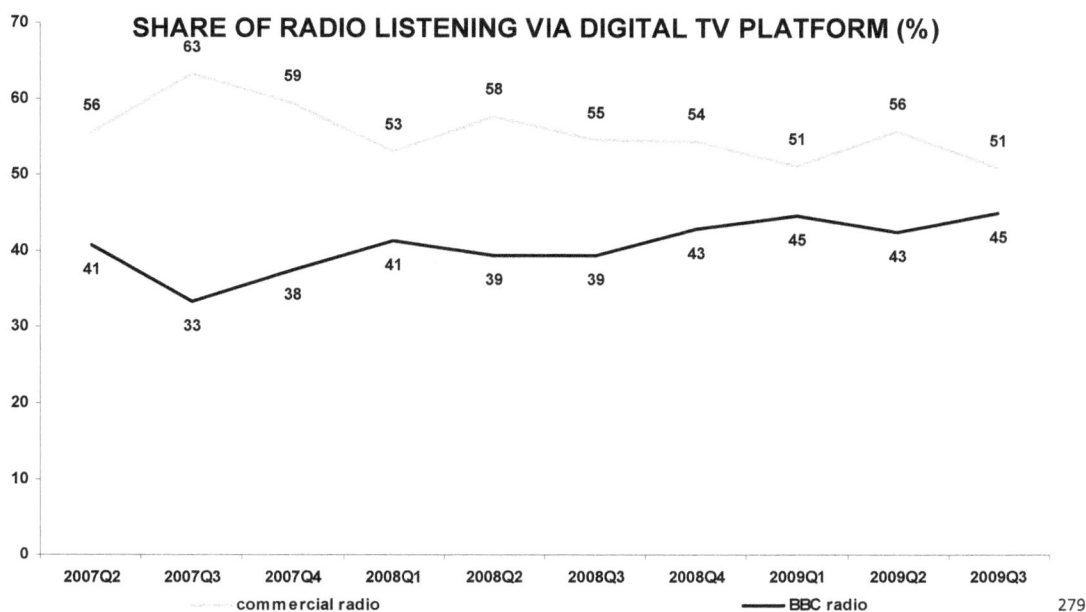

commercial radio   BBC radio   279

143

commercial radio's access to Freeview spectrum even more so. Before long, it is likely that the BBC will dominate the digital TV platform, just as it already does on DAB. Presently, the BBC has a 45% share, compared to commercial radio's 51%.[280]

As might be expected, the BBC's strong online presence has already put it in the commanding position in terms of its share of listening via the internet platform. The integration of BBC radio into the iPlayer has no doubt helped as well, whereas commercial radio's offerings are relatively more fractured and less heavily marketed, despite the excellent innovation of the RadioCentre Player. The BBC has a 50% share of listening on the internet platform, compared to commercial radio's 37%.[281]

SHARE OF RADIO LISTENING VIA INTERNET PLATFORM (%)

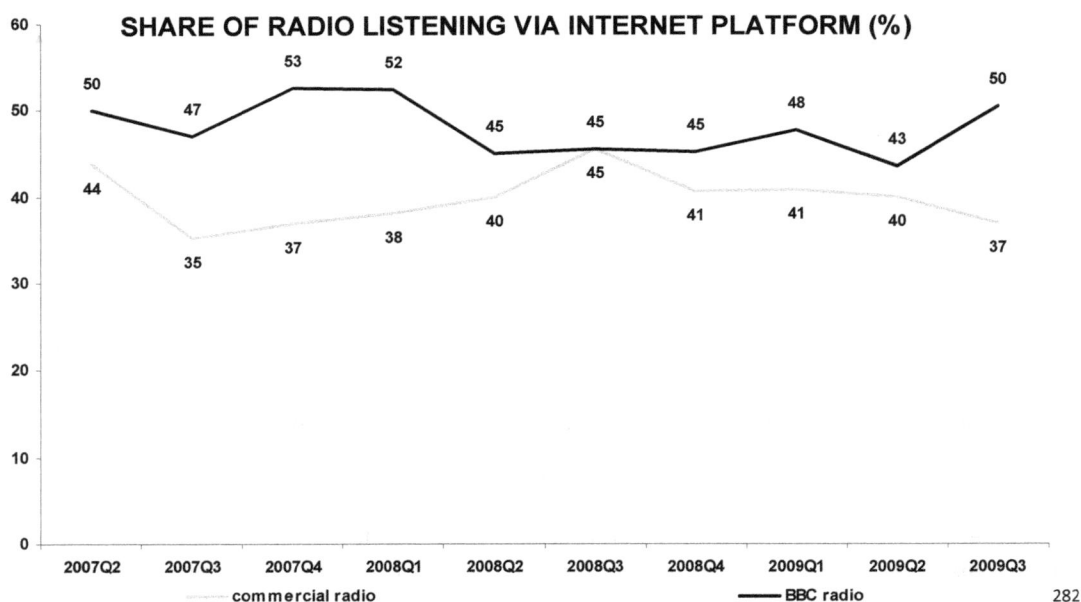

commercial radio   BBC radio   282

The significance of commercial radio's diminishing share of these three digital platforms is demonstrated when we look at the two sectors' listening shares achieved on the analogue platform alone. Once one removes the digital platforms from the picture, it is evident that the shares of both the BBC and commercial radio have remained relatively stable in recent years. In other words, it is commercial radio's declining share of listening on digital platforms that is effectively pulling the sector's total share of listening (analogue + digital) down, particularly as digital platforms are growing as a proportion of total radio listening (21.1% in Q3 2009).[283]

There is a paradox here. The commercial sector invested heavily in the DAB platform, believing that the new technologies would help it **INCREASE** its overall share of radio listening versus the BBC. In fact, that investment has recently helped to **DIMINISH** commercial radio's overall share of listening. Digital television remains the only platform in which commercial radio dominates, and yet this is the very platform where commercial radio will be forced to cede spectrum and face, once more, losing out to the BBC whose spectrum for radio is guaranteed.

It is important to emphasise that these graphs show only the **SHARE** of listening on these platforms. The volumes of listening on each of these platforms have demonstrated absolute growth for both commercial radio and for the BBC over the same time period. But, more than any other digital platform, it is significant that the DAB platform is dominated by the

144

BBC which now accounts for almost two-thirds of its usage. Such data are important when making decisions about the potential returns

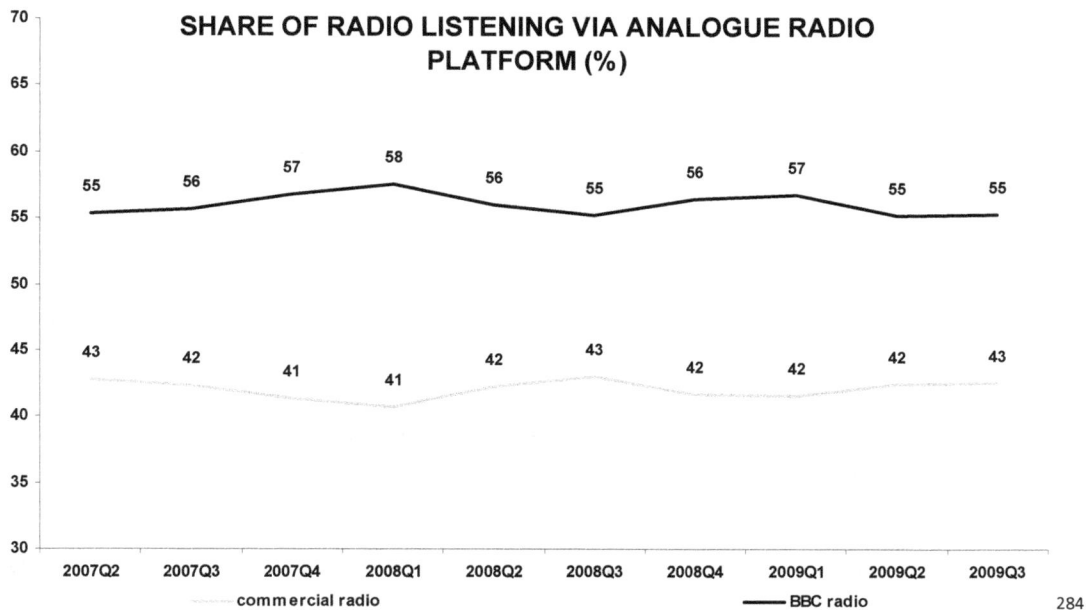

SHARE OF RADIO LISTENING VIA ANALOGUE RADIO PLATFORM (%)

	2007Q2	2007Q3	2007Q4	2008Q1	2008Q2	2008Q3	2008Q4	2009Q1	2009Q2	2009Q3
BBC radio	55	56	57	58	56	55	56	57	55	55
commercial radio	43	42	41	41	42	43	42	42	42	43

284

on further investments in DAB infrastructure. Will further investment simply maintain the existing imbalance, or will it really improve commercial radio's share? Does investment in infrastructure also require parallel investment in new content that will appeal directly to the older age groups who own DAB radios?

Some possible reasons for commercial radio's diminishing share of listening on digital platforms include:

- commercial radio's tendency to invest in DAB infrastructure more significantly than in original digital-only content
- recent closures of many digital-only radio stations in the commercial sector
- the BBC's relatively stable resource base, at a time when commercial radio revenues are falling precipitously
- the BBC's long-held policy to invest simultaneously in multiple platforms, whereas commercial radio has focused on DAB and, to a lesser extent, Freeview
- the BBC's focus on creating exclusive digital-only content unavailable on the analogue platform
- the BBC's 360-degree music royalty agreements which allow it to use diverse platforms, whereas commercial radio requires separate (and more restrictive) agreements for time-shifted content and podcasts
- the BBC's long-term, consistent promotion of content and digital platforms across TV, radio and the internet whereas commercial radio is less willing to cross-promote content or digital platforms that migrate listeners away from its core analogue offerings
- frequent management changes and ownership changes in some parts of commercial radio, where substantial consolidation has often translated into short-term 'slash and burn' rather than 'invest and build' policies.

Whatever the reasons, we are not where we were meant to be – that is, we are not where it had been anticipated more than a decade ago commercial radio would be when investment in digital platforms, notably DAB, was expected to produce a beneficial outcome for

145

commercial radio audiences versus the BBC. To put it plainly, the strategy conceived in the 1990's has not worked. Commercial radio offerings do not dominate digital platforms (yes, they are more numerous, but they do not attract more hours listened than the BBC). DAB has become a largely BBC platform.

So, what can be done? Some of the issues noted above require a more level playing field to be established between commercial radio and the BBC. One such example of a practical solution is the Radio Council plan for a new UK Radio Player that will offer BBC and commercial radio content from a single aggregated access point. Other issues remain mostly in the lap of the gods (revenues, for example). Some issues require the BBC to be less predatory (or more regulated) and for the commercial sector to be more focused on strategic, long-term objectives (such as an online strategy that is more than simulcasting).

There is no single answer to this complex problem, though the commercial radio sector is hobbled by both its present lack of profitability and the regulatory strings that are attached to the majority of its analogue radio licences. What is desperately needed in these difficult times is not minor regulatory tinkering (such as adjusting how many hours of local content a local station is required to broadcast) but a wholesale change in strategy to maintain a commercial radio sector that can thrive in the digital marketplace we now inhabit. Will the imminent Digital Economy Bill prove sufficiently forward-thinking in its radio policy proposals?

[Statistical note: The graphs above to do not sum to 100% because the minimal amount of platform data released by RAJAR are 'rounded' (hours listened to 1,000,000; listening shares to 0.1%) and the listening apportioned to the BBC and commercial radio sometimes does not add up to the total for a platform. Some of this shortfall may be accounted for by 'other' listening (neither the BBC nor commercial radio) which is not itemised by platform. Data for individual quarters are therefore somewhat inconsistent, though the trend over several quarters is likely to be indicative. Additionally, there is an element of radio listening unattributed to any platform, 12.8% of the total in Q3 2009, but which is roughly equally applicable to BBC radio and commercial radio.]

---

[275] source: RAJAR, Q3 2009
[276] source:RAJAR
[277] source:RAJAR
[278] source:RAJAR
[279] source:RAJAR
[280] source:RAJAR
[281] source:RAJAR
[282] source:RAJAR
[283] source:RAJAR
[284] source:RAJAR

47.

30 October 2009

Digital Radio Upgrade and the Digital Economy Bill

Westminster eForum Parliamentary Reception[285]
Terrace Pavilion, House of Commons, London
28 October 2009 @ 1600

"The informal discussion that takes place can be expected to cross a range of current policy issues but the chosen theme is *digital switchover and DAB*."

John Whittingdale MP, Chairman, House of Commons Culture, Media & Sport Select Committee:

The future of radio is very much a topic under debate. My Select Committee is currently conducting an inquiry into the future of local and regional media, of which radio is an absolutely critical part. So, yesterday, we were hearing evidence from Andrew Harrison of RadioCentre, Travis Baxter [of Bauer Media] who is here somewhere today, and Steve Fountain from KM Group. And we are very much aware of the pressures on commercial radio and the difficulties faced. But, at the same time, there are opportunities. And when Digital Britain came out, much of it had been trailed in advance, a lot of it quite controversial – things like top-slicing and file-sharing legislation – but the one bit which came as something of a surprise, I think, was the announcement of the date for Digital Radio Upgrade. Certainly, when I saw that in the Report, my immediate reaction was rather like the 'Yes Minister' Permanent Secretary who said: "That is a very brave decision, Minister."

It is going to be challenging. It is slightly controversial. Not everybody in the industry is 100% yet signed up to it. Equally, there is a cost attached and we can have interesting debates about who is going to pick up the bill for it. And there will be quite a task to persuade people. In the same way that we had to work hard to persuade people that analogue switch-off of television was going to be beneficial, I think the task to persuade people in the case of radio is going to be even greater, particularly whilst we still have the overwhelming majority of cars with analogue radios in them. So there are challenges, but equally there are going to be benefits.

We heard yesterday about the costs to radio of having to transmit simultaneously in both analogue and digital and, clearly, that is something which would be reduced if we managed to get switchover. So this is a very important debate and I am keen that, when we come to debate the Digital Economy Bill when it is introduced, we should not overlook radio. There is always a danger that everybody focuses on television and there will be a huge argument about whether or not the BBC should be the exclusive recipient of the Licence Fee, and whether or not we should be trying to stop teenagers in bedrooms file-sharing, but it is important we should also debate radio and, certainly, that is something which I will try and do my best to ensure happens. But I think this afternoon is a good start to that and it is good to see so many people from the industry assembled in one room. So that's enough from me, just to say welcome to the reception this afternoon …..

Paul Eaton, Head of Radio, Arqiva:

I would like to welcome you all as well on behalf of Arqiva and Digital Radio UK. Arqiva is part of Digital Radio UK, with the BBC and commercial radio, and I am very pleased to be joined today by Andrew Harrison, chief executive of RadioCentre, and Tim Davie, director of Audio & Music at the BBC.

Digital Radio UK has been formed by the radio industry to get the UK ready for the Digital Radio Upgrade. That upgrade is vital because radio faces a stark choice – we can either stay in the analogue world or we can move forward into the digital one. Both need considerable investment from all of the players but only one, digital, can give radio that exciting future that listeners deserve. Digital radio will mean more choice, a better quality listening experience and the kind of interactivity that we can only dream about today.

We all know that the road ahead is a difficult one. We know that the coverage is not good enough yet, we know that we haven't got digital radio in enough cars, and we know that we need to get converters onto the market to turn analogue radios into digital ones – set-top boxes for radio, if you like. We know that there is new content and new services that need to go digital. So there's a lot to do. But, in creating Digital Radio UK, the radio industry is demonstrating that it is serious about the digital future and is determined to address the issues and, in doing so, give the digital future that listeners deserve.…..

Simon Mayo, Presenter, BBC Five Live:

I had one of those 'blimey, you're old' moments this morning. I was talking about radio with my son – my eldest son is eighteen – and I asked him what he listened to and what his friends listen to. He thought for a moment and then he said "none of my friends have got a radio." I thought that was quite an astonishing moment. Now, obviously, he is an unrepresentative sample of one, that is true. They kind of know about radio and they might listen online, and it's on in the kitchen and they hear it in the car and they have an opinion of [BBC Radio One breakfast presenter] Chris Moyles, but that was it. It occurred to me that, really, radio has got a bit of a fight on its hands, which is where the kit here [points to display of DAB radio receivers] comes in, I think.

My parents' generation didn't need to be told that radio was fantastic. My father, if he was here, would talk about listening to Richard Dimbleby and Wynford Vaughan-Thomas and The Goons. The Goons generation didn't need to be told that radio was great. The 60s generation didn't need to be told that radio was great – they had the pirates, then they had Radio One. My generation fell asleep listening to the Radio Luxembourg Top 40 on a Tuesday night. It finished at 11 o'clock and that was quite daring – I see a few people nodding. That was quite daring staying up to 11 o'clock, and the fact that is was sponsored by Peter Stuyvesant cigarettes was even more dangerous. But we remembered it and we fell in love with radio, and I think there is a job to be done to make future generations fall in love with radio.

So enter digital. Partly that has to be done by the broadcasters in coming up with exciting new stations filling gaps that don't exist. BBC7 is wonderful. Everybody will have their own particular favourites. Absolute Classic Rock is really rather good. If you want Supertramp and Led Zeppelin any time of the day, that's the place to go. Really good stuff. There are some really big gaps that need to be filled, but that's exciting. Analogue is full, so digital is the place to be.

But the kit is really exciting. If you have a radio when you are listening to a piece of music .... and you're listening to the radio and an Angelic Upstarts track comes on, you press a button and it sends you an e-mail that tells you that they have reformed, you can buy their records and this is where they are playing. Or someone is listening to 'Yesterday In Parliament' and they hear a speech from a parliamentarian that they like, and they think "he's interesting, she's interesting", press a button, you get sent an e-mail and it tells you who it is, how you can contact them – this sounds quite exciting. If you are listening to one of [presenter] Mark Kermode's film reviews on Five Live, and you like the sound of the film, you press a button, and its sends you an e-mail, you go to your in-box and it's got an e-mail telling you where that film is on, how you can go to see it, maybe a link to the trailer. All of that kind of information means that radio has got an exciting future, but it just means that we have to go out and explain it a bit more because people might not get it the way they used to.

Hopefully, there is still a role for the humble presenter. So you do a little bit as well. Thank you very much indeed for coming.....

---

[285] author's recording

48.

31 October 2009

Catalonia: DAB radio in a "coma"

A new 176-page report documenting the current state of radio broadcasting in Catalonia (the autonomous region in Northeast Spain with 7.4m population) says that DAB radio "is in a state that could qualify as a technical coma", according to 'infoperiodistas.'[286] Of 48 licences awarded by the media regulator for DAB radio, only 23 stations are presently broadcasting and are reported to have no impact on radio audiences.

---

[286] http://www.l-obsradio.org/files/informe_radio_08.pdf
http://www.infoperiodistas.info/busqueda/noticia/resnot.jsp?idNoticia=8835

49.

5 November 2009

DAB radio: "let us get on this horse or get off it"

House of Commons Culture, Media & Sport Committee [excerpts][287]
'The future for local and regional media'
27 October 2009 in the Thatcher Room, Portcullis House

Andrew Harrison, chief executive, RadioCentre
Travis Baxter, managing director, Bauer Radio
Steve Fountain, head of radio, KM Group

Mr Tom Watson: *Can I ask you about Digital Britain and the Digital Britain Report? Do you think the report gave a good way forward for the commercial sector to journey out of its current troubles?*

Mr Baxter: *Perhaps I could ask Andrew to give an overview on that and then maybe we can give our respective views?*

Mr Harrison: *To give an overview, I think the short answer to that is 'yes.' One of the fundamental issues the sector faces right now is the appalling cost of dual transmission. Ultimately, right now, this is a small sector and very many of our stations are simultaneously paying for the cost of analogue and digital transmission. That clearly does not make any financial sense. What we advocated for in Digital Britain was a pathway for all stations to end up with a very clear plan of what is the single transmission platform for them. That led, as I said in my opening remarks, to three very complementary tiers of the commercial radio offer. The first tier is a strong national offer on digital to compete with the BBC, and that is critical for the sector because the truth is that the FM spectrum is full. I am sure all of you will know from some of the other conversations we have had before that the BBC dominates the gift of analogue spectrum. It has four national FM stations; we only have one with Classic FM. For the sector to compete and capture its share of national advertising revenue, the ability to have a national digital platform I think is critical. As we then had the conversations with Digital Britain, I think it became very clear to all of us that you cannot just migrate national stations to digital and leave all of the large metropolitan local stations, like City in Liverpool for example or Metro in Newcastle, all the BBC's local stations, as analogue only. The listeners to those stations will want the functionality, experience and benefits that come with digital. It is then very important that we have a second tier of the large local and regional stations which also migrate to digital. Critically, however, that nevertheless leaves an important third tier, which are the smaller or the rural stations for which either DAB coverage is currently not present – there is just not the transmitter build-out in some of the rural areas – or for which it is likely to be prohibitively expensive going forward. That sector equally needs clarity and that sector being able to stay on FM alongside community radio we feel gives a very balanced ecology where the sector has the most opportunity to compete and the lowest cost base because each station can ultimately choose whether it is on one* *transmission methodology, i.e. digital, or another, analogue. At the moment, we are in limbo where stations are paying for both but the profitability of the sector is fragile and there is not a plan. So we absolutely welcome the beginnings of that plan, which we recognise is the start of what is going to be a long and difficult journey as stations migrate and decide if their future is on digital only or their future is on analogue. The quicker we can move the industry there, clearly the better for the fragile economics of the sector.*

Mr Baxter: *Perhaps I can encapsulate some of the things we sent in to the Carter Review. Our business view generally is that the future is digital. There is hardly the need for me to make that clear to you. Our view has been for the last ten years that we will look at all platforms as we develop our business. We have successful radio stations, primarily operating for example off the audio channels on the Freeview digital television system. However, within that we think it is of real value for radio to have a bespoke platform and the one that is available to us that is a bespoke broadcast platform is DAB. It has, however, taken 12 to 13 years of very slow development for that platform to get to its current state. Therefore, our proposition to Carter's Review was: let us get on this horse or get off it. We think we should get on it and put every possible energy we can over the next view years into getting consensus, direction and pace into the whole process of take-up, like there has not been during the last 12 years. If that can be achieved, it will produce a new resonance for commercial radio as a whole, indeed for the whole of radio. It will help position radio more effectively in the fragmenting media landscape we all have to deal with and give us an opportunity, as Andrew said, of clarifying our investment levels around platforms where currently we are having to pay for two when, in a future where either one is successful, we would only have to pay for one, thereby allowing resource to be put into developing content and other things around our business.*

Mr Fountain: *KM Group does have a digital platform. It is currently costing us over £100,000 a year and we get absolutely nothing back from it. I think the company at the time, six years ago, took the view that they wanted to be a part of the future. Circumstances since have not really helped them to be able to develop that particular medium. I think we too take the view that we would want to be part of a digital platform going forward, but there are a number of issues that would need to be overcome, not least of all the cost of entry and also in our particular case our DAB coverage and the coverage of our FM stations is not mirrored. We have better coverage right now on our FM platforms than we do on our one single DAB coverage. The problem around the coast, if you take that from Medway right the way round perhaps as far down as Rye, around the Kent coast and just touching into Sussex, is such that DAB does not actually reach into large parts of that coastal area.*

Mr Watson: *Would DAB+?*

Mr Fountain: *I could not answer that because I do not actually know.*

Mr Harrison: *No, there is no difference in terms of the coverage for DAB or DAB+. DAB+ is just a different method of compressing the signal so you can actually get more signal down the pipe, if you like; you tend to get more stations, but it does not actually affect the coverage.*

Mr Fountain: *You can see that in order for us to extend the coverage of DAB, there is clearly a cost involved, and there is also a conversation to be had between Ofcom and the French communication authorities as well.*

Mr Watson: *Presumably you are all relatively happy with what is quite a demanding timetable outlined in Digital Britain if your view is that we should just get on with it and do it?*

Mr Harrison: *I think you have expressed it exactly right. The timetable is demanding. I think it is set deliberately as being demanding. Digital Britain does not set a date for switchover. What it sets are two criteria that it says are axiomatic to be hit before switchover can be contemplated: one on listener levels and one on coverage, both of which we support. The aspiration in Digital Britain is to try and hit those two gates, if you like, by the end of 2013. On what Travis was saying earlier on, we think that is absolutely right, that the industry now works terrifically hard together, alongside the BBC and alongside the Government and the regulator to do our very best to hit those criteria. Once we then hit the criteria, the Digital Britain report identifies that it will probably take a couple of years from the criteria being hit before we could actually contemplate switchover. That is aggressive but we think it is appropriately aggressive against the context of an industry that is clearly struggling financially now, and the vast majority of my members are highlighting the cost of dual transmission as the single biggest cost issue that they face and self-evidently one that could be eliminated the quicker we can get to a decision one way or the other.*

Mr Watson: *May I ask you a bit of a left field question? You are quite confident that we should move to digital radio quite quickly. How confident are you that consumers will want to make that journey and that they will not migrate to internet, radio or choose to listen to live streaming sites like Spotify?*

Mr Harrison: *There are two different points there. We are quite confident, as you say, about the movement to digital, but purely because what the Digital Britain Report sets up are consumer-led criteria to drive that change. The criteria are absolutely that we will not move until coverage is built out to match FM. It would be absolutely suicidal for the industry to switch people off who currently listen and enjoy radio services, so it is axiomatic that we have to build coverage out. Secondly, the criterion is that listenership to digital has to be that the majority of all listening has to be to digital before you would contemplate switchover. We are not going to rush into this without being led by the consumer. What we are trying to do, as Travis said earlier, is inject some pace, momentum and energy into the process. If we wait for the natural replacement of sets and the natural progression of DAB – it has taken a long time to get to the listener levels we have right now, we still have all of the BBC's services for example available on analogue – it is going to be very difficult to kick start the progression. We are very comfortable but we are comfortable because it is led by the consumer. The second part of your question is: are we worried about competing services? We are absolutely. I think there is a whole generation of new entrants into the market – Spotify, Last.fm, Pandora – available on-line, all of which are unregulated and against which we are competing for listeners and for advertising revenue. When you have a small, heavily regulated, constrained local radio sector competing with an unregulated world-wide series of music offerings, that is one of the challenges we have to face. We are, however, absolutely committed to the importance of a broadcast transmission methodology for digital. That is not to say that the internet will not be an important complement to that but our business model is based on a broadcast signal of one signal to a wider audience. There is very little evidence so far that on-line music offerings are in themselves profitable business models. For UK citizens and consumers, for our listeners, we think it is absolutely critical that radio remains free at the point of delivery. That has been one of its great strengths ever since the BBC was founded in the 1920s. Of course at the moment, although as I heard this morning the cost of broadband is potentially down to £6 a month, nevertheless, to access any* *internet-delivered service, you have to pay an ISP connection. That may change but I suspect we are a long way away from that.*

[edited]

Mr Watson: *Do you think the car industry is sufficiently prepared for the digital revolution?*

Mr Baxter: *I think we have had some very encouraging conversations with the motor industry over the last six months. The response to Carter's work during the beginning of this year has helped galvanise interest in that area quite significantly, so I think there is a very different aura around those discussions than there was 12 months ago.*

Mr John Whittingdale, Chairman: *Just on the cost of the digital upgrade, what is your best estimate of how much it is going to cost?*

Mr Harrison: *I was on the working party, the Digital Radio Working Group, that was the forerunner for Digital Britain. That working group identified the cost of build-out, the one-off capital cost, as between £100 million and £150 million. That is quite a spread. The reason for the spread ultimately depends on what degree of coverage build-out you get to from equalling FM to universality and at what signal strength. Of course, you get real diminishing returns as you go to the very rural areas. That is the reason for the spread. There has been a lot of debate about that number. In reality, the way we have tended to look at it is that if you take that spread of £100-£150 million over the 12 year period of a licence, which is typically when a radio station is licensed or a multiplex is licensed, and if you said for round figures it is £120 million, that is £10 million a year for the licence period. I think it was £10 million a year that the Secretary of State quoted for example last week. Funding that we have always felt is actually absolutely critical to the build-out and conversation to Digital Britain. The commercial sector is absolutely happy to pay its way to the extent that the build-out is commercially viable but, after that, there is a clear public policy imperative. If the Government and Parliament decide that it is important to have a dedicated transmission structure for radio, that will be a public policy decision and it will need funding. That said, we believe that funding is very affordable. If you take that £100 million number, we believe that, for example, the BBC would save much more than that over the period of the 12-year licence just on what it will save on FM transmission alone, so there is a straightforward business proposition. Another way to think about the £100 million over a 12-year licence with the current Licence Fee settlement for the BBC at around about £3.5-£3.6 billion a year is that over 12 years that is £43 billion. The £100 million infrastructure cost for DAB radio is less than a quarter of one per cent of what the BBC's income will likely be over the next 12 years. So it is eminently affordable if there is a public policy decision that it is important to do that build-out.*

Chairman: *Those two arguments suggest that you are looking for the BBC to pay for this.*

Mr Harrison: *We have said very clearly and very fairly that we are absolutely happy to pay our fair share in our way to what is commercially viable.*

Chairman: *What does that mean?*

Mr Harrison: *That means that we have already put our hands in our pockets substantially to build out coverage on a local and a national basis as far as we judge is affordable. I think realistically, given the state of the sector, the vast majority of the cost going forward, which* *is primarily designed to meet the BBC's obligations of universality rather than the commercial sector's obligations of viability, should rest with the BBC.*

Chairman: *So whilst RadioCentre is keen to move ahead with the digital upgrade, the economics of your sector at the moment means that you cannot really afford to put any more money into it?*

Mr Harrison: *We believe that transmission coverage build-out is axiomatic; it is one of the criteria to effect switchover. We cannot afford it but we absolutely believe the BBC can.*

Philip Davies: *Andrew, on this part can I ask you about how representative your view is of the industry as a whole? It was over this issue it seems more than any other that UTV Radio quit the RadioCentre and said that it felt that it was no longer representing the interests of the wider industry and gave too much power to its biggest member.*

Mr Harrison: *Yes, UTV did say that. Scott Taunton, the UTV Radio managing director, actually represented the commercial radio industry with me on the Digital Radio Working Group through all the per-work that was done for Digital Britain, and so they have been intimately involved. To be fair to UTV's position, they have a particular reservation over the date and the timing for digital, but to be fair to the Digital Britain Report, and indeed we await the clauses of any potential Bill because it is not yet written, there has never been a formal switchover date actually agreed. Although, for example, I think Scott in his Guardian article yesterday talked about a 2015 date being farcical, that date has never been set. What have been set are two consumer-led criteria that have to be hit and then a transition period after that before we all migrate. As Travis said earlier, the majority of opinion across the sector, and certainly across my members and representing my board, is that we need now to put our foot on the gas and work hard to deliver the criteria. Inevitably, there is going to be a spectrum of views with different businesses in different places in terms of their own business models as to the urgency or not they see behind that. UTV are absolutely right to have their own position. They are more at the tail end of the timing.*

Philip Davies: *UTV did not just say that they had a different position to you. They said something a bit more fundamental than that that they felt that you were no longer representing the interests of the wider industry. It was not just as if they had a disagreement. They were indicating that there were others in the sector who shared their view. Do you accept that there are many others or some others in the sector that would share their view?*

Mr Harrison: *I would absolutely accept that we are a broad church and there is a breadth of opinion. I represent large and small stations, local and national, rural and metropolitan, so there is a breadth of opinion. To give you an example of that, our other major national station member that is on AM is Absolute Radio and they believe that the timing for digital should be sooner rather than later. They already have over 50% of their listening on digital platforms, one way or another, so they would move sooner. I have a number of digital-only stations in membership, stations like Jazz and Planet Rock, which clearly are already digital-only and would like to be in the vanguard. Inevitably, there is a spectrum of opinion and we try our best to reflect the overall views. The truth is that it is very unfortunate that UTV have left membership but we continue to represent the vast majority of the sector and its stations and will continue to try to steer a path, helping Government and helping the regulator through this tension.*

---

[287] http://www.publications.parliament.uk/pa/cm200809/cmselect/cmcumeds/uc699-iii/uc69902.htm

50.

7 November 2009

Austria: media regulator puts DAB radio on hold

On 3 November 2009, the Austrian media regulator RTR presented the results of a study that had been commissioned in December 2008 on the potential for digital radio in Austria. Dr Alfred Grinschgl, managing director of RTR, commented:

"The reason we in Austria are not presently digitising radio is that, in neighbouring Germany, there is no unanimous agreement on adopting DAB or DAB+ and therefore no successful large-scale launch."[288]

ORF, Austrian state radio, technical director Peter Moosmann commented that the time was not yet *"ripe"* for the introduction of digital radio and he rejected the notion of planned FM switch-off.

"In every Austrian household, there are four or five radio sets that would need to be replaced with one blow," he said. *"We do not want to force the listener to switch, but want to entice them to digital radio with the appeal of new radio formats."*[289]

The chairman of the association of Austrian private broadcasters, Christian Stogmuller, said that the successful launch of digital radio in Austria could only happen under a single European-wide technical standard, the existence of sufficient digital radio devices in the market and a significant financial commitment from public funds to launch digital radio.[290]

The Austrian Association of Free Radios, VFRO, said it was sceptical because DAB/DAB+ is designed for large-scale radio services, whereas it suggested alternative technologies such as the DRM+ system be used for expanding local radio.[291]

Michael Wagenhofer, managing director of transmission company ORS (60% owned by ORF), said:

"The introduction of a digital radio transmission standard will require simulcasting [broadcasting content on both analogue and digital] for about 15 years because the car industry alone has a six-year implementation cycle."[292]

He estimated the cost of building DAB multiplex infrastructure in Austria would be around 8m Euros.

Dr Grinschgl of RTR said that the anticipated carriage cost of DAB transmission for a local station would be around 100,000 Euros per annum, and that nationwide coverage on DAB would cost 800,000 Euros per annum. He warned that, if individual stations had to bear the costs of broadcasting on both FM and DAB, then DAB would develop only *"very slowly."*[293]

Back in June 2008, an ORF article on digital radio had noted:

"The reason for the lack of consumer interest in digital radio is that RTR [the regulator] licenses a good range of conventional analogue radio services. The consumer is more than satisfied with the existing diversity of content and the quality of reception."[294]

Meanwhile, at the Media Days event in Munich this week, Jurgen Doetz, president of VPRT, the German association of private broadcasters, was reported to have proclaimed:

"Digital radio is dead, dead, dead."[295]

---

[288] http://www.rtr.at/en/pr/PI03112009RF
[289] http://www.rtr.at/en/pr/PI03112009RF
[290] http://www.rtr.at/en/pr/PI03112009RF
[291] http://www.rtr.at/en/pr/PI03112009RF
[292] http://derstandard.at/1256743925467/Bedarfserhebung-Ja-zur-Radiodigitalisierung---aber-noch-nicht-jetzt---ORF-will-Spartenkanaele
[293] http://diepresse.com/home/kultur/medien/519348/index.do?_vl_backlink=/home/kultur/index.do
[294] http://futurezone.orf.at/stories/1631041
[295] http://derstandard.at/1256743879271/Digitales-Radio-ist-tot-tot-tot

51.

8 November 2009

Digital radio in France: T-DMB or not T-DMB?

The question of which digital radio broadcast standard should be adopted in France continues to be an issue of growing concern, even as the proposed launch in 2010 nears. According to l'Humanite, Socialist Party deputy Jean-Marc Ayrault has written to Culture Minister Frederic Mitterand, questioning the risks for local radio owners involved in the Ministry of Culture & Communications choosing a single digital radio standard.[296]

According to Ayrault, the decision in 2007 by then Culture Minister Christine Albanel of *"T-DMB in Band III as the sole [digital radio] standard is worrying stakeholders more and more of the value to have chosen only one standard between T-DMB and DAB+."*[297]

Ayrault noted that tests by local radio groups in Nantes had demonstrated that *"the joint implementation of both standards is technically possible."*[298]

L'Humanite commented that *"the launch of digital terrestrial radio is being complicated by a real scepticism, almost mistrust, in the notion that it can be started in early 2010."*[299]

The Ministry of Culture & Communications is due to present to the government a delayed second report on the proposed rollout of digital radio.[300]

---

[296] http://www.humanite.fr/2009-11-02_Medias_Les-critiques-sur-la-norme-choisie-s-amplifient
[297] http://www.humanite.fr/2009-11-02_Medias_Les-critiques-sur-la-norme-choisie-s-amplifient
[298] http://www.humanite.fr/2009-11-02_Medias_Les-critiques-sur-la-norme-choisie-s-amplifient
[299] http://www.humanite.fr/2009-11-02_Medias_Les-critiques-sur-la-norme-choisie-s-amplifient
[300] http://www.itespresso.fr/frederic-mitterrand-instaure-une-commission-sur-la-numerisation-des-fonds-des-bibliotheques-32145.html

52.

9 November 2009

Radio in the Digital Economy Bill: the tail wagging the dog

The government's forthcoming Digital Economy Bill will be the most significant legislation for the UK radio industry since the passage of the Communications Bill in 2002.[301] Published at the end of November 2009, the Digital Economy Bill will propose 'primary' legislation that sets out a new regime for the licensing and regulation of commercial radio in all its forms – national analogue stations, local analogue stations and local DAB multiplexes.

The main thrust of the new legislation for commercial radio was contained in the Digital Britain final report published in June 2009.[302] According to the Department of Culture Media & Sport, Lord Carter's almost year-long consultation was intended to set out *"the Government's strategic vision for ensuring that the UK is at the leading edge of the global digital economy"* and would introduce *"policies to maximise the social and economic benefits from digital technologies."*[303]

Indeed, some of the changes proposed for the radio industry are forward-looking and designed to place the sector in a multimedia future in which it could survive and thrive.

However, some of the recommended changes to existing radio legislation are there only because parts of the commercial radio industry have lobbied for them to be there. At the time, these interested parties might have claimed that such changes would be beneficial to the commercial radio industry as a whole. Increasingly, other parts of that industry have realised that some Digital Britain proposals were lobbied for inclusion only because they suit the interests of a particular player, offering little or no benefit to the wider industry.

Worse, one proposal ties the future of the whole industry to a dangerous poker game with the government which commercial radio is unlikely to win. This is the Digital Britain proposal [page 102, paragraph 44] to automatically extend the existing licenses of the three national commercial radio stations for a further seven years. Why is this proposal there, and what does it have to do with the UK's digital future? What price is the commercial radio industry being forced to pay for its inclusion?

During the Digital Britain consultation period, Global Radio had lobbied intensively to have the licence of its national analogue station, Classic FM, automatically renewed beyond its 2011 expiry date. In January 2009, I had written:

Classic FM's licence expires on 30 September 2011 and it cannot be automatically renewed. This is a big problem. Whereas local commercial radio licences are still awarded (and re-awarded) by Ofcom under a 'beauty contest' system, national commercial radio licences are not. The system for national commercial radio licences is simple. Sealed bids are placed in envelopes. Ofcom opens the envelopes. The bidder willing to pay the highest price wins the licence. That's it. This system is enshrined in legislation. Even if Ofcom wants a different system, it cannot change it without legislation.

As Classic FM's new owner, Global Radio definitely wants a different system that will enable it to hang on to this most valuable asset. Global has been busy bending the ears of anybody and everybody who it might be able to persuade to interpret the broadcasting rules in a way that lets it keep Classic FM after 2011. Even Ofcom has had its lawyers busy examining the legislation to see what flexibility it has to interpret the rules in a way that might maintain the status quo.

Unfortunately, the legislation in the Broadcasting Act 1990 is quite specific:
"[Ofcom] shall, after considering all the cash bids submitted by the applicants for a national licence, award the licence to the applicant who submitted the highest bid."

The solution for Global Radio was to lobby, lobby and lobby some more for the current legislation detailing the licensing system for national commercial radio to be revoked, changed, amended – whatever needed to be done to ensure that Global could hang on to its valuable Classic FM licence. When Digital Britain was published, it was evident that the phone calls and meetings had paid off handsomely. Lord Carter had listened and offered a solution – a significant change to primary legislation that would allow Global Radio to retain its Classic FM licence for a further seven years, replacing the existing legal requirement that it be re-awarded by Ofcom to the highest bidder in an auction in 2010.

Why exactly is Global Radio so desperate to hang on to Classic FM?

Firstly, Classic FM is a 'cash cow' and has always been the most successful of the UK's three national commercial radio stations launched in the early 1990s. It attracts 40m hours listening per week which, at current sector yields, would earn it around £50m per annum revenues. However, its earning power is further enhanced by the affluence of its audience. Of its hours listened, 66% derive from ABC1 adults, 85% from 'housewives', and 68% from adults aged 55+, a target age group that very little commercial radio reaches. As a result, Classic FM is likely to be attracting more than 10% of total UK commercial radio revenues, significant for a single player out of 300 commercial stations.[304]

Global Radio overpaid to acquire GCap Media for £375m in 2008. The challenge for Global is that the radio business is dominated by fixed costs. In other words, however many listeners an individual station has within its service area, that station's costs are relatively static. Many of the stations in Global's portfolio are medium-sized local operations, whereas Classic FM is a 'giant' with national coverage. Its profit margin probably far outstrips every other commercial station in the UK. Classic FM alone probably generates more operating profit than all Global's other radio stations added together.

Classic FM occupies a unique position in the radio market (the only competitor in the classical music format is BBC Radio Three) and its market power has proven relatively stable over time, with a current listening share of 3.7%, only slightly down from 4.1% a decade ago. By comparison, GCap Media's prime local radio assets also acquired by Global Radio have lost immense market power over the same period – the market share of London's Capital FM down from 13.0% to 6.2%, and Birmingham's BRMB down from 17.1% to 4.8%, for example. Thus, Classic FM is very much a 'rock' at a time many local commercial stations occupy a 'hard place'.[305]

Global Radio desperately does not want to partake in an auction for the Classic FM licence. It might under-bid and lose. It might over-bid and win. Either outcome would be a disaster, the former losing it the 'crown jewels', the latter allowing it to keep the licence but at a price that could lose the station its 'cash cow' status. Because there has been no auction of a national commercial radio licence auction since the early 1990s, nobody knows what the winning bid price might be. Worse, in the 1990s, the field had been open only to European Union companies. Legislation since then has opened up the bidding to the global market. Thus, a licence auction would be an extremely dangerous game for Global to play and, if it lost, would force it to write off its entire Classic FM balance sheet valuation only two years after it acquired the station.

Global Radio has a bargain on its hands in the current Classic FM licence. Not only does this one radio station attract more than a tenth of all commercial radio revenues, but its Ofcom-issued broadcast licence costs very little by market standards. The present cost is fixed at £50,000 per annum + 6% of revenues, probably amounting to around £3m per annum, not a huge expense for a station that generates around £50m. Why is the licence fee so little?

It is the regulator (initially the Radio Authority, now Ofcom) that sets the price of the licence, in the first instance according to the amount that the applicant has bid in its licence application to win the right to broadcast. The price of the licence is collected by the regulator but remitted directly to the Treasury in payment for the scarce FM radio spectrum used by the station.

In 1991, when it won the licence at auction, Classic FM had bid £670,000 per annum plus 14% of its revenues. In 1999, the Radio Authority increased this to £1m per annum plus 14% of revenues. However, in 2006, Ofcom reviewed the Classic FM licence payment and slashed it to £50,000 per annum plus only 6% of revenues. As the table below shows (using estimated amounts because the advertising revenues generated by Classic FM are not published), Global Radio purchased Classic FM just at the time when its licence started to cost significantly less than in previous years.

CLASSIC FM ESTIMATED LICENCE FEE			
year	estimated revenues (£m)	licence fee formula	estimated licence fee (£m)
2000/1	55	£1,000,000 + 14%	8.7
2001/2	56	£1,000,000 + 14%	8.8
2002/3	58	£1,000,000 + 14%	9.2
2003/4	62	£1,000,000 + 14%	9.7
2004/5	61	£1,000,000 + 14%	9.5
2005/6	59	£1,000,000 + 14%	9.3
2006/7	56	£1,000,000 + 14%	8.8
2007/8	56	£50,000 + 6%	3.4
2008/9	47	£50,000 + 6%	2.9
2009/10		£50,000 + 6%	
2010/1		£50,000 + 6%	

306

Why did Ofcom decide to reduce the cost of Classic FM's licence so substantially? Because Ofcom believed that the analogue FM spectrum used by Classic FM would become less and less important with time, as listening via digital platforms, mostly DAB, rapidly replaced FM listening.[307] Ofcom's own forecast, made in November 2006, anticipated that digital platforms would account for 60% of all radio listening by 2011, the date when Classic FM's licence expires. Quite how this justified a 95% cut in the licence fee, alongside a 57% cut in the revenue charge, was not explained by Ofcom. Essentially, Ofcom offered Classic FM's owner the bargain analogue radio licence deal of a lifetime.

RADIO LISTENING VIA DIGITAL PLATFORMS (% share of total listening)

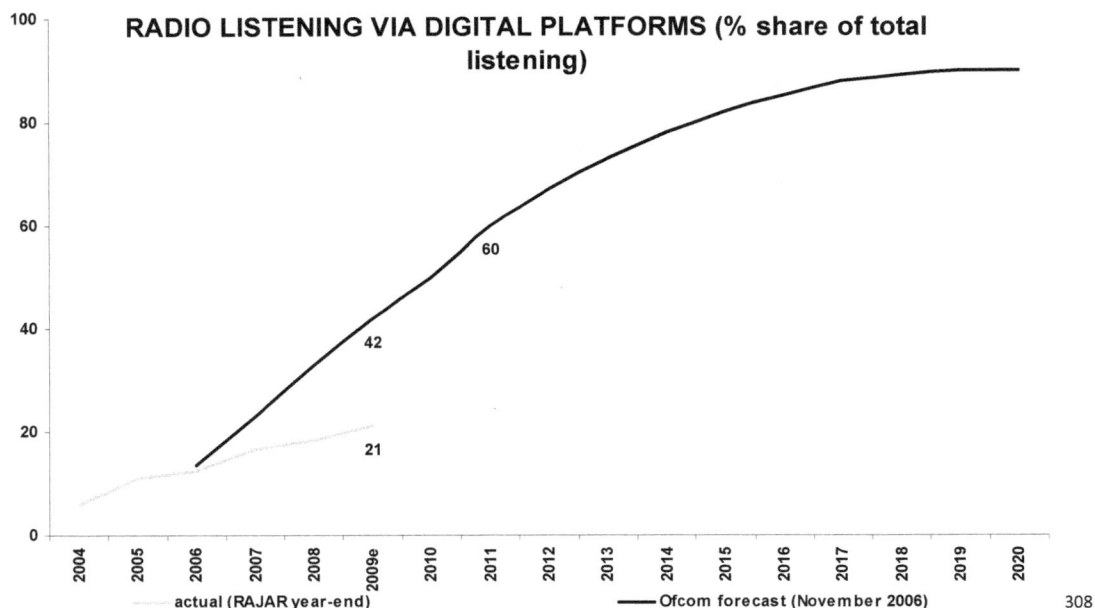

actual (RAJAR year-end) ———— Ofcom forecast (November 2006)   308

Ofcom's forecast of digital radio listening turned out to be wildly over-optimistic, appearing to be based more on wishful thinking than on available evidence. Whilst Ofcom had forecast that digital platforms would account for 42% of radio listening by year-end 2009, industry data show the present outcome to be 21% for all radio and 20% for commercial radio.[309]

The inaccurate Ofcom forecast for consumer uptake of digital radio (never subsequently updated publicly) merely confirmed the belief within a large part of the radio industry that digital radio was about to exhibit exponential growth. This Ofcom forecast, accompanied by supporting comments from the regulator (for example, six months later, Ofcom director of radio Peter Davies said: *"we are potentially at a Freeview moment with digital radio"*), proved significant in misleading stakeholders into believing that the death of analogue radio was just around the corner.[310] The regulator could not have got it more wrong.

Ofcom's inability to forecast the radio market it regulated has resulted in a loss of millions of pounds of potential commercial radio licence fees for the Treasury, not only from Classic FM, but from the other two national commercial stations whose licence fees were also reduced. By Ofcom's own estimate, under the previous formula the three stations combined had paid £7m per annum, but were now being charged less than £1.5m per annum.[311] Over the four-year period until the three stations' licences expire in 2011/2, the total revenue foregone to the Treasury will be around £22m. The Digital Britain proposal to extend these national radio licences for a further seven years, if the present licensing payment scheme is continued, would increase the total potential revenue lost to the Treasury to more than £50m.

Neither RAJAR nor Classic FM release data publicly showing the proportion of the station's listening derived from digital platforms, but it presently seems unlikely that the station would voluntarily give up using FM for broadcasts after 2011 (when the present licence expires), and probably not even after 2018 (the revised expiry date if Digital Britain's proposed seven-year licence extension were legislated). Effectively, the Digital Economy Bill would merely enable the largest player in the commercial radio sector not only to hang on to its 'cash cow', but to continue paying its present low licence payments to the Treasury for the FM radio spectrum it uses.

The losers from this arrangement are:

- taxpayers who, thanks to Ofcom's poor forecasting, are now effectively subsidising the FM spectrum used by the commercial radio sector's single most profitable asset
- the rest of the commercial radio sector who will never be able to match Classic FM's operating margin because their own costs and revenues are considerably more constrained
- new entrants to the radio sector who wish to bid for the Classic FM licence when it expires in 2011 and are willing to pay a realistic, market price for the licence, but will be denied the opportunity by the government's offer of an automatic licence renewal.

Politically, the proposals in the Digital Britain Final Report could not have isolated Classic FM as the sole commercial radio station to have its licence automatically renewed through new legislation. So the renewal proposal was extended not only to all three national commercial stations, but also to all local analogue stations that are broadcasting on the DAB platform. In July 2009, I suggested that this Digital Britain proposal was still iniquitous to the remaining local commercial stations that cannot or will not broadcast on DAB. It appears now that the Digital Economy Bill is likely to extend the proposed licence extension to all analogue commercial radio stations (whether or not they simulcast on DAB).

So every analogue commercial radio station will now be offered an automatic licence extension! Is that not a universal 'good thing'? Well, no, because there is rarely a 'free lunch'. Lord Carter was determined to extract a price from the entire commercial radio sector for bowing to persistent demands from Global Radio for new legislation to renew its Classic FM licence. The strings he attached are related to the government's insistence that the whole radio industry use DAB as its main broadcast platform. This is why two entirely unrelated issues – Classic FM's licence and DAB consumer uptake – have now become so intertwined in the proposed legislation.

In the seven-year renewal offered to every commercial radio licence, the government proposes to insert a clause that will allow it (via Ofcom) to terminate that licence extension with two years' notice if the radio industry as a whole (commercial radio and the BBC) does not achieve these goals:

- 50% of radio listening to be via digital platforms by 2013
- DAB transmission infrastructure to be upgraded significantly.

It is a 'carrot and stick' approach: 'We the government will give you all a free licence extension if you collectively promise to make DAB work. But, if we find you do not succeed in making DAB work, we will take your licences (and hence your businesses) away altogether'. The problem here is that the buck has been passed on to a wide and varied constituency of 300 commercial radio stations, many of whom have very little or no control over whether DAB can be turned into a successful delivery platform.

It is the entire commercial radio industry that will be expected to potentially pay the price with its own lives in exchange for changes to primary legislation that allow Global Radio to hang on to its 'cash cow' Classic FM licence. What seems even more unfair is that the entire DAB platform is owned and controlled by a mere handful of the largest UK commercial radio companies who, between them (and the BBC and transmission company Arqiva), wield the power to make DAB a success or failure.

If the largest commercial radio owner, Global Radio, had demonstrated incredible confidence in the DAB platform, maybe it might instil confidence in the rest of the radio sector that DAB could be made a consumer success by 2013. However, although Global Radio has regularly talked the DAB talk, it has hardly walked the DAB walk. Global had been the largest owner of commercial DAB infrastructure until, in April 2009, it sold its 63% stake in the national DAB multiplex and its wholly owned group of local DAB multiplexes.[312] At the same time, it has sold or closed all but two of its digital-only radio stations, which exist now only as music jukeboxes.

Of course, for Global Radio, none of the DAB 'strings' really matter. It thinks it has got exactly what it wanted in the forthcoming Digital Economy Bill – to keep its valuable Classic FM licence. This is its significant short-term goal and may be the only thing that can keep the group afloat financially. Who knows? If the media ownership rules are relaxed, Global might be able to sell its entire radio business to Murdoch or RTL or MTG before the 2013 date of judgement on DAB is even reached.

For a while, many in the industry had seemingly been happy to line up behind Global Radio, uncertain of their own futures and relatively uninformed on these complex regulatory and legislative issues. But the truth is dawning on many – what is good for Global Radio is not necessarily good for the rest of the commercial radio industry. The future of commercial radio should remain in the collective hands of the industry itself, not be determined by one individual owner. And the issue of radio licence renewals should not have to be linked to the future performance of the DAB platform.

Digital Britain and the Digital Economy Bill offer a rare opportunity to update the regulatory regime for the entire commercial radio sector, rather than merely to offer one company a 'phone a friend' millionaire lifeline.

[Note to the table: the estimated costs of the Classic FM licence fee are simplified. Firstly, the cash amount paid increases annually from £1,000,000 in 1999 to £1,161,000 in 2006 and subsequently, in line with the Retail Price Index. The £50,000 cash payment will similarly be adjusted. Secondly, the revenue percentage paid is applied only to "advertising and sponsorship revenue attributable to national analogue listening hours", but these data are not published, so 100% of estimated revenues have been assumed to derive from the FM platform.]

---

[301] http://www.parliament.the-stationery-office.co.uk/pa/cm200203/cmbills/006/2003006.htm

[302] http://www.culture.gov.uk/images/publications/digitalbritain-finalreport-jun09.pdf

[303] http://www.culture.gov.uk/what_we_do/broadcasting/6216.aspx

[304] source: RAJAR, Q3 2009

[305] source: RAJAR, Q3 2009 & Q3 1999

[306] source: Ofcom and Grant Goddard

[307] http://www.ofcom.org.uk/consult/condocs/methodology/financialterms/financialterms.pdf

[308] source: RAJAR and Ofcom

[309] source: RAJAR, Q3 2009

[310]

http://pqasb.pqarchiver.com/smgpubs/access/1259491541.html?dids=1259491541:1259491541&FMT=ABS&FMTS=ABS:FT&type=current&date=Apr+22%2C+2007&author=Steven+Vass&pub=Sunday+Herald&desc=Analogue+radio+could+be+history+by+2020+.+.+.if+Ofcom+wins+the+day+RADIO%3A+SWITCHOVER+RADIO%3A+SWITCHOVER+Optimistic+projections+on+digital+listening+suggest+end+of+MW+and+FM&pqatl=google

[311] http://www.ofcom.org.uk/consult/condocs/methodology/financialterms/financialterms.pdf

[312] http://www.arqiva.com/press-office/press-releases/press-releases-2009/arqiva-to-take-full-ownership-of-digital-one-commercial-dab-mult

53.

11 November 2009

France: digital radio roll-out hangs in the balance

The much delayed report on digital radio commissioned by the French government from Marc Tessier, former CEO of France Televisions, was published on Monday 9 November 2009. It suggested that the planned launch of France's first digital radio stations in mid-2010 was *"implausible"* and it proposed that an economic model for digital radio needed to be identified before an estimated 600m to 1bn Euros is spent over the next ten years on the rollout of digital radio in France.[313]

The 54-page report raised queries concerning almost all the various aspects of digital radio implementation: the funding, the T-DMB standard adopted in France, the date for FM switch-off, and the cost of simulcasting on both analogue and digital spectrum for a ten-year period. It concluded:

"There is still time to consider the appropriateness of pursuing digital terrestrial radio, at the point when several players are unwilling to endorse its prerequisites [coverage, reception quality] or to pay for the additional transmission costs for ten years."[314]

Interviewed in Le Figaro, Tessier was asked if digital radio should be halted altogether:

"That's for the radio bosses to decide. But I believe there is serious doubt over the desirability of a project that will take ten years, at a time when a new radio platform is evolving at great speed via mobile phone networks. Digital terrestrial radio has less to offer now than it did three years ago. Where will we be in five years time?"[315]

Asked what the crucial issues are, Tessier said:

"The coverage area and the number of stations available to every citizen would be the main benefits of the project. That is why each radio station owner must commit to covering at least 90% of the population, which involves huge costs at a time when radio advertising revenues are declining under pressure from new media. If we reduced the coverage area to cut the costs, the project would attract little interest."[316]

Also interviewed in Le Figaro, Rachid Arhab, president of the CSA [France's broadcast regulator] digital radio working group, was asked if the report threatened the future of digital radio.

"I am not bothered", he responded. *"This report is but one part of the digital radio project, and we are now awaiting the Hamelin report on the funding for radio groups. I recall that the letter to Tessier commissioning this report explicitly required it 'to map out the successful path for digital terrestrial radio' based on the notion of public funding."*[317]

Asked whether digital terrestrial radio would be overtaken by other radio platforms, Arhab said:

"Technology is evolving very quickly. But the longer we wait, the more difficult it will be. In ten years time, perhaps a significant portion of listening will no longer be delivered by radio waves. But that would pose a major problem in terms of platform neutrality. If there is no digital terrestrial radio platform, then public radio will be obliged to negotiate with the internet service providers for distribution. The CSA does not want to take digital radio away from radio receivers. For each new problem, we find a solution."[318]

Asked when digital radio will launch, Arhab said: *"I can no longer give a precise date."*[319]

The commercial radio owners started meeting the evening the report was published to draft their response, which they will deliver to the CSA by 23 November.[320]

L'Express newspaper commented: *"Without a huge effort from the radio industry, which right now does not believe in it, digital terrestrial radio is doomed to failure even before it starts."*[321]

---

[313] http://www.electronique.biz/editorial/409385/coup-de-froid-sur-la-radio-numerique/

[314] http://www.lefigaro.fr/assets/pdf/Rapport_RNT_031109.pdf

[315] http://www.lefigaro.fr/medias/2009/11/10/04002-20091110ARTFIG00360-un-doute-serieux-sur-l-opportunite-de-la-rnt-.php

[316] http://www.lefigaro.fr/medias/2009/11/10/04002-20091110ARTFIG00360-un-doute-serieux-sur-l-opportunite-de-la-rnt-.php

[317] http://www.lefigaro.fr/medias/2009/11/10/04002-20091110ARTFIG00366-radio-numerique-plus-on-attend-plus-ce-sera-difficile-.php

[318] http://www.lefigaro.fr/medias/2009/11/10/04002-20091110ARTFIG00366-radio-numerique-plus-on-attend-plus-ce-sera-difficile-.php

[319] http://www.lefigaro.fr/medias/2009/11/10/04002-20091110ARTFIG00366-radio-numerique-plus-on-attend-plus-ce-sera-difficile-.php

[320] http://www.lefigaro.fr/medias/2009/11/10/04002-20091110ARTFIG00357-la-radio-numerique-attendra-.php

[321] http://blogs.lexpress.fr/media/2009/11/la-radio-numerique-cest-pour-2.php

54.

13 November 2009

FM radio in mobile phones: the universal standard

Although some politicians and civil servants might try to convince us that the UK can lead Europe and the world in technological innovation, new broadcast standards and electronic hardware, the reality is that the sun set on the British Empire a long time ago. Almost none of the gadgets we use are manufactured in the UK, and even those that have British corporate logos glued on the front are inevitably assembled in China or Korea. When global commercial forces make a decision on the adoption of a new consumer mass technology, the best Britain can do is follow in the slipstream and make the most of innovations that the rest of the world is pioneering.

Right now, the new broadcast standard for mobile radio reception is being decided in the corridors of power in Washington DC and in the boardrooms of the mobile phone manufacturers. That standard will be FM radio. This inevitably means that FM radio delivered to on-the-go consumers via mobile devices will become the universal standard for years to come. Please, Ofcom and the Department of Culture, Media & Sport [DCMS], do not bother getting uppity just because you were not consulted by Congress, Nokia, Samsung or Apple. Neither were several hundred other countries around the globe. And please, DCMS and Ofcom, do not even think about committing the UK to going its own sweet way unilaterally on this issue. All it will do is create more embarrassment.

Recall the DCMS-led Digital Radio Working Group which spent a year deliberating on the digital radio issue and included in its Final Report published in December 2008 a note that *"consumer groups believe that, once an announcement [of digital radio switchover] is made, no equipment should be sold that does not deliver both DAB and FM."*[322] In the margin, at the time, I had scribbled *"in your dreams!"* After ten years of DAB broadcasting, there is still not one mobile phone currently on sale in the UK that incorporates DAB radio.

Recall the report from Ingenious Consulting published in January 2009 which suggested that, in order for the DAB platform to succeed for commercial radio, it would need a *"commitment [from radio stakeholders] not to pursue alternative technologies to DAB."*[323] So, is the only way to drive consumers' use of DAB to prevent them from listening to radio on other competing platforms? Will 'DAB police' be storming some West Country town next year and taking all the residents' analogue radios away from them?

Whereas the UK has too often pursued these sort of fundamentally impractical strategies to achieve its aims (and thus usually fails), the US is adopting a much more practical and sensible approach. Almost everyone in the US carries a mobile phone. Therefore, mobile phones should all have FM radios in them. An FM chip costs next to nothing for a mobile phone manufacturer. The benefit to the consumer is that FM radio is free at the point of access and its usage is only limited by the battery power of the phone.

This week, Homeland Security Secretary Janet Napolitano and FCC chairman Julius Genachowski received a cross-party letter, signed by 60 members of the House of Representatives, encouraging FM radio capability to be included in mobile phones sold in the US. The letter noted that the Warning Alert & Response Network Act of 2006 requires the mobile phone industry to create an emergency alerting system in the US, and it stated:

"There are well over seven hundred million cell phones with FM radios globally. Currently, only a handful of FM radio enabled cell phones are in the U.S. market. There is no excuse for American consumers' access to advanced technology to lag behind that available worldwide. "[324]

In June 2008, the commercial broadcasters' trade body, NAB, had published a report which outlined the potential benefits from including FM radios in mobile phones.

"Radio is a service that already reaches 235 million American listeners every week," said NAB president & CEO David Rehr. *"With 257 million cell phones currently in service, we're confident that implementation of a new FM-radio feature would result in rapid penetration, benefiting not only the radio business and American consumers, but the cell phone, electronics manufacturing, and music industries as well."*[325]

The NAB report included a graph (see below) which displays data supplied by iSuppli Corporation forecasting that, by 2011, 45% of all mobile handsets globally will incorporate FM radio.

Exhibit IV
Global FM-capable Handset Market
(2005-2011)

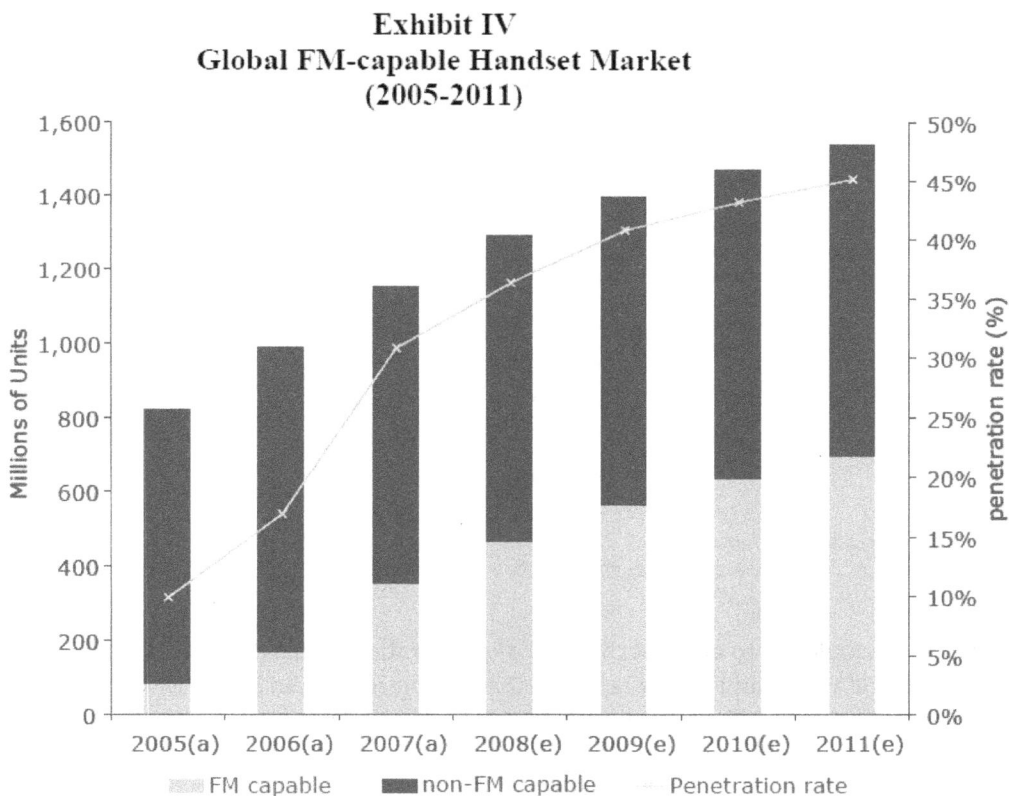

326

It is noteworthy that the US, in this case, seems totally happy to 'follow' the rest of the world in incorporating FM radios in its mobile phones, a feature that is already widely available in many other countries (including the UK). The US is not trying to argue that some new proprietary broadcast standard (such as HD Radio) be adopted in phones to further the objectives of a particular commercial US business.

In the UK, we are in a somewhat different position. Mobile phones with FM radios are already out there and being purchased by most consumers. My survey earlier this year of mobile phones available in the UK found that more than half the available models included FM radio (see table below).

UK MOBILE PHONE MARKET					
	2008 UK market share	no. of current models			% with FM radio
		total	with FM radio	with internet	
Nokia	43%	97	73	66	75%
Samsung	21%	37	18		49%
Sony Ericsson	18%	72	53	9	74%
LG	7%	40	13	27	33%
Motorola	5%	14	9	12	64% [327]

It is remarkable that the hardware is already sitting in millions of UK citizens' pockets with the capability to listen to FM radio. And it costs them nothing (but battery power) to listen. The only disappointment is that people do not seem to be using their phones much to listen to the radio, according to Ofcom data (see Ofcom graph below):

Figure 5.49   Use of mobile phone functions

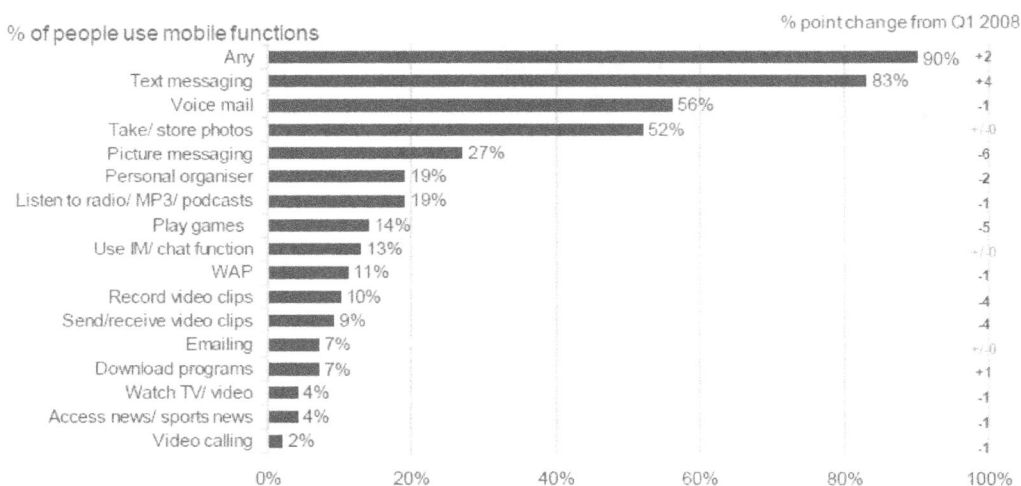

% of people use mobile functions / % point change from Q1 2008

Function	%	change
Any	90%	+2
Text messaging	83%	+4
Voice mail	56%	-1
Take/ store photos	52%	+/-0
Picture messaging	27%	-6
Personal organiser	19%	-2
Listen to radio/ MP3/ podcasts	19%	-1
Play games	14%	-5
Use IM/ chat function	13%	+/-0
WAP	11%	-1
Record video clips	10%	-4
Send/receive video clips	9%	-4
Emailing	7%	+/-0
Download programs	7%	+1
Watch TV/ video	4%	-1
Access news/ sports news	4%	-1
Video calling	2%	-1

Source: Ofcom research, Quarter 1 2009
Base: Adults aged 15+ who personally use a mobile phone (n= 5273)   328

Most industries would kill to achieve the kind of penetration levels that FM radio has already achieved in the UK with mobile handsets. Yet the commercial radio industry in the UK, unlike in the US, appears to see little advantage to directing listeners to the mobile phone platform. Why?

Maybe because:
- RAJAR, the radio audience metric, does not publish listening data separately for the mobile phone platform in its quarterly survey [confusingly, it presently seems to lump

respondents' reported mobile phone listening to live radio into its 'digital unspecified' platform category, even though FM radio received via mobiles is, in fact, analogue]

- DAB is the platform of choice for the commercial radio industry because it (like the BBC) has invested so heavily over a decade in building its expensive infrastructure, so why persuade listeners to go elsewhere? The questions to be asked are: What is your radio company primarily – a content provider or a platform operator? Are 'hours listened via DAB' really more important to you than 'TOTAL hours listened'?
- DAB (like FM) restricts consumers' listening to BBC and UK commercial radio stations, whereas mobile devices increasingly offer a much wider choice of content (not on FM, but via G3 or broadband). So there is reluctance to promote a mobile platform that could potentially attract a previously loyal listener to, say, Last.fm.

As a result, a drive to encourage FM radio listening on mobile phones does not figure in UK commercial radio's overall strategy, even though it might help maintain the sector's audiences and revenues (admittedly, some companies such as Global Radio and Absolute Radio have individual initiatives that do push the point). You cannot help but think that opportunities are being lost here because:

- all the industry's platform eggs have been placed in the DAB basket
- the DAB campaign in the UK seeks to persuade consumers to **PURCHASE** a new radio receiver, whereas almost everyone already owns a mobile phone, so a campaign to persuade consumers to use its FM radio will involve no additional purchase
- the UK industry wants to maintain its 'walled garden' that shields consumers from experiencing non-BBC/non-UK commercial radio content, thus maintaining the cosy content duopoly.

A parallel might be Tesco not wanting to tell customers about its 'Metro' stores within petrol stations because it was worried that they might spend their disposable income on forecourt petrol rather than Tesco items. That would be crazy. Tesco simply wants consumers to be offered as many opportunities as possible to buy Tesco goods, wherever that opportunity might arise.

The incongruity is that the US radio industry desperately wants to be at a place where we, in the UK, **already are** (lots of mobile phones incorporating FM radio). Yet, what are we ourselves doing to promote FM radio listening on mobile phones? Almost nothing.

---

[322] http://www.culture.gov.uk/images/publications/DRWG_Final_Report.pdf
[323] unpublished report
[324] http://news.radio-online.com/cgi-bin/rol.exe/headline_id=b12020
[325] http://www.nabfastroad.org/fmradiofeaturecellularhandsets052808.pdf
[326] http://www.nabfastroad.org/fmradiofeaturecellularhandsets052808.pdf
[327] source: Mintel and Grant Goddard
[328] http://www.ofcom.org.uk/research/cm/cmr09/cmr09.pdf

55.

16 November 2009

Switzerland: five of eight DAB+ radio licences expire unused

Yesterday, five out of eight broadcast licences issued in Switzerland for DAB+ radio expired without their owners having launched the promised digital stations. According to the Klein Report, only three DAB+ stations – Open Broadcast, Radio Eviva and Swiss Mountain Holiday Radio – are now on-air, the latter having launched on yesterday's deadline.[329]

Two stations, Radio.ch and SoundCity, had already handed back their unused licences to Swiss media regulator Bakom. Now, Energy Zurich's licence to provide a service named Radio For Youngsters, aimed at 10 to 14 year olds, has also expired. Egon Blatter, owner of Radio 105, said he had hoped that Energy Zurich's proposed switch from FM to DAB in 2010 would have changed the whole audience for digital radio. He has *"reluctantly"* let the DAB+ licence expire of his proposed station DJ Radio Deluxe, as has Oliver Fluckiger's RadioJay.

---

[329] http://www.kleinreport.ch/meld.phtml?id=54023

56.

18 November 2009

"It is only in recent months … that digital radio has been a real focus and priority for the [UK radio] industry"

Campaign magazine, 6 November 2009 [excerpts][330]

Karen Stacey, director of broadcast sales and brand solutions, Bauer Radio [KS]
Mike Gordon, group commercial director, Global Radio [MG]

Q: *Why has the take-up of digital radio been so slow?*

KS: *I don't think take-up necessarily has been slow given that it is only in recent months, since the publication of the Digital Britain report [June 2009], that digital radio has been a real focus and priority for the industry.*

MG: *Until recently, digital has been viewed as a complementary, rather than a primary, platform. Now that the industry is unified and the Government has given clear targets, I'm absolutely convinced take-up will accelerate.*

Q: *How realistic is the government's analogue radio switch-off target?*

KS: *As an industry, we're aware that we've got a lot of work to do - coverage needs to be improved, the cost of sets needs to come down and more cars need digital radios installed. But the vision and commitment is now in place to make that happen.*

MG: *2015 is a target that we are focused on and working towards, and the fact of the matter is that, without it, a lot less would be happening. There's been a huge amount of activity since the Government gave us a target. Earlier this month, for instance, we called a summit with the motoring industry and other stakeholders to progress take-up in cars. If we keep up this kind of momentum, then I believe it is achievable. It's still more than six years away.*

Q: *How has your company invested in digital radio?*

MG: *We have invested more in digital radio than any other radio group. Global also took the lead in supporting the call for a target date to galvanise the industry and will continue to be a pioneer in commercial radio on all distribution. [sic]*

---

[330] http://www.campaignlive.co.uk/news/rss/966849/Media-Double-Standards---adland-prepared-retune-digital-radio

57.

20 November 2009

Commercial radio and DAB: turkeys voting for Christmas

Significant players in the UK commercial radio industry, along with Digital Radio UK, the Digital Radio Development Bureau, DCMS and Ofcom are all lobbying for DAB receiver take-up to be accelerated and for consumers to migrate their radio listening to DAB as quickly as possible. However, the industry's own data suggest that the pursuit of these strategies will simply reduce even further commercial radio's already declining share of radio listening versus the BBC.

The commercial radio sector's diminishing success in competing for listeners against the BBC remains one of its most pressing problems. In 1998, commercial radio's share of listening was 51.1%, but that figure is now down to 42.4%.[331] Conversely, the BBC's share has increased from 46.8% to 55.0% over the same period. The long-term decline in commercial radio's market power looks like this in recent quarters (see graph below):

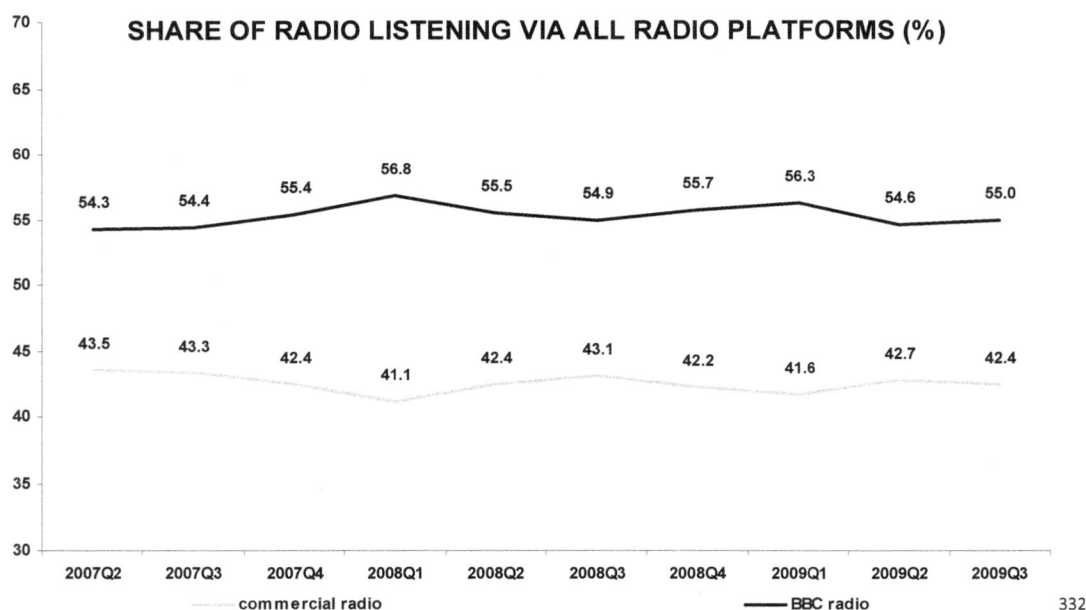

SHARE OF RADIO LISTENING VIA ALL RADIO PLATFORMS (%)

BBC radio: 54.3, 54.4, 55.4, 56.8, 55.5, 54.9, 55.7, 56.3, 54.6, 55.0

commercial radio: 43.5, 43.3, 42.4, 41.1, 42.4, 43.1, 42.2, 41.6, 42.7, 42.4

Quarters: 2007Q2, 2007Q3, 2007Q4, 2008Q1, 2008Q2, 2008Q3, 2008Q4, 2009Q1, 2009Q2, 2009Q3

········ commercial radio ——— BBC radio   332

However, if we examine listening solely on digital radio platforms, we see that commercial radio is losing listening share much more sharply (see graph below). In 2007, commercial radio's share of listening via digital platforms had been above the average for all platforms and so was 'helping' the overall fight against the BBC for market share. However, in two of the last three quarters, commercial radio's share via digital platforms has been lower than for all platforms, and so is now dragging down the sector's overall market share.

SHARE OF RADIO LISTENING VIA DIGITAL RADIO PLATFORMS (%)

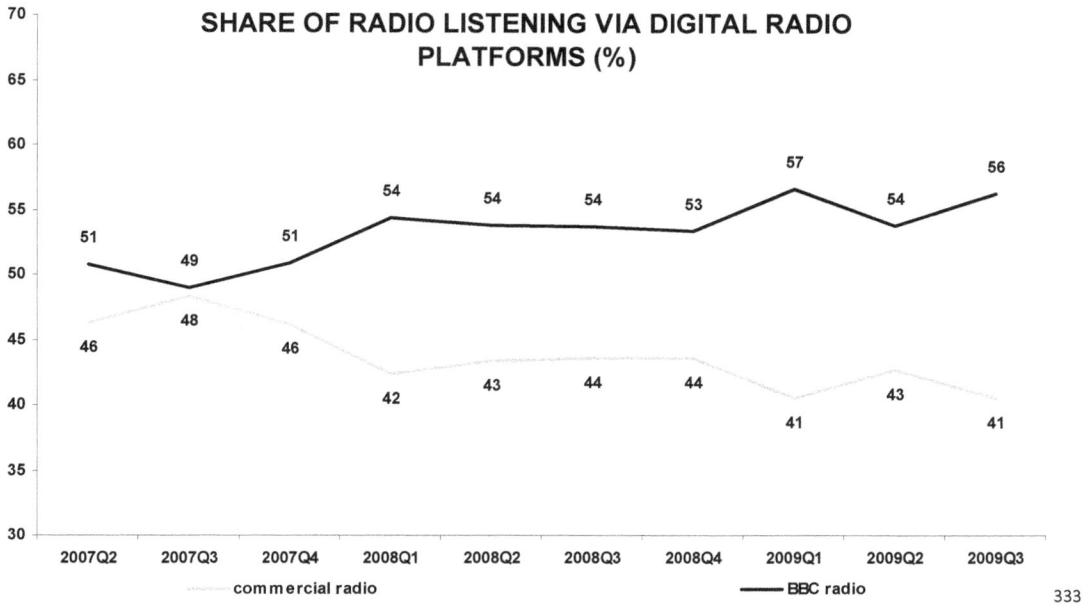

	2007Q2	2007Q3	2007Q4	2008Q1	2008Q2	2008Q3	2008Q4	2009Q1	2009Q2	2009Q3
BBC radio	51	49	51	54	54	54	53	57	54	56
commercial radio	46	48	46	42	43	44	44	41	43	41

333

Worse, with each new quarter, radio listening via digital platforms is growing as a proportion of total radio listening, so that the 'contribution' of digital platforms to the overall picture is becoming greater. In Q2 2007 (the earliest point on the timescale of these graphs), digital platforms accounted for only 12.9% of total listening. In the latest quarter, that proportion has increased to 21.1%.[334]

Now, if we extract listening via DAB from the total for all digital platforms, we observe two phenomena (see graph below). Firstly, commercial radio is badly losing the battle for DAB platform usage to the BBC by a ratio of 1:2. Secondly, commercial radio's performance on the DAB platform is worsening over time. It is the combination of these two trends which is dragging down not only the commercial sector's share of digital platforms, but also its overall competitive performance against the BBC.

SHARE OF RADIO LISTENING VIA DAB PLATFORM (%)

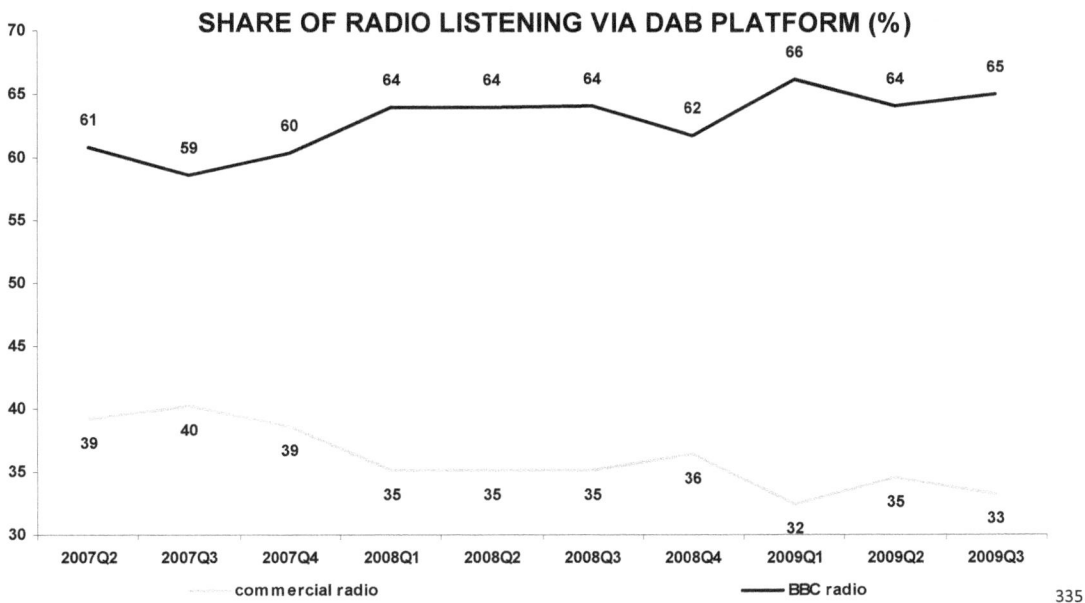

	2007Q2	2007Q3	2007Q4	2008Q1	2008Q2	2008Q3	2008Q4	2009Q1	2009Q2	2009Q3
BBC radio	61	59	60	64	64	64	62	66	64	65
commercial radio	39	40	39	35	35	35	36	32	35	33

335

175

To make matters worse, DAB is the largest element of radio listening via digital platforms (up from 54.4% in Q2 2007 to 62.9% in Q3 2009 of listening via all digital platforms), and the DAB platform's contribution to total radio listening is similarly growing (up from 7.0% in Q2 2007 to 13.3% in Q3 2009).[336] DAB is the focus of the radio industry's digital platform marketing campaigns, so the commercial sector's current poor performance on this platform is disastrous.

The data suggest that, far from the DAB platform helping the commercial radio sector compete more effectively against the BBC, the absolute opposite holds true:
- the average adult with a newly acquired DAB radio uses it for listening in a way that effectively reduces the commercial radio sector's overall share of listening versus the BBC
- acceleration of DAB usage will only serve to accelerate the decline in commercial radio's share of radio listening versus the BBC.

These outcomes are hardly surprising when one considers industry data which show that:
- DAB radios are purchased predominantly by older people (the average age of a DAB radio receiver owner is 46, according to RAJAR)
- older people listen to BBC radio much more than to commercial radio (BBC radio accounts for 63% of radio listening amongst over-45s).[337]

The paradox is that stakeholders in commercial radio continue to push for DAB to be adopted by consumers as quickly as possible, even though the inevitable outcome will be to reduce further the commercial sector's listening share, handing the BBC even more of a competitive advantage.

So why exactly does the notion continue to be voiced by significant players in commercial radio that the DAB platform is itself the answer to the sector's present lack of competitiveness with the BBC?

[Data source: RAJAR. Statistical note: The graphs above to do not sum to 100% because the minimal amount of platform data released by RAJAR is 'rounded' (hours listened to 1,000,000 per week; listening shares to 0.1%) and the listening apportioned to the BBC and commercial radio sometimes does not sum to the total for a platform. Part of this shortfall may be accounted for by 'other' listening (neither the BBC nor commercial radio) which is not itemised by platform. Data for individual quarters are therefore somewhat inconsistent, though the trend over several quarters is likely to be indicative.]

---

[331] source: RAJAR Q2 1998 & Q3 2009
[332] source: RAJAR
[333] source: RAJAR
[334] source: RAJAR
[335] source: RAJAR
[336] source: RAJAR
[337] source: RAJAR

58.

26 November 2009

Digital Economy Bill: radio industry reaction

The Digital Economy Bill was published by the government last Friday.[338] Press headlines about the radio provisions in the Bill included:

Media Week: **'Radio Bill creates deep divisions among industry'**[339]
Radio Today: **'Digital Bill inflames industry'**[340]
The Telegraph: **'Digital Economy Bill: no date for radio digital switchover'**[341]

Below is a selection of initial reactions from stakeholders in the commercial radio sector:

Scott Taunton, Managing Director, UTV Media, said: *"What radio really needs is some meaningful deregulation and licensing reform to enable it to thrive and adapt in the multimedia age. The Digital Economy Bill shows that government is out of tune with listeners, who are delighted with the broad choice and accessibility that radio already offers."*[342]

William Rogers, Chief Executive, UKRD and The Local Radio Company, said: *"Where is the fairness in a proposal to permit 100 per cent of the BBC's local radio stations a digital future and deny over 100 local commercial radio stations the same opportunities? The Bill is fundamentally unfair and dishonest and I hope the Peers give it the going over it thoroughly deserves."*[343]

He added: *"We need to inject some honesty and fairness into this debate. What is honest about suggesting to millions of listeners in the country that they will be delivered a wonderful digital listening future when the results will be to deny them the opportunity to listen to their favourite local radio station?"*[344]

Andrew Harrison, Chief Executive of RadioCentre, said: *"There was no permanently set date in the Digital Britain report – 2015 was just a target. This is a clever and enabling piece of legislation as it allows the Secretary of State to set a date for switchover without having to create more legislation. Based on annual progress reports from Ofcom and BBC, they can set a date when the circumstances are rights for which at least two years notice must be given."*[345]

Travis Baxter, Managing Director of radio at Bauer, said: *"There is always a difference between setting out objectives and actually having government's support. This piece of legislation has created a mechanism to create a date when consumer demand for digital radio is great enough."*[346]

Two days before the Bill was published, a spokesperson for Digital Radio UK said: *"Whilst the target date of 2015 may be ambitious, the consumer-led criteria are achievable for operators who are committed to a digital future. Without an ambitious target date, the alternative is to* condemn the industry to an indefinite period of dual transmission, a financial burden that commercial radio cannot continue to bear and that represents a poor use of public funds for the BBC."[347]

A few weeks earlier, a spokeswoman for Digital Radio UK had said: "The Digital Britain report does not set annual targets and we do not have a specific target to meet in terms of total share beyond the two criteria for upgrade. It has only been a matter of months since the publication of the Digital Britain report and all stakeholders in digital radio have galvanised their activity."

"There are numerous ways that a graph could plot a path to growth and we anticipate that take-up will accelerate as a result of the excellent foundations we are currently building. There is clearly some way to go but the industry is confident that with the formation of Digital Radio UK, the provisions in the Digital Economy Bill and the collective will of all stakeholders, we will successfully deliver the benefits of digital radio upgrade to listeners." [348]

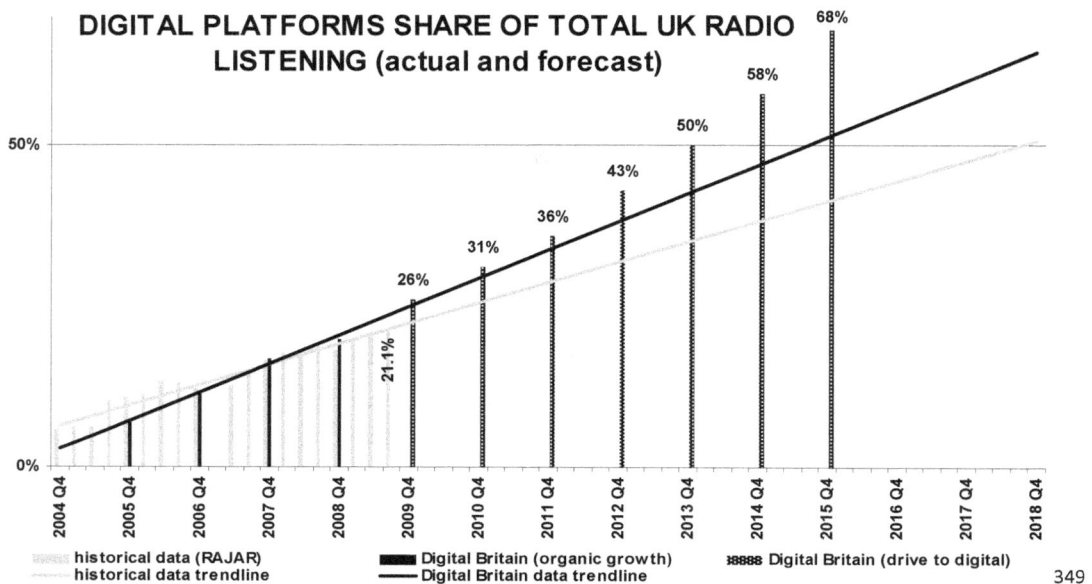

349

Andrew Harrison, Chief Executive, RadioCentre [AH]
Interviewed on Today, BBC Radio 4, 20 November 2009 @ 0837[350]

Q: Everybody is really worried about this and can't quite see the point of why we are pushing ahead [with digital switchover], even if it is a target, by 2015.

AH: Well, I think that why there is a need is fairly straightforward. The truth is that Britain loves its radio services. About 40 million of us tune into the radio every week. But the way we receive our news and entertainment is changing all around us in a digital world. It's changing in television, it's changing for films and music and newspapers. So it's very important that radio, which is at the heart of British daily life, has its own chance to look forward and face the future in a digital sense, rather than being trapped in a sort of analogue environment while the rest of the entertainment world goes digital.

Q: But we love to hear our radio in decent quality and the problem with digital [DAB] at the moment is it doesn't work very well in cars, indoors or in rural Britain.

AH: *That's absolutely right. That's why we need some time as an industry to roll up our sleeves and improve the coverage for digital and improve the listener experience so that, when we actually get to the stage of contemplating actually switching over services, it will be a service for listeners with more choice and better functionality, but with the transmission quality that would be absolutely imperative.*

Q: *And you can guarantee that you can solve the problems that currently exist by 2015?*

AH: *Well, the Bill doesn't set out 2015 as a date. What the Bill sets out is the two key criteria which will be important to set a switchover date. That is that consumers or listeners can hear the service, so that they can get a digital transmission where they can currently get an FM transmission, and that they are actually listening to those services, so one of the criteria will be that over 50% of listening is via digital. Once those two thresholds have been crossed, then I think we'll be very confident that we can deliver the services fit for a digital age.*

Q: *When do you think that those thresholds can be crossed?*

AH: *Well, our target is to try to hit the thresholds by the end of 2013 to then start the two-year window to switchover. It remains to be seen whether the industry collectively can work together – that's the commercial sector, the BBC and Arqiva, our transmission provider – very hard across the next four years to get on with the job.*

---

[338] http://www.publications.parliament.uk/pa/ld200910/ldbills/001/2010001.pdf
[339] http://www.mediaweek.co.uk/News/MostEmailed/968765/Radio-Bill-creates-deep-divisions-among-industry
[340] http://radiotoday.co.uk/news.php?extend.5387
[341] http://www.telegraph.co.uk/technology/6616161/Digital-Economy-Bill-No-date-for-radio-digital-switchover.html
[342] http://radiotoday.co.uk/news.php?extend.5387
[343] http://radiotoday.co.uk/news.php?extend.5387
[344] http://www.mediaweek.co.uk/news/968765/Radio-Bill-creates-deep-divisions-among-industry/
[345] http://www.telegraph.co.uk/technology/6616161/Digital-Economy-Bill-No-date-for-radio-digital-switchover.html
[346] http://www.telegraph.co.uk/technology/6616161/Digital-Economy-Bill-No-date-for-radio-digital-switchover.html
[347] http://www.techradar.com/news/audio/hi-fi-radio/digital-radio-uk-hits-back-at-critics-652167
[348] http://business.timesonline.co.uk/tol/business/industry_sectors/media/article6896093.ece
[349] source: RAJAR and Digital Britain
[350] http://www.bbc.co.uk/programmes/b00nsnvm

59.

26 November 2009

France: digital radio "already dead"?

'Digital terrestrial radio: already dead?' enquired the headline of a French news web site early on Thursday morning 26 November 2009.[351] It proved to be prophetic.

Only hours later, a press release was published by the Bureau de la Radio (a radio trade group comprising the largest commercial radio owners) which effectively hammered another nail into the digital radio coffin in France. It stated:

"As it stands, the Bureau de la Radio estimates that the cost of the [digital radio] project is not compatible with the economics of the radio medium and does not allow plans for the launch of digital radio to proceed under positive conditions."[352]

On Monday 23 November, more than 70 stakeholders from the radio sector had met at the offices of the CSA [France's media regulator] to discuss digital radio. Afterwards, the CSA had issued a press statement which claimed that, during the meeting, all those present had endorsed the launch of digital terrestrial radio in France.[353] The meeting had decided to establish four working groups to examine: resource planning, the rollout timetable, transmission signals and data content. A further meeting was planned for February 2010.

However, the Bureau de la Radio has now stated its *"regret that the question of the [digital radio] economic model is not being addressed centrally"* and has called on the CSA to pursue more detailed work on this issue with stakeholders.[354]

Reflecting the seriousness of this announcement, Agence France-Presse headlined its news story 'Digital Radio: commercial radio opposes the launch of digital terrestrial radio.'[355]

There are some interesting parallels between developments in France and the situation in the UK, even though we are already ten years further down the digital terrestrial radio road. Just as in France, the UK government (through its DCMS department) and media regulator (Ofcom) have both been insistent that digital terrestrial radio must replace FM/AM radio broadcasting, without seeming to pay sufficient attention to the pre-requisite for an economic model to make it work.

In the UK:
- industry data point to minimal consumer demand for DAB, which I documented two months ago
- industry data point to minimal revenues from digital commercial radio stations, which I documented three months ago
- industry data suggest that the faster the take-up of DAB radio is accelerated, the faster commercial radio will lose more listening share to the BBC, which I documented recently
- the development of DAB has already proven disastrous for the financial health of the UK commercial radio industry.

After ten years of DAB radio development in the UK, precisely the same question needs to be answered here as is being asked in France this week:

Why has nobody published a realistic economic model for digital terrestrial radio which demonstrates convincingly that it is financially worthwhile?

Perhaps because one does not exist?

[351] http://www.ouest-france.fr/actu/actuDet_-Deja-depassee-la-radio-numerique-terrestre-_3639-1168143_actu.Htm
[352] http://www.google.com/hostednews/afp/article/ALeqM5g5R5vFv6Wfs7Tus1RU52ZfYFgfjA
[353] http://www.csa.fr/actualite/communiques/communiques_detail.php?id=129731
[354] http://www.google.com/hostednews/afp/article/ALeqM5g5R5vFv6Wfs7Tus1RU52ZfYFgfjA
[355] http://www.google.com/hostednews/afp/article/ALeqM5g5R5vFv6Wfs7Tus1RU52ZfYFgfjA

60.

28 November 2009

DAB radio: the customer is always wrong?

Politicians, government, civil servants, regulators. We pay their wages. They work for us, don't they? So why does the voice of the consumer, the citizen, the customer so often seem to be ignored or become lost when the government makes new policies or passes new legislation. DAB radio seems to be a case in point.

The government had convened the Digital Radio Working Group [DRWG] in 2007 to consider:
- what conditions would need to be achieved before digital platforms could become the predominant means of delivering radio?
- what are the current barriers to the growth of digital radio?
- what are the possible remedies to those barriers?

The Group met for a year and published its Final Report six days before Christmas 2008.[356] It had created a number of sub-groups to examine specific aspects of digital radio. One of these, the Consumer Impact Group, submitted its own report to the Working Group in November 2008 to inform its Final Report.

The Consumer Impact Group's recommendations about DAB radio make sober reading and carry as much gravitas, maybe more, now as when they were written a year ago. To quote directly and extensively from its report:

"The group is concerned that the case for digital [radio] migration has not been made clearly enough from the point of view of the consumer. While it is clear what the rationale is for the radio industry, the group would like to see a compelling argument as to why digital migration is desirable for consumers and what its benefits would be for consumers."

"The group also considers that the proposed migration criteria of 50% of all listening through digitally enabled devices is too low, and disproportionately affects disadvantaged groups who are less likely to be represented in the first 50% to take up digital radio. The group would therefore like to see the 50% figure analysed in more detail and a stronger case made for it, before it is adopted by the full DRWG, to ensure this is not the case."

"The group notes that neither the market nor consumers are currently prepared for migration at this stage. Information provided to the group shows that take-up varies from region to region and amongst demographic groups. Therefore, the group recommends that if digital migration proceeds, a help scheme will be essential to assist those where the cost of migration is significantly greater than the benefit. The information provided by the cost benefit analysis for the more vulnerable social groups will be an essential element in considering where and how a help scheme is best delivered."

"The group believes that further research should be undertaken to examine the extent of ownership and usage of analogue and digital radio particularly amongst disabled people, older people, people whose first language is not English and consumers from low income households. The research must be structured and use appropriate methodology to capture information on those over 65 and those over 75. The findings should be fed into plans to protect the consumer interests, i.e. for a help scheme, for effective labelling, for information and education campaigns and for the development of easy-to use products."

"The group urges caution with migration to digital radio should the uptake amongst older people, disabled people and low-income households be found to be low or should the costs be found to be prohibitive for these groups."[357]

Commenting on DAB radio take-up and the proposed digital migration criteria, the report said:

"The RadioCentre was asked to present figures, drawn from the existing Rajar and DRDB figures, setting out the current information on the number of DAB sales, household penetration and listening, defined by region, age and social class."

"The figures, which are annexed to this report at B [but excluded from the published version], show a number of interesting trends. For example sales, penetration and listening to DAB vary across the UK. Generally speaking, listening and awareness of DAB is highest in London and the South East, and the English Midlands. These have been the areas of longest DAB broadcasts and the widest choice of stations."

"When awareness and penetration are broken down by Socio-Economic Group and age, there does appear to be a divide. The figures show that consumers in lower income groups are considerably less likely to own a DAB set than other social groups. Even when owning a DAB, in some areas weekly listening to DAB by the over 65s is very low at less than 10%."

"The main conclusions to be drawn from this research is the general low level of ownership and listening by the over 65s compared to other age groups, and the low listening figures for consumers in the lower socio economic groups. This perhaps reflects that financially lower income groups are finding the price of sets a barrier, whilst for older groups, despite having sets, over 65's may find DAB radio's less easy to use than analogue sets, or perhaps prefer the traditional use of their analogue sets."

"Whilst recognising that universal DAB coverage is not achievable, the group considered that after migration, DAB coverage for UK-wide stations and stations for the nations should be equivalent or better than that available for analogue radio at present."

"The group stressed the importance of encouraging availability and use in cars, and noted that it would be virtually impossible to meet any listening criteria without addressing the issue of take up in cars. The group feels this should be a priority for the full DRWG."[358]

On the topic of research, the Consumer Impact Group commented:

"More and wider research is required, particularly about the ownership and usage of analogue and digital radio amongst those people with disabilities, people whose first language is not English, older people (both over 65s and over 75s) and those in low income groups. This additional research, when used together with the RadioCentre research and

Rajar figures should be used to guide future work in this area, particularly around take-up, equipment features, programming and a help scheme. The group feels that there is an opportunity here to ensure that future work is based on comprehensive and reliable evidence and analysis. The findings should be fed into plans for any help scheme, for effective labelling, for information and education campaigns and for developing easy-to use products. Where it doesn't already, this research should also take into account ways of listening to digital radio other than through a DAB enabled set, for example via the internet, digital terrestrial and satellite television, which may provide a significant proportion of the growth in the future."[359]

The Consumer Impact Group's recommendations included:

- *"We believe, that before migration could begin, additional research into radio users who are disabled, older people (both over 65 and over 75) and consumers from low income households is essential, since these people are likely to require particular assistance with migrating to DAB. This research should inform the development of plans for a help scheme, for effective labelling, for information and education campaigns and for developing easy to use products."*
- *"In the absence of the finalised cost benefit analysis at this point in time, the group recommends that the cost of converting to digital radio for the average household, as well as the affordability for low income groups should be investigated. In addition, the current take-up amongst older people, disabled people and low-income households needs to be investigated. The group urges extreme caution with migration to digital radio should the uptake in these groups be found to be low or should the costs be considered to be prohibitive by any of these groups, unless an appropriate help scheme is in place.*"[360]

Analysis. Research. Cost benefit analysis. Comprehensive and reliable evidence. All were considered to be very important by the Consumer Impact Group.

However, when the 26-page Final Report of the Digital Radio Working Group was published in December 2008, it did not include a single graph, a single numerical table or the results of any commissioned consumer research. Neither were such data attached in appendices.

The Final Report of the Digital Radio Working Group did recommend that *"the government should conduct a cost benefit analysis of digital migration."*[361] The government accepted this recommendation. One might think that this would be an urgent imperative, given that proposed legislation on DAB radio in the Digital Economy Bill is about to be debated in Parliament.

Wrong! The government has stated explicitly that it is *"committed to a full cost benefit analysis of the Digital Radio Upgrade programme before any Digital Radio Upgrade is set"* which would include *"the timings and costs to consumers."* But the government has stated that *"**this is likely to begin in 2011**."*[362]

What? The government wants a huge (some would say impossible) commitment from the UK radio sector and from the British public to forge ahead with migration of radio listening to DAB, even though its own full cost benefit analysis of pursuing that policy will not be **STARTED** until 2011.

Is this not mad? Are our public servants working for us? Does the consumer viewpoint on these issues count for nothing?

[356] http://www.culture.gov.uk/images/publications/DRWG_Final_Report.pdf
[357] http://www.culture.gov.uk/images/publications/Consumer_Impact_Group_Report_to_DRWG.pdf
[358] http://www.culture.gov.uk/images/publications/Consumer_Impact_Group_Report_to_DRWG.pdf
[359] http://www.culture.gov.uk/images/publications/Consumer_Impact_Group_Report_to_DRWG.pdf
[360] http://www.culture.gov.uk/images/publications/Consumer_Impact_Group_Report_to_DRWG.pdf
[361] http://www.culture.gov.uk/images/publications/DRWG_Final_Report.pdf
[362] http://interactive.bis.gov.uk/digitalbritain/wp-content/themes/clean-home/DEB-Impact-Assessments.doc

61.

1 December 2009

DAB radio UK sales: 10 million in 10 years is "an incredible achievement"?

The Digital Radio Development Bureau [DRDB] published a press release yesterday trumpeting the "incredible achievement" that 10 million DAB receivers had been sold to date in the UK which, it said, "proves that digital radio is here to stay."[363] The press release was notable not for what it said, but for what it omitted.

Ten million radios sounds like a big number until you realise that this has been achieved over more than a decade of DAB product sales in the UK. There are 51.3 million adults (aged 15+) in the UK. So, averaged over the decade, roughly one out of every fifty adults bought a DAB radio each year. Not so impressive.

Revisit the DRDB's own forecasts for DAB receiver sales. In 2004, it forecast 13.15m DAB radios would be sold by year-end 2008 (the reality was 8.53m). In 2005, it forecast 19.96m to be sold by year-end 2009. In 2006, it forecast 17.2m sold by year-end 2009. In 2007, it was too embarrassed to revise its year-end 2009 forecast (but even its year-end 2008 forecast of 9.16m was overstated, as the reality was 8.53m).[364] In 2008 and 2009, understandably, the DRDB did not publish its forecasts. The DRDB forecasts of the very consumer market in which it is specialising have consistently been shown to be wildly inaccurate.

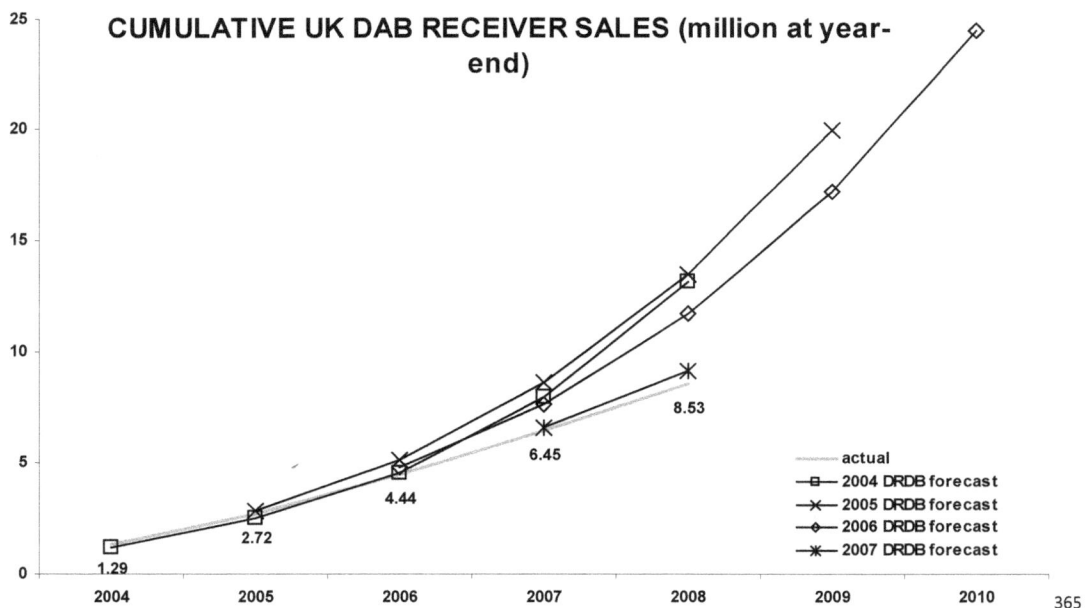

CUMULATIVE UK DAB RECEIVER SALES (million at year-end)

The DRDB press release also claimed that "for the past three years, sales of digital radio sets have remained solid."[366] 'Solid' is an interesting choice of word to describe the present situation of declining sales. Sales in Q2 2009 were the lowest in two years and were down

186

6% year-on-year. Sales in the previous two quarters, Q1 2009 and Q4 2008, were also down 1% and 10% respectively year-on-year.[367] Three consecutive quarters of negative sales growth can hardly be described as 'solid'.

UK QUARTERLY DAB RECEIVER SALES ('000)

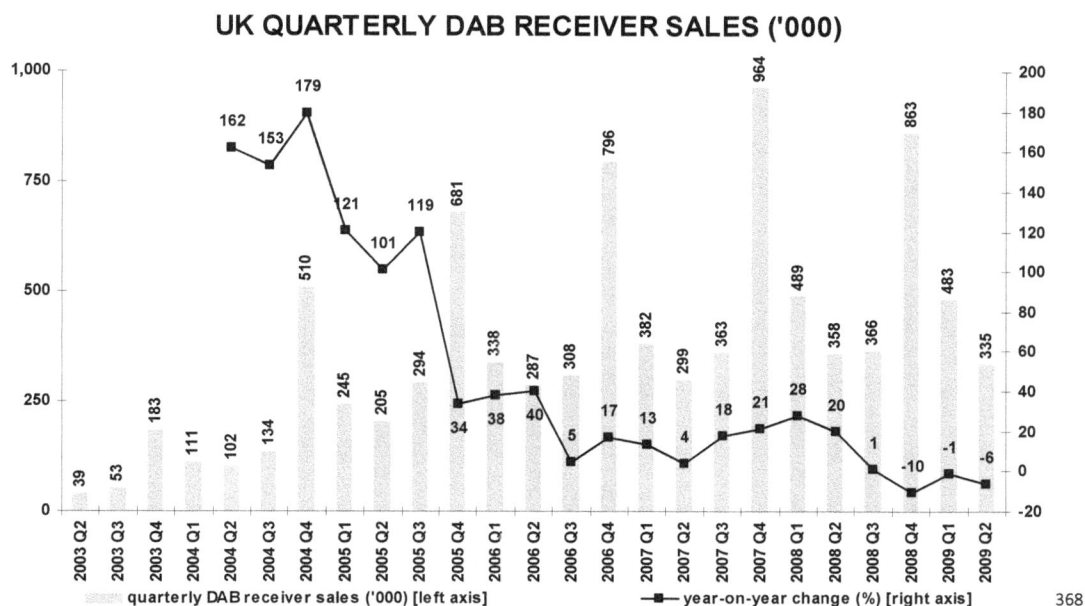

| | quarterly DAB receiver sales ('000) [left axis] | ─■─ year-on-year change (%) [right axis] | 368 |

As the graph above shows, the rot set in at the end of 2005, when year-on-year DAB radio sales growth fell from triple to double digit figures. Both 2006 and 2007 included quarters of single digit growth. Now, in 2009, growth has been negative all the way. This is no temporary blip caused by the recession. The writing was already on the wall by 2006 – the DAB party is over. Now we are merely waiting for the last few guests to leave.

The other remarkable statement in the DRDB press release is its satisfaction that sales of "all categories of analogue radio showed significant decline."[369] As I have pointed out previously (see graph below), sales of radio receivers generally are in long-term decline in the UK. Is this a fact that a stakeholder within the radio broadcast industry should be crowing about? It's like two passengers on the Titanic fighting over which has the bigger cabin – does it really matter if the whole ship is slowly going down?

It should be pointed out that the DRDB data exclude sales of mobile phones, despite the fact that the majority of current models sold in the UK include FM radios, whilst not one model includes a DAB radio. More than 30m mobile phones were sold in the UK in 2008, which puts the 2m DAB radios sold in stark perspective.

Also, it should be pointed out that the vast majority of what the DRDB calls 'DAB radios' on sale in the UK also incorporate analogue FM. It is increasingly difficult to find a DAB-only radio to purchase in UK shops. This renders the DRDB's proclaimed digital versus analogue victory completely hollow. For every 'DAB radio' sold that the DRDB hopes will automatically lead us to some kind of digital heaven, in probably 90%+ of purchases, yet another FM radio is also being added to the millions already in UK households.

187

UK radio receiver sales - four quarter moving average - '000 units per annum
DAB receivers as percentage of total receiver sales

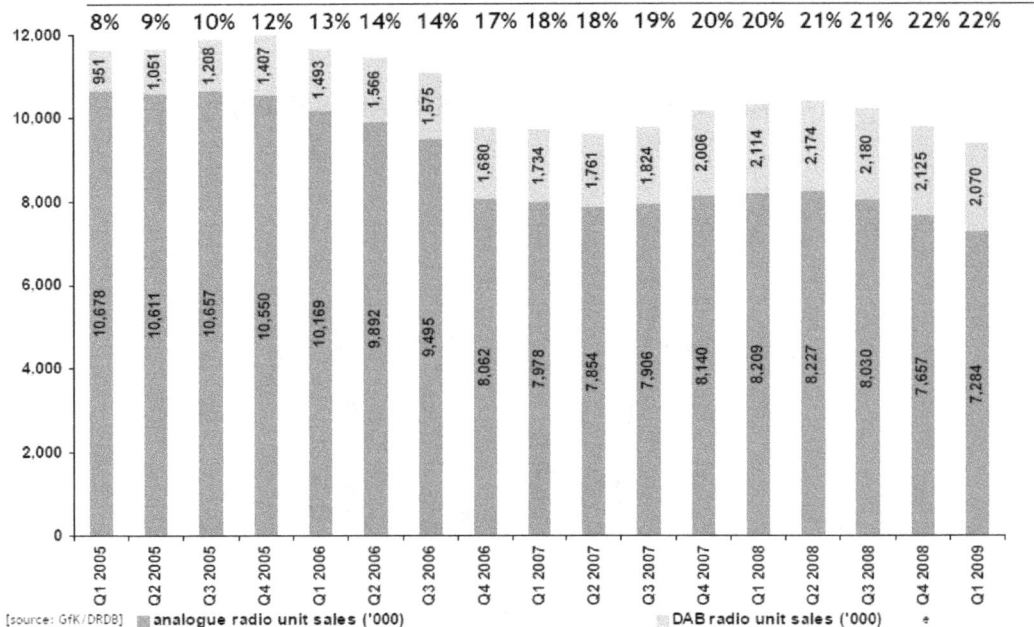

	Q1 2005	Q2 2005	Q3 2005	Q4 2005	Q1 2006	Q2 2006	Q3 2006	Q4 2006	Q1 2007	Q2 2007	Q3 2007	Q4 2007	Q1 2008	Q2 2008	Q3 2008	Q4 2008	Q1 2009
%	8%	9%	10%	12%	13%	14%	14%	17%	18%	18%	19%	20%	20%	21%	21%	22%	22%
DAB	951	1,051	1,208	1,407	1,493	1,566	1,575	1,680	1,734	1,761	1,824	2,006	2,114	2,174	2,180	2,125	2,070
analogue	10,678	10,611	10,657	10,550	10,169	9,892	9,495	8,062	7,978	7,854	7,906	8,140	8,209	8,227	8,030	7,657	7,284

[source: GfK/DRDB] ■ analogue radio unit sales ('000)   ▨ DAB radio unit sales ('000)   ◆   370

Finally, recall that 8m analogue radios (without DAB) are still being sold annually in the UK. Now add to that the 30m mobile phones purchased, most of which include FM radio. Then compare it with the "incredible achievement" of 2m DAB radios sold per year, most of which include analogue radio anyway. The future of radio is looking less and less like a DAB world. Rather, analogue radios are probably multiplying faster in the UK marketplace than they have ever done, thanks to mobile phone manufacturers. This is good news for radio, bad news for investors in DAB.

These facts might not conveniently fit the DRDB 'story'. But they are the facts.

---

[363] http://www.drdb.org/article.php?id=832&from=lat
[364] source: Digital Radio Development Bureau
[365] source: Digital Radio Development Bureau
[366] http://www.drdb.org/article.php?id=832&from=lat
[367] source: Digital Radio Development Bureau
[368] source: Digital Radio Development Bureau
[369] http://www.drdb.org/article.php?id=832&from=lat
[370] source: Digital Radio Development Bureau

62.

3 December 2009

Radio in the Digital Economy Bill: House of Lords Second Reading

Digital Economy Bill
2 Dec 2009 @ 1539
Second Reading, House of Lords [excerpts][371]

The First Secretary of State, Secretary of State for Business, Innovation and Skills and Lord President of the Council (Lord Mandelson):
We have also set out our vision for the future of digital radio, which will see the country shift to digital, when transmission coverage and audience numbers are wide enough, by the end of 2015.

The Lord Bishop of Manchester:
The switchover to digital radio may produce more problems than expected. Of course there is much to welcome in the creation of platforms for new content to meet the needs of specialist audiences. I think, for example, of Premier Christian Radio's recent acquisition of a national DAB licence. However, there may be much to be concerned about over the plan to cut off national stations and many local services as early as 2015. While the Government have indicated that that will not be finalised until digital services account for 50 per cent of all radio listening and can reach 90 per cent of the population, it is also clear that, without an early deadline, sufficient pressure may not build on radio manufacturers and retailers to shift to selling DAB sets only for cars as well as homes. The radio switchover again underlines the risk of creating another two-tier system where significant swathes of the country could lose their favourite national stations from the FM dial, including the BBC stations they pay for through the Licence Fee. Surely that cannot be right.

What government support will there be for the switchover to digital radio, which is likely to be not only more problematic but, generally, more expensive across the population than the TV switchover has been? Will the Minister accept that over-rushing towards analogue switch-off will not allow proper time for the Government, this House and the other place to think through the unintended consequences? Is there anything that the Government can learn from the German Government's experience and their postponements of switchover plans?

..... On voluntary supported broadcasting, do the Government intend to keep some of the analogue spectrum going, for example, for hospital radio?

This country must, of course, embrace the opportunities offered by a digital economy, but the advantages must be shared by the widest possible number of citizens. Some, if not all, of the unintended consequences that could unfairly disadvantage people might be avoided by not being trapped in too rigid a timetable. If that happens, I fear that this country will not benefit from the best rewards that a digital economy offers.

189

Lord Carter of Barnes:
Secondly, in the critical areas of investment, infrastructure, spectrum liberalisation and the digitalisation and deregulation of sound radio, it provides a framework for innovation, development and investment.

Baroness Howe of Idlicote:
My Lords, when the noble Lord, Lord Carter of Barnes, introduced Digital Britain a little while ago, we all recognised that things were beginning to happen and there were some very welcome realisations, for example, on the need to move forward with digital radio.......

I welcome those parts of the Bill which incorporate the Digital Britain promise to speed up delivery of a fully operational DAB digital radio platform. I spend a lot of time in cars and have had hearing difficulties since the arrival of my first child, so it is a real pleasure to enjoy the quality and clarity of digital sound, especially when listening to music, whether it is Radio 3 or Classic FM, both of which are excellent stations. The plank for Ofcom to be able either to terminate analogue licences without consent, subject to a minimum two years' notice, or where appropriate to extend analogue licences up to and beyond switchover, on condition that digital services are also provided, will no doubt help to build in the much-needed flexibility to enable radio switchover. I very much hope and have confidence in the plans that have been outlined that it will happen by 2015. It is important that it does.

Lord Roberts of Llandudno:
Today, I looked at the figures for radio listeners in Wales who have ever listened to Digital Audio Broadcasting [DAB]. I shall not go through the whole list, but in Cardiff, it was 27 per cent, while in the Valleys, it was only 4 per cent. That is the difference. The most needy areas will not have the opportunity to benefit from these new high-tech developments. There is a pressing need for an extension of broadband, not least because of the commitment already made by the Government that fibre optic broadband should be prioritised in 'notspots', where other technologies have also failed.

Lord Clement-Jones:
I move on again, to independent radio services. We broadly welcome the provisions for digital switchover. Of course, full switchover will only happen on a specified date if certain criteria for uptake are met, and the only way that one will get further adoption is by setting a firm date. I hope that the Minister will confirm that we are currently working off a 2015 date, but there are concerns among smaller radio stations that the digital multiplex regions that have been defined are too large. Small, local stations will be broadcast across the whole of a large region covered by a multiplex, and may be expected to pay a rental reflecting that. That would be unfair on some of those small stations. Many of them are arguing for DAB+, a technology which would be, I believe, much more in tune with their requirements. I would be grateful to hear what the Minister says in that respect.

Lord Howard of Rising:
While we on these Benches support the switch from analogue to digital radio, it is a sensitive area. It would be good if the Government could give some assurances of what criteria will be used to decide when will be the appropriate time for the changeover. Will the Government be guided by the criteria set out in the Digital Britain White Paper, referred to by the right reverend Prelate the Bishop of Manchester? If so, we remain unconvinced that the 2015 target date is realistic and worry that millions of listeners and hundreds of local stations will be disadvantaged.

There are many for whom the digital switchover will cause problems: the elderly or the lonely, who may have had a wireless for many years which has become almost a companion; the blind person who will not be able to work the digital radio because the instructions are on a screen that they will not be able to see. I hope that the Secretary of State can reassure the House that proper care and attention will be paid to the needs of those who will encounter difficulties with the transition.

Lord Davies of Oldham:

The right reverend Prelate the Bishop of Manchester indicated the issues that arise with the digital switchover. I emphasise that we will not make the switchover for radio until there is already 90 per cent coverage in the United Kingdom and until 50 per cent of hours of radio are listened to via digital stations. We have criteria before we actually make the move. This follows on from points about the switch from analogue to digital television. I take on board his point that it is important that any changes that are made benefit people and do not shock them with a possible loss of services and extra cost. That point has to be addressed.

[next stage: House of Lords Committee, 6 January 2010]

---

[371] http://www.publications.parliament.uk/pa/ld200910/ldhansrd/text/91202-0004.htm

63.

5 December 2009

France: digital radio postponed until at least year-end 2010

After a period of uncertainty about a timetable for the launch of digital terrestrial radio in France, the regulator has finally admitted that the first transmissions, which had been scheduled to take place this month, will be postponed until at least year-end 2010.

During an online chat yesterday, Michel Boyon, president of the CSA [France's media regulator], said:

"While everyone recognises the need to act quickly, despite the current economic challenges, it will be year-end 2010." He argued that *"if radio does not go digital, It will slowly decline"* and noted that *"internet radio is very good, but it is totally inadequate to meet the demands of listeners."*[372]

Nevertheless, the French press seems unconvinced that digital radio will ever happen.

*'**Digital radio silence is delayed!**'* said the headline in trade magazine SatMag, which commented:

"After having been delayed for years in favour of digital television, digital radio is taking too long and is being overtaken by other technologies."[373]

*'**Too expensive, digital radio postponed indefinitely**,'* said the headline of Agence France-Presse the day before Boyon's announcement. It added:

"The latest figures from Mediametrie confirm the change in radio listening habits: almost 50% of listening takes place on the move, and a quarter of the population has already listened to radio via the internet. During the last year, the numbers listening to radio via mobile phones has increased by 50%."[374]

*'**Has the internet killed the digital radio proposal?**'* asked the headline in rue89. Francoise Benhamou, professor of economics at the University of Paris 13, commented:

"Consider that a cost of 600m to 1bn Euros [to implement digital terrestrial radio] over ten years is viable only if [radio] advertising revenues increase by 20 to 25%. Such a forecast would be very risky given the uncertain economic background and the competition from the internet for advertising revenues." She added: *"Many of us already receive radio broadcasts, live and on-demand covering a wide range of content, as well as associated interactive services, by connecting via broadband. Do we really have a need for digital terrestrial radio?"*[375]

Professor Benhamou concluded: *"This situation does not please everyone, particularly the CSA [media regulator] who saw [digital radio] as an opportunity to extend their domain ..."*[376]

It sounds all too familiar to us in the UK.

[372] http://www.latribune.fr/entreprises/communication/publicite-medias/20091204trib000449887/la-radio-numerique-attendue-fin-2010.html

[373] http://www.satmag.fr/affichage_module.php?no_theme=1&no_news=11626&id_mod=50

[374] http://www.google.com/hostednews/afp/article/ALeqM5hXXSU8J3Gj_AadFl4KewR1VlnLmA

[375] http://www.rue89.com/en-pleine-culture/2009/12/01/internet-a-t-il-tue-le-projet-de-radio-numerique-terrestre

[376] http://www.rue89.com/en-pleine-culture/2009/12/01/internet-a-t-il-tue-le-projet-de-radio-numerique-terrestre

64.

7 December 2009

Digital radio switchover: amendment to "consider the needs" of listeners and small stations

Clause 30 of the government's Digital Economy Bill sets out the process for determining the date for radio 'digital switchover':

97A: Date for digital switchover
(1)   The Secretary of State may give notice to OFCOM nominating a date for digital switchover for the post-commencement services specified or described in the notice.
(2)   When nominating a date, or considering whether to nominate a date, the Secretary of State must have regard to any report submitted by OFCOM or the BBC under section 67(1)(b) of the Broadcasting Act 1996 (review of digital radio broadcasting).[377]

An amendment has been tabled by Lord Howard of Rising and Lord de Mauley which would require the government additionally to consider:

- the needs of local and community radio stations
- the needs of analogue listeners

as well as any reports submitted by Ofcom and the BBC. This amendment will be considered, along with many others not concerned with radio, when the Bill is debated by a House of Lords committee on 6 January 2010.

Although this amendment does not suggest a specific mechanism for canvassing the opinions of listeners or local radio stations, it nevertheless acknowledges implicitly that the consumer and small commercial/community radio stations need to have a voice in the process. It is about time.

From its earliest formulation, the proposal for radio broadcasting to be switched from FM/AM to DAB seemed to have been intended to create:
- a 'walled garden' under the control of the UK's largest commercial radio owners and the BBC who, between them and transmission provider Arqiva, not only own the entire DAB infrastructure but also act as 'gatekeeper', deciding which station has access to the platform
- a 'walled garden' on DAB that would hopefully stop consumers listening to content not produced or approved by the BBC or the largest commercial radio companies, such as online radio (most of which originates or is owned overseas), pirate radio, community radio and small independent stations.

Massive consolidation in commercial radio since then has resulted in a more divided industry than ever, in which the biggest commercial players are eager to 'nationalise' or 'regionalise' what had been licensed as local radio stations, whereas most of the smaller commercial and community owners want to keep local radio as local as they can.

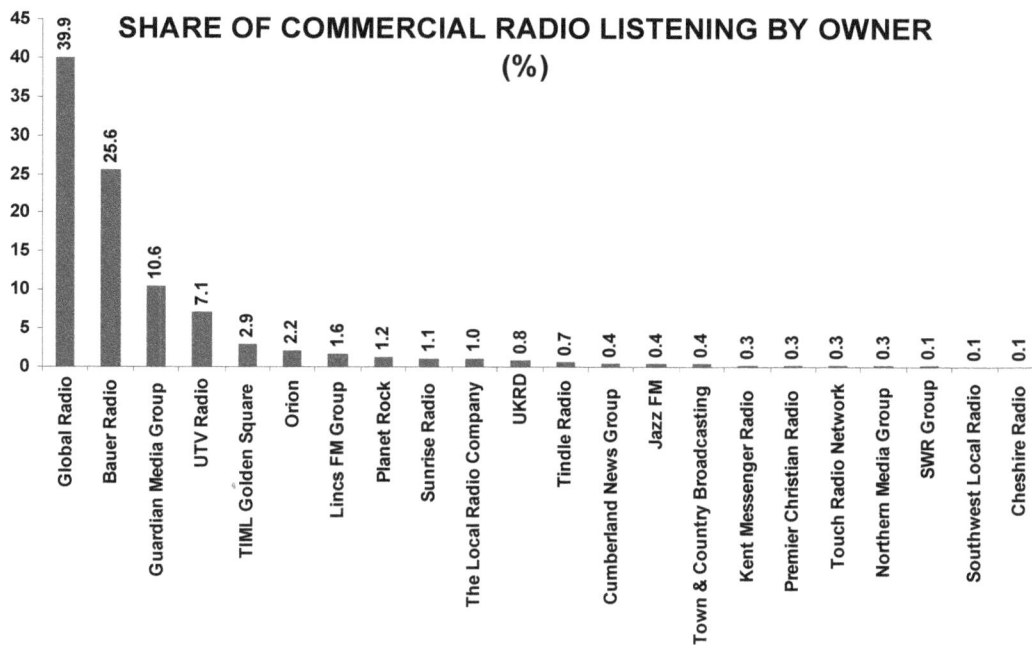

SHARE OF COMMERCIAL RADIO LISTENING BY OWNER (%)

Owner	%
Global Radio	39.9
Bauer Radio	25.6
Guardian Media Group	10.6
UTV Radio	7.1
TIML Golden Square	2.9
Orion	2.2
Lincs FM Group	1.6
Planet Rock	1.2
Sunrise Radio	1.1
The Local Radio Company	1.0
UKRD	0.8
Tindle Radio	0.7
Cumberland News Group	0.4
Jazz FM	0.4
Town & Country Broadcasting	0.4
Kent Messenger Radio	0.3
Premier Christian Radio	0.3
Touch Radio Network	0.3
Northern Media Group	0.3
SWR Group	0.1
Southwest Local Radio	0.1
Cheshire Radio	0.1

378

There is no longer likely to be a single organisation that can embrace the full range of stakeholders in the radio sector. Even government agencies such as Ofcom and DCMS seem wilfully to be ignoring the wider picture, as if seduced by notions that 'DAB must happen', 'bigger must be better', 'Britain must lead the way' and 'consumers don't know what's good for them'.

Inevitably, it will end in tears. You can pass all the laws you want but, if you cannot get the consumer interested in DAB, it will fail. And, to date, the consumer seems largely disinterested and could not care less that manufacturers of DAB radios are mostly British (though they manufacture outside the UK) or whether they listen to British radio content.

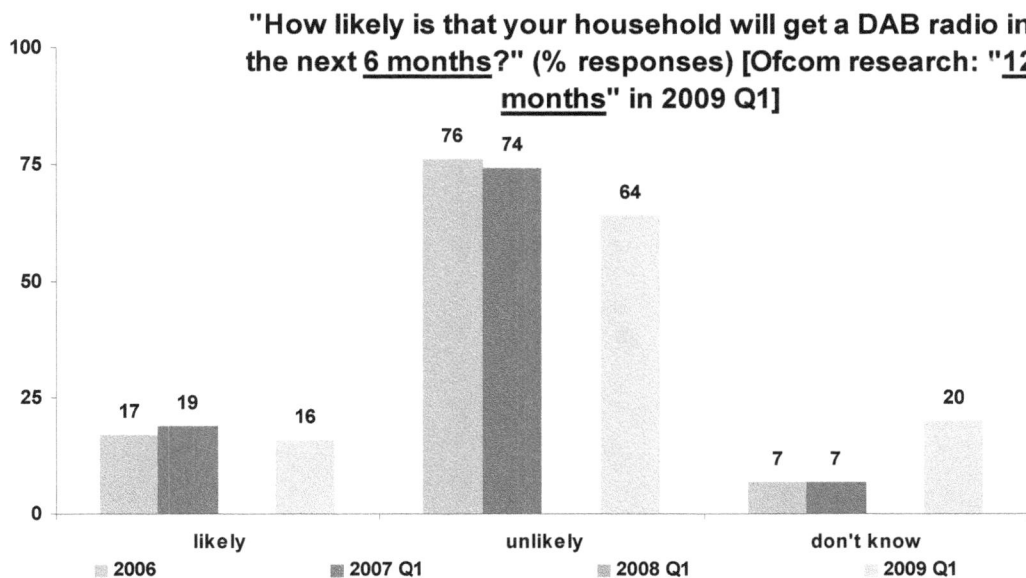

"How likely is that your household will get a DAB radio in the next 6 months?" (% responses) [Ofcom research: "12 months" in 2009 Q1]

	likely	unlikely	don't know
2006	17	76	7
2007 Q1	19	74	7
2008 Q1	16	64	
2009 Q1			20

379

195

Ofcom's most recent market research shows the stark reality: 64% of households say they are unlikely to buy a DAB radio in the next 12 months, and a further 20% say they don't know.[380]

You ignore consumer opinion at your peril.

---

[377] http://www.publications.parliament.uk/pa/ld200910/ldbills/001/2010001.pdf
[378] source: RAJAR
[379] source: Ofcom
[380] source: Ofcom

65.

11 December 2009

Local government unhappy about digital radio switchover

The Local Government Association, representing 424 local government authorities in England and Wales, is backing a campaign to lobby the government to re-think its proposal for digital radio switchover.

"I'm urging the Government not to confirm the 2015 switchover date from analogue to digital radio until proposals have been properly rural proofed," said Peter Phillips, Liberal Democrat councillor for Bishop's Castle in Shropshire. *"The proposed switchover will also have significant carbon footprint implications, as DAB radios consume more power than transistor sets. Waste authorities will be affected in having to dispose of analogue radio sets."*[381]

Phillips presented a report to a September 2009 board meeting of the Local Government Association, at which he *"raised a number of important issues for both the Association and Local Authorities to consider in preparing for any switchover to digital radio"*, according to the minutes.[382]

The Association is reported to be contacting the government, Ofcom and Digital Radio UK to express its concerns about the proposal in the Digital Economy Bill for digital radio switchover.

---

[381] http://www.ludlowadvertiser.co.uk/news/4759576.Radio_plan_will_be_rural_switch_off_in_South_Shropshire/
[382] http://www.lga.gov.uk/lga/aio/4577439

66.

13 December 2009

Internet radio: denigrate it, ignore it, marginalise it … consumers will still listen

It was a surprise to find that the entire front page of the most recent issue of the World DMB Forum's global newsletter ('Eureka!') was filled with an article that did not extol the virtues of the DAB/DMB platform, but instead tackled the online radio platform and drew the conclusion that the internet *"will NOT replace traditional broadcasting."* The article, entitled 'The Future of Radio', sought to debunk the assertion that *"the internet is the future of radio."*[383]

It stated that the BBC iPlayer *"allows the UK public to access almost all of its radio and TV programmes broadcast during the previous seven days."*[384] This is inaccurate. The iPlayer offers nothing like *"almost all"* the BBC's radio and TV output. Indeed, for some of the BBC's radio and TV networks, the selection of content remains remarkably thin (mostly due to rights issues).

The article continued: *"Given the outstanding success of the BBC's iPlayer, it is surprising to learn from RAJAR's latest audience figures that 'radio via the Internet' (in all its forms: live streaming; on-demand services and podcasting) accounts for only 2.2% of radio listening in the UK."*[385]

This is untrue. The RAJAR 2.2% share figure ONLY includes simulcast live streams of the BBC and UK commercial broadcasters. It does not include on-demand services; it does not include podcasts; it does not include listening to online radio services such as Last.fm, Spotify and Rhapsody; and it does not include listening to audio from overseas broadcasters. There is a detailed section on the RAJAR web site that explains these facts.[386] RAJAR has never claimed that its data for 'internet' listening include anything other than simulcast live streams of BBC and UK commercial radio stations.

The article then drew the conclusion: *"Taking these differences in penetration into account shows that DAB listening in the UK is 10 times more popular than listening via digital TV or via the internet."*[387] However, it is unclear what the phrase *"10 times more popular"* is trying to imply. Is that '10 times more listening'? Or maybe '10 times more reach'?

Interestingly, exploring the latter metric, RAJAR's own research (as part of its MIDAS survey, rather than the main diary survey) found in December 2008 that the weekly reach of all internet-delivered radio content in the UK was 14%, compared to the DAB platform's weekly reach of 17.8% during the same quarter (see graph below).[388] Ten times more popular? The platforms were almost neck-and-neck in the 'reach' metric. I wrote about this research a year ago. It is the closest we have for now to a like-for-like comparison that includes all forms of audio delivered by the internet.

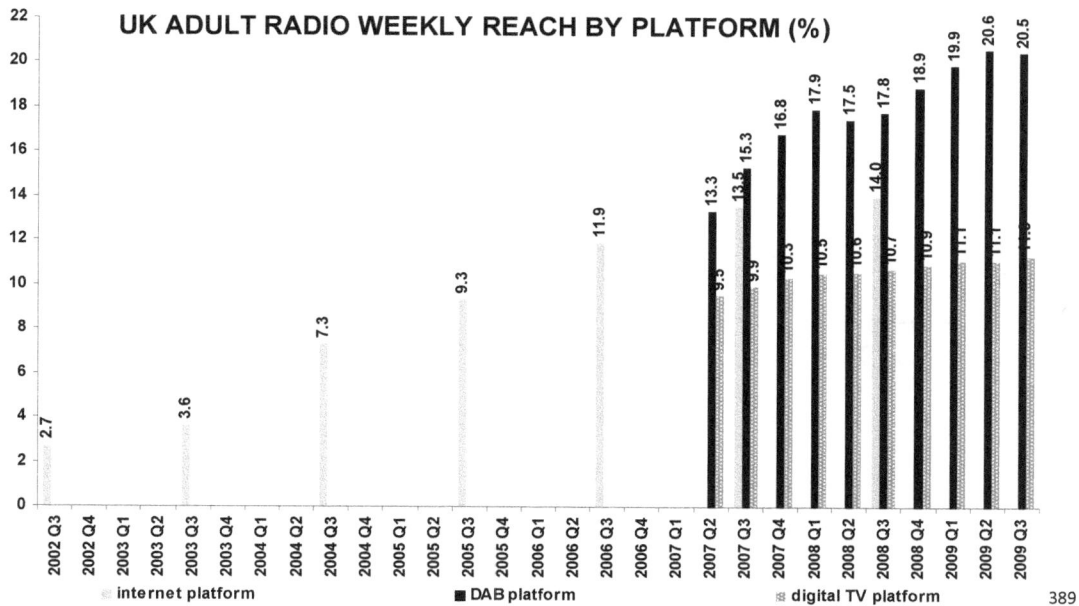

UK ADULT RADIO WEEKLY REACH BY PLATFORM (%)

	2002 Q3	2003 Q3	2004 Q3	2005 Q3	2006 Q3	2007 Q2	2007 Q3	2007 Q4	2008 Q1	2008 Q2	2008 Q3	2008 Q4	2009 Q1	2009 Q2	2009 Q3
internet platform	2.7	3.6	7.3	9.3	11.9										
DAB platform						13.3	15.3	16.8	17.9	17.5	17.8	18.9	19.9	20.6	20.5
digital TV platform						9.5	9.9	10.3	10.5	10.6	10.7	10.9	11.1	11.1	

389

The most recent reach data for the internet platform in the above graph derive from Q3 2008 because RAJAR has not publicly released comparative data derived from its two subsequent MIDAS surveys (which are now only available on subscription).

RAJAR was keen to stress in its press release accompanying this week's latest MIDAS 5 survey that:

"74% of those Listen Again [audio on-demand] listeners said the service has no impact on the amount of live radio to which they listen, while half said they are now listening to radio programmes to which they did not listen previously."[390]

Somehow, the Daily Mail managed to mangle this factual statement into something that, yet again, portrayed the internet platform as an aggressor against DAB:

"Rajar says the figures do not mean people are abandoning traditional or DAB radio sets but that more Britons are trying and using online stations as well."[391]

The problem the radio industry faces with the RAJAR audience metric is that it cannot have its cake and eat it. Either it chooses:

- to restrict RAJAR to measuring 'traditional', live radio and accepts that, as a result, the data will inevitably show that listening to 'traditional' radio is in continuing decline (which is RAJAR today, see graph above); or
- to expand the RAJAR metric to measure 'audio' consumption that includes on-demand and podcast content, as well as non-traditional radio such as Spotify and Last.fm, thus demonstrating that total listening is not at all in decline but, on the contrary, has been enhanced by audio content increasingly consumed via non-broadcast platforms and 'on the go'.

For the BBC, Director of Audio & Music Tim Davie hinted at the last RadioCentre conference that he would be interested to see RAJAR extended to encompass time-shifted and downloaded audio, both of which account for an increasing proportion of BBC radio listening.

15-44 YEAR OLDS RADIO LISTENING ('000 hrs/wk)

Quarter	all radio: hours listened ('000/wk)	commercial radio: hours listened ('000/wk)
Q2 1999	499,777	302,019
Q2 2000	503,851	304,802
Q2 2001	535,921	318,209
Q2 2002	494,373	290,589
Q2 2003	484,452	285,098
Q2 2004	497,863	291,367
Q2 2005	489,061	273,200
Q2 2006	480,950	258,911
Q2 2007	484,436	260,285
Q2 2008	451,449	239,533
Q2 2009	455,850	241,273

392

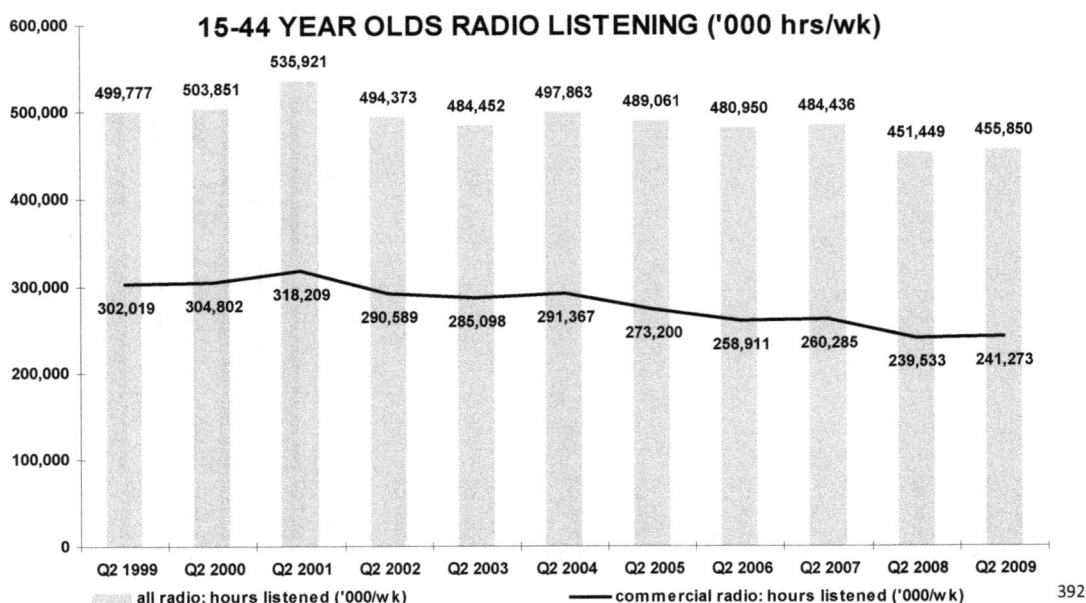

For its part, commercial radio has shown no interest in advocating such a re-definition of the RAJAR metric. Not only do its offerings of time-shifted and downloadable audio remain miniscule compared to the BBC, but it is locked into a strategy to maintain its 'walled garden'. Understandably, it has no desire to demonstrate to the world that it is losing listening to competitors' time-shifted audio and online 'radio'. UK commercial radio has enjoyed a nice little over-the-air duopoly from 1973 until recently – best just to pretend that it remains one of only two games in town.

The paradox here is that commercial radio is busy presenting advertising agencies and potential advertisers with RAJAR data that only tell part of the story of how and what audio people are listening to in 2009. However, once their meetings with commercial radio people are over, those same advertisers and agencies will inevitably be busy booking advertising with all sorts of online media, including Last.fm and Spotify. They know precisely what opportunities are out there in the wide world beyond traditional broadcasting.

Simply ignoring new businesses that are competing for your listeners' attentions is not going to make them go away. Sticking your head in the sand can only have the effect of devaluing RAJAR as a useful and accurate metric in the long term.

Remember King Canute.

---

[383] http://www.worlddab.org/news/document/925/Eureka__issue_10_FINAL_72dpi.pdf
[384] http://www.worlddab.org/news/document/925/Eureka__issue_10_FINAL_72dpi.pdf
[385] http://www.worlddab.org/news/document/925/Eureka__issue_10_FINAL_72dpi.pdf
[386] http://www.rajar.co.uk/content.php?page=about_process
[387] http://www.worlddab.org/news/document/925/Eureka__issue_10_FINAL_72dpi.pdf
[388] http://www.rajar.co.uk/docs/news/MIDAS3_report.pdf
[389] source: RAJAR
[390] http://www.rajar.co.uk/docs/news/MIDAS5_news_release.pdf
[391] http://www.dailymail.co.uk/news/article-1234860/Popularity-online-radio-soars-UKs-adult-population-tunes-internet.html
[392] source: RAJAR

67.

16 December 2009

Radio in the Digital Economy Bill: three more amendments tabled

The following amendments to the Digital Economy Bill will be considered at Committee Stage in the House of Lords, scheduled for 6, 12, 18 and 20 January 2010.[393]

CLAUSE 30: DIGITAL [RADIO] SWITCHOVER

What does Clause 30 do? According to the government's Explanatory Notes:

"Clause 30 allows the Secretary of State to give notice to OFCOM of a date by which digital switchover must occur for services specified in the notice. In making a decision to nominate a switchover date, the Secretary of State must take account of any reports by the BBC and OFCOM about the future of analogue broadcasting.
The date for digital switchover is the date after which it will no longer be appropriate for the service in question to be broadcast in analogue form.
The Secretary of State may nominate different switchover dates for different types of radio services and may withdraw a nomination of a switchover date.
After a switchover date has been set, OFCOM are required to vary the licence periods of all licences for the services specified by the Secretary of State so that they end on or before that date. However, OFCOM cannot shorten the duration of a licence so that it would end less than 2 years from the date on which OFCOM give notice of the variation, unless the licence-holder consents.
OFCOM may not vary a licence period so that it ends after the switchover date."[394]

A.   Lord Clement-Jones and Lord Razzall have proposed an amendment to Clause 30:

Page 33, line 19, at end insert —
"(2A) The Secretary of State may not nominate a date for switchover —
(a)   unless it can be established that all local commercial radio stations will have the opportunity to move to digital audio broadcasting,
(b)   until the proportion of homes in each of the four nations of the UK able to receive —
    (i) national BBC services,
    (ii) national commercial radio services,
    (iii) local BBC services, and
    (iv) local commercial radio radio services [sic],
via digital audio broadcasting is equal to the proportion able to receive them via analogue broadcasting.
(c)   until digital audio broadcasting accounts for at least 67 per cent of all radio listening, and
(d)   until digital audio broadcasting receivers are installed in 50 per cent of private and commercial vehicles."

Page 33, leave out line 21 and insert —
"(a)   must ensure that all commercial and BBC radio services broadcasting in the UK have the opportunity to switchover on the same date,"

Page 34, line 1, leave out *"2"* and insert *"4"*

In (my) plain English, this amendment would prevent the government from announcing a single digital radio switchover date until:
- Listening via DAB accounts for two-thirds of all radio listening
- DAB radios are installed in 50% of cars
- 50% of households in each nation have access to a DAB radio
- All BBC and commercial stations, large and small, have been offered the opportunity to migrate from analogue to DAB.

In practice, none of these criteria could possibly be met within the next decade, which would effectively scupper the notion of a digital switchover date. Additionally, a four-year termination notice period would be inserted into renewed commercial radio licences (instead of the government's proposed two-year period).

B.   Lord Cotter has proposed a separate amendment to Clause 30:

Page 33, line 33, at end insert —
"97AA Disposal and recycling of domestic analogue radios
(1)   Following a decision to give notice to OFCOM under section 97A of a date for digital switchover, the Secretary of State must devise a scheme for the disposal and recycling of domestically owned analogue radios.
(2)   The scheme must include provision for a financial incentive for domestic owners of analogue radios to purchase a radio suitable for digital audio broadcasting following disposal and recycling of their analogue radios.
(3)   The financial incentive must be based on any profit made from the disposal and recycling of analogue radios and must not be derived from public funds."

In plain English, consumers will have to be paid something for all those analogue radios they will be expected to no longer use.

CLAUSE 31: RENEWAL OF NATIONAL RADIO LICENCES

What does Clause 31 do? According to the government's Explanatory Notes:

"Clause 31 allows the further renewal of national analogue licences for a period of up to seven years. All of these licences have already been granted a renewal of 12 years under the powers in section 103A of the Broadcasting Act 1990 ('the 1990 Act'). ......"[395]

Lord Clement-Jones and Lord Razzall have proposed an amendment to Clause 31:

"The above-named Lords give notice of their intention to oppose the Question that Clause 31 stand part of the Bill."

In plain English, this amendment would delete the proposal in the Bill to automatically renew the three national commercial radio licences for a further seven years. Instead, the licences would have to be auctioned individually to the highest bidder, as required by existing legislation. The greatest impact would be on Classic FM, whose licence would have to be advertised by Ofcom in 2010, if this amendment were passed. Its owner, Global Radio, would be faced with Hobson's choice – either to bid a significantly higher amount (maybe £10m+ per annum rather than the present £2m+ per annum) to win/retain the licence, thus diminishing its 'cash cow' status, or to lose the single most profitable licence in commercial radio. Either option might seriously undermine Global Radio's ability to trade profitably and to service its debt.

---

[393] http://www.publications.parliament.uk/pa/ld200910/ldbills/001/2010001.pdf
[394] http://www.publications.parliament.uk/pa/ld200910/ldbills/001/en/2010001en.pdf
[395] http://www.publications.parliament.uk/pa/ld200910/ldbills/001/en/2010001en.pdf

68.

18 December 2009

Germany: government proposes to re-launch DAB in 2011, if sufficient interest

In Germany, the Commission for the Approval & Supervision of State Media Authorities, ZAK, has published a new directive today that attempts to stir interest in resuscitating the country's DAB radio system. It requires the media authority of each German state to issue a common tender by 22 January 2010, calling for applications by 12 March 2010 from those who want to provide national radio services on DAB.

This new plan involves re-launching DAB radio in Germany in early 2011, but only *"if sufficient qualified commercial applicants"* show interest in acquiring licences, according to ZAK.[396] Two-thirds of national DAB multiplex capacity has been allocated by the government to commercial radio, with the remainder for state broadcaster Deutschlandradio.

ZAK says it is seeking proposals for new digital stations that *"strengthen the diversity of viewpoints in Germany"* by offering information, business, sport, religion and specialist music formats.[397] However, this suggestion flies in the face of evidence from other countries where it has not proven commercially viable to offer specialist radio formats on a national DAB platform, even after many years of consumer hardware take-up. For example, in the UK, many radio formats have come and gone on the DAB platform over the last ten years, including:

- news (ITN News 2000-2002)
- business (Talkmoney 2000-2003)
- 50s/60s music (PrimeTime 2000-2006)
- country music (3C 2000-2007)
- teenagers (Capital Disney 2002-2007)
- contemporary pop music (Core 1999-2008)
- soul music oldies (Virgin Groove (2000-2008)
- pop music for young women (Capital Life 1999-2008)
- book readings & talk (OneWord 2000-2008)
- jazz music (TheJazz 2006-2008)
- extreme rock music (Absolute Xtreme 2005-2009)

The worst thing a nascent or potential business can do is fail to learn from the experiences of those who have attempted the same proposition previously and failed. Continually re-inventing the wheel is a waste of human and financial capital. The agencies that are charged with promoting DAB broadcasting could do the global broadcast industry a massive favour by analysing and documenting why each of these stations, and others like them in other countries, failed. There is much more to be learnt from the 95% of business failures than from the 5% of successes.

Propagating the notion globally that DAB radio has been nothing other than a huge success in the UK is horribly irresponsible. More than £600m has been sunk into DAB in the UK over the last decade, but not one content provider has yet generated an operating profit from the platform. Actively encouraging and promoting implementation of DAB radio overseas as a means for broadcast entrepreneurs to emulate the 'success' achieved in the UK is as immoral an export as selling cigarettes to developing countries as a 'luxury' good.

Judged by their previous rejections of the DAB platform, radio stakeholders in Germany have demonstrated that they are not so easily duped. .

---

[396] http://www.alm.de/34.html?&tx_ttnews%5btt_news%5d=563&cHash=3f7f39b44e
[397] http://www.alm.de/34.html?&tx_ttnews%5btt_news%5d=563&cHash=3f7f39b44e

69.

11 January 2010

Criteria and a date for digital radio switchover: where'd they go?

When will the UK government's proposed 'digital radio switchover' happen? For a long time, we had always been told that the pre-requisites were:
- market criteria that had to be reached before switchover could be announced;
- a fixed, single date for switchover to happen.

So both of these must be in the Digital Economy Bill somewhere, surely? Well, it seems that everything (except the Bill itself) points to 2015 as the switchover date. But as for the criteria?

The government's press release of 20 November 2009 announcing the Digital Economy Bill stated:
- *"Digital radio: update the regulatory framework to prepare for moves to digital switchover for radio by 2015."*[398]

The government's accompanying Factsheet of 20 November 2009 stated:
- *"At the centre of our ambition is the delivery of a Digital Radio Upgrade programme by the end of 2015."*[399]

The government's accompanying Impact Assessments of 20 November 2009 referred to:
- *"a switchover to digital radio by 2015"*
- *"a switchover to digital only radio by 2015"*
- *"a Digital Radio Upgrade programme, which should be completed by the end of 2015".*[400]

However, the government's Explanatory Notes to the Digital Economy Bill said:
- nothing about criteria that have to be met;
- nothing explicitly about a switchover date.[401]

Published on 20 November 2009, the Digital Economy Bill itself contained nothing about:
- criteria that have to be met;
- an explicit date for digital radio switchover.[402]

What? Is this not strange? Somewhere along the way, it seems as if the agreed criteria and the switchover date just vanished into thin air. So what happened? Let's go back and follow the timeline of how we got to where we are now.

JUNE 2008
The Interim Report of the government's Digital Radio Working Group recommended:
- *"Government should agree a set of criteria and timetable for the migration to digital.*
- *These criteria should include an assessment of:*

    ○ *The percentage of listening to DAB enabled devices;*

    ○ *Current and planned coverage of DAB and FM; and*

- *In considering the case for migration we expect the Government will also want to consider the take-up of digital radio in cars, affordability, functionality, and an environmental impact plan."*[403]

DECEMBER 2008

The Final Report of the Digital Radio Working Group recommended:

- *"Three broad criteria that must be met in order to trigger the digital migration process:*
  - ○ *That at least 50% of total radio listening is to digital platforms;*
  - ○ *That national multiplex coverage will be comparable to FM coverage by time of digital migration;*
  - ○ *That local multiplexes will cover at least 90% of the population and, where practical, all major roads ...."*
- *"Government should announce a date for digital migration, ideally two years after the criteria have been met".*[404]

JANUARY 2009

The Interim Report of the government's Digital Britain recommended:

- *"We will create a plan for digital migration of radio, which the Government intends to put in place once the following criteria have been met:*
  - ○ *When 50% of radio listening is digital;*
  - ○ *When national DAB coverage is comparable to FM coverage, and local DAB reaches 90% of population and all major roads."*[405]

JUNE 2009

The Final Report of Digital Britain recommended:

- *"The delivery of a Digital Radio Upgrade programme by 2015"*
- *"Included within the Digital Radio Upgrade timetable is our intention that the criteria should be met by the end of 2013":*
  - ○ *"When 50% of listening is to digital; and*
  - ○ *When national DAB coverage is comparable to FM coverage, and local DAB reaches 90% of the population and all major roads"*[406]

This Report also included a critically important graph (see below) which, it said, *"shows the projected digital share of listening under two scenarios: organic growth and with a concerted drive to digital".*[407]

Shockingly, the historical data in this graph had been 'doctored' to make it look as if the faster growth path advocated by Digital Britain was easily achievable [confusingly, the key on this graph labels the lines round the wrong way]. When I queried the source of this false data, the government told me it had been supplied by another party, which I later found to be a report produced by the Digital Radio Development Bureau, but not made public.

Digital Britain's graph sought to demonstrate that continuation of the current growth trend in digital listening would lead to the 50% criterion being achieved in early 2015, whereas the actual data (from RAJAR) in my graph show the 50% criterion not being reached until the end of 2018 [the trend line here is automatically generated by Microsoft Excel from all available quarterly data].

% Share of All Radio
Listening (at year end)   Total Digital Share of Listening

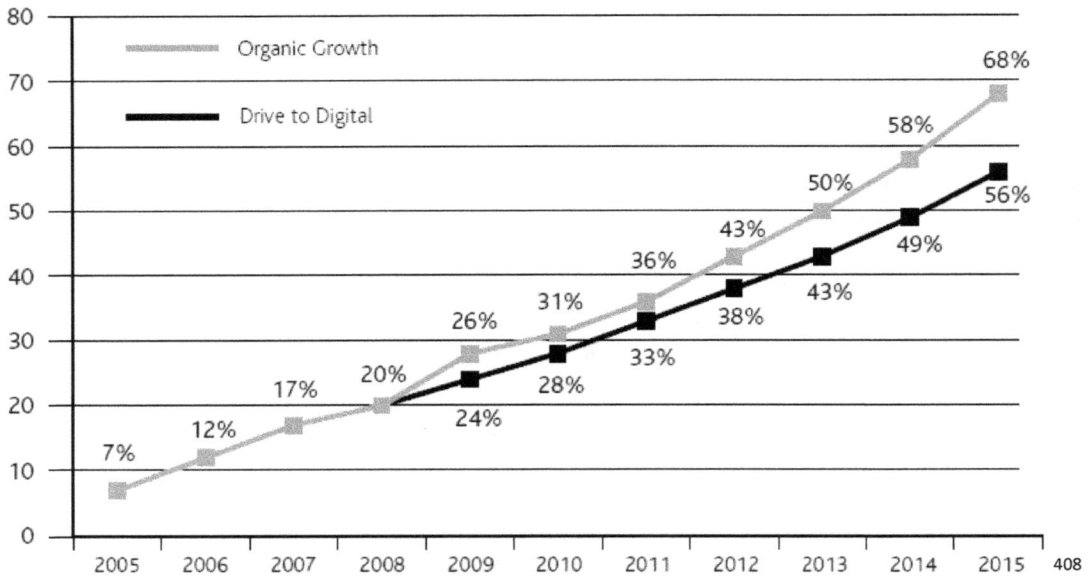

Organic Growth

Drive to Digital

7% 12% 17% 20% 26% 31% 36% 43% 50% 58% 68%
24% 28% 33% 38% 43% 49% 56%

2005 2006 2007 2008 2009 2010 2011 2012 2013 2014 2015   408

DIGITAL PLATFORMS SHARE OF TOTAL UK RADIO
LISTENING (actual and Digital Britain forecast)

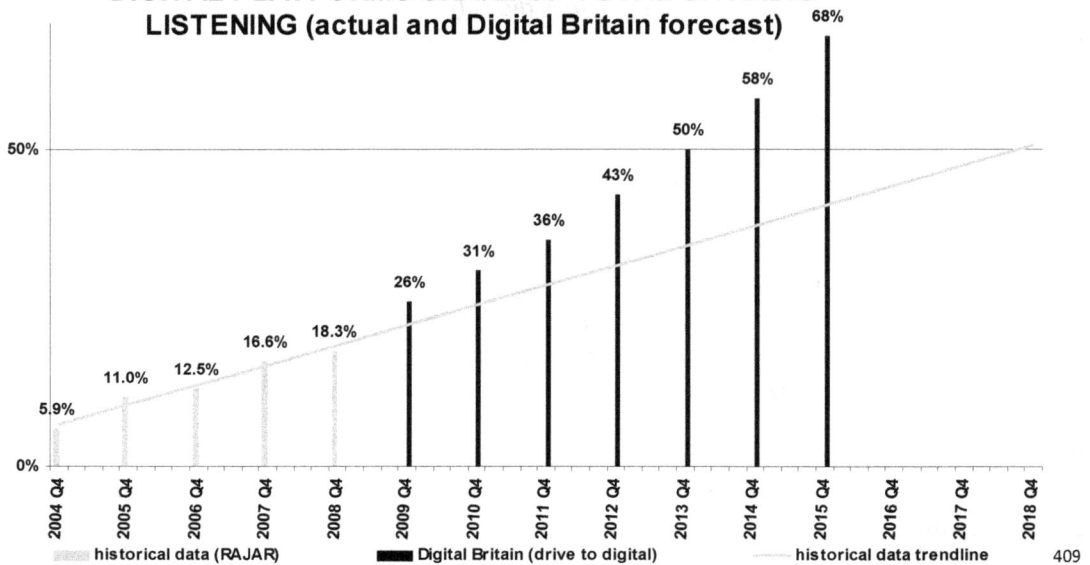

5.9% 11.0% 12.5% 16.6% 18.3% 26% 31% 36% 43% 50% 58% 68%

2004 Q4 2005 Q4 2006 Q4 2007 Q4 2008 Q4 2009 Q4 2010 Q4 2011 Q4 2012 Q4 2013 Q4 2014 Q4 2015 Q4 2016 Q4 2017 Q4 2018 Q4

historical data (RAJAR)   Digital Britain (drive to digital)   historical data trendline   409

Digital Britain proposed policies to accelerate DAB take-up which, it said, would ensure that the 50% criterion would be achieved by year-end 2013, a gain of a little over one year from its natural trend. However, in my graph that uses RAJAR data, the acceleration necessary is shown to be five years, not one year, which would prove an almost impossible task to achieve [I wrote about the false data in June 2009].

JUNE TO DECEMBER 2009
Between the publication of the Digital Britain final report in June 2009 and today, it has slowly dawned on some of radio's stakeholders that the agreed criteria necessary for digital radio switchover stand zero chance of being achieved by 2013. Neither do they stand a chance of being achieved by 2014 or 2015, nor probably by 2016. It always was pie in the

sky, wishful thinking, fiction rather than fact. The manipulation of key data in a significant government report only demonstrates the duplicity.

So, what to do about it now? Admit you were wrong? Admit your culpability? Best to simply pretend that the criteria and the proposed switchover date never really mattered. Botched data – ignore it. Unrealistic targets – lose them. Perhaps nobody will notice the whole, sorry deception.

In the here and now, Digital Radio UK (the new organisation responsible for implementing DAB) explains the current thinking:
- *"The [Digital Economy] Bill does not set a definite date for digital radio switchover …"*
- *"The Government has stated that switchover will not happen until the majority of radio listening is to digital, and until anyone who can currently receive FM is able to receive digital radio"* [but fails to address why these criteria are not included in the Bill].

In the here and now, RadioCentre (the commercial radio trade body) explains:
- *"[Digital Economy Bill Clause 30] allows the Secretary of State to set a [digital switchover] date, but does not require one to be set, or indicate when the date might be"*.
- *"The objective that switchover should not occur until certain thresholds have been reached for listening … appears sensible on first reading. However, RadioCentre does not believe it is appropriate for the industry to be tied to any figures in primary legislation. This is a very inflexible mechanism against which to manage our industry going forwards"*.

Figures. Numbers. Dates. Criteria. This kind of factual evidence or hard data might obstruct a future decision to force consumers to switch to DAB radio.

So to answer the original question – the criteria and the switchover date that had been agreed upon by stakeholders, over two years of deliberations, have now quietly been relegated to oblivion.

When would digital radio switchover have happened if the agreed criteria had been implemented in law? Probably never.

When will digital radio switchover happen now? Whenever those in power want it to.

[398] http://www.culture.gov.uk/reference_library/media_releases/6447.aspx
[399] http://interactive.bis.gov.uk/digitalbritain/wp-content/uploads/2009/11/Factsheet-DigitalRadio.pdf
[400] http://interactive.bis.gov.uk/digitalbritain/wp-content/uploads/2009/11/DEB-Impact-Assessments.pdf
[401] http://www.publications.parliament.uk/pa/ld200910/ldbills/001/en/2010001en.pdf
[402] http://www.publications.parliament.uk/pa/ld200910/ldbills/001/2010001.pdf
[403] http://www.culture.gov.uk/images/publications/DRWG2008-interimreport.pdf
[404] http://www.culture.gov.uk/images/publications/DRWG_Final_Report.pdf
[405] http://www.culture.gov.uk/images/publications/digital_britain_interimreportjan09.pdf
[406] http://www.culture.gov.uk/images/publications/digitalbritain-finalreport-jun09.pdf
[407] http://www.culture.gov.uk/images/publications/digitalbritain-finalreport-jun09.pdf
[408] http://www.culture.gov.uk/images/publications/digitalbritain-finalreport-jun09.pdf
[409] source: RAJAR and Digital Britain

70.

13 January 2010

Parliamentary debate on local radio: Minister reads from the government DAB script

"The Future Of Local Radio" [excerpts][410]
Private Members' Debate
Westminster Hall, House of Commons
12 January 2010 @ 1330

The Parliamentary Under-Secretary of State for Culture, Media and Sport (Mr. Siôn Simon): *Local radio is, without question, important to the Government and to communities, playing an important role in binding together the social fabric. We take it very seriously.*

Dan Rogerson (North Cornwall) (Liberal Democrat): *On the point about the importance the Government place on local radio, it seems that local radio stations, and certainly those in my constituency, Pirate FM and Atlantic FM, do not necessarily feel that they have had the opportunity to get their points across at an early stage. That is why they are now contacting local Members to look at some of the issues when the Digital Economy Bill is debated on the Floor of the House. What sort of consultations are taking place with local radio stations?*

Mr. Simon: *The hon. Gentleman is quite right; there is undoubtedly some concern in the industry. There has been a bit of a campaign, led by UTV. I recently met, at RadioCentre, representatives of many local commercial local radio stations from across the country, and some of them will have been those he mentioned from his constituency. There was extensive consultation when the Bill was drafted, so we do take it seriously. During my remarks, I hope to allay some of the fears which may have emerged through misunderstanding.*

Bob Spink (Castle Point) (Independent): *There are genuine fears that the Bill will lead to a two-tier system, so would the Minister address a couple of those fears? Will Clause 34 genuinely lead to deregulation for smaller local radio? Will digital be affordable for smaller local radio, and how can we achieve that? Will smaller local radio get more access to higher-quality FM while it is still around?*

Mr. Simon: *I am pretty confident that I shall address all those points in my brief remarks. Let me make some progress before I take any more questions. Digital switchover provides new opportunities and increases functionality. It is an essential part of securing the long-term future. The total revenue of the commercial sector has fallen from £750 million in 2000 to £560 million now. At the same time, transmission costs have gone up, with stations now bearing the cost of carriage on FM, DAB, online and digital TV. A market facing such rising costs and falling revenue is unsustainable and puts the health of the entire sector under threat. Although the path to digital may not be easy, we are convinced that it is the only route for securing the long-term future of radio, and that is a view shared by the vast majority of the sector, notwithstanding some of the reservations raised by hon. Members.*

Therefore, rather than a catalyst for decline, the changes set out in the Digital Economy Bill are essential to secure the survival of local radio. For the first time, we will have three distinct tiers. First, there will be a tier of national services, both commercial and BBC, with a wide range of content. It will allow the commercial sector to compete more effectively with the BBC, employ high-profile presenters and attract high value national advertising and sponsorship. Secondly, a regional or large local tier, again comprising commercial and BBC services, will provide a wide range of programmes, including regional news, traffic and travel. The tier will increase the coverage size and potential revenue of many large local stations which, in turn, will increase the opportunity for linked advertising between regions so that regional commercial operators can benefit from quasi-national advertising. The hon. Member for Orkney and Shetland mentioned the issue of advertising being badly commissioned by the Scottish Government, which I understand. None the less, the benefits of linked advertising for regional radio can be very great if commissioned sensitively. Most important in the context of today's debate, there will be a tier of local and community radio stations with the specific focus of informing and reflecting the communities they serve. They will be distinct from the national and regional tiers because of the very local nature of their content and they will benefit from less competition for local advertising funding.

Mr. Oliver Letwin (West Dorset) (Conservative): *People in my constituency and elsewhere who depend on radios will not be able to get local radio if it is purely digitised.*

Mr. Simon: *Local radio will not be purely digitised. That tier will stay on FM for the foreseeable future, but it will not be an FM ghetto; it will be an accessible FM, as I shall explain.*

Mr. Brian H. Donohoe (Central Ayrshire) (Labour): *Given the time constraints, will the Minister agree to meet Members who are interested in the subject?*

Mr. Simon: *Yes, I am happy to meet Members who are interested. I have another meeting scheduled with local radio operators from all over the country, which will be under the same auspices as my recent meeting with them. (I am not sure whether I have enough time to continue. I do.) So, let me be clear: we see a digital future for all radio eventually. However, with more than 50 BBC services, nearly 350 commercial stations, 200 licensed community stations, the current infrastructure will not support a move to digital for everybody. For small commercial and community stations, the coverage area and the cost of carriage of a digital multiplex are too great. That is one reason why, for the time being, we believe that those stations are best served by continuing to broadcast on FM.*

Malcolm Bruce (Gordon) (Liberal Democrat) rose —

Mr. Simon: *I am nearly coming to my point, but I give way to the right hon. Gentleman.*

Malcolm Bruce: *Some of the small stations have already invested in being on digital. Are they not in danger of being kicked off to FM, having made that investment, and would that be a fair outcome?*

Mr. Simon: *No, small stations are not in such danger. Stations that are already on digital are not in danger of being kicked off digital, but they are suffering the extra cost of running on two platforms. That is one of the reasons why we need an orderly, managed and reasonably speedy transition to an affordable single platform for as many people as can afford to be on it. The idea of stations on more than one platform is not new, which moves us to a key point* *that has not been widely understood — it is really important. Listeners have for decades moved between FM and Medium Wave, and historically also to Long Wave. The current generation of DAB sets has tended to make that move a rather sharp distinction, which has led to the fear that FM will end up being a second-class ghetto tier. To avoid that, we are committed to ensuring the implementation of a combined station guide, which is similar to an electronic programme guide, that will allow listeners to access all sets will simply have a list of station names. The listener will not distinguish between FM and digital stations, but will simply select the station by name. We are already working with the industry on that system and encouraging its development and introduction as quickly as possible. That is a crucial difference that has not been widely promulgated or understood. It means that people can stay on FM and the new sets can service the same market. Only 5 per cent of the digital radio receivers currently on sale cannot receive FM. It is our intention that all digital receivers should be able to receive FM as well as complying with the World DMB profile, which will ensure that they can support other technologies to accommodate future changes. That crucial distinction has not been widely understood. When I explained it to people in the industry, it made a big difference. The hon. Member for Orkney and Shetland asked whether we could upgrade to DAB+ from the beginning. I understand why he says that, but we are not right at the beginning. There are 10 million DAB sets out there for which people have laid out large amounts of money. The BBC completed a study into the issue last year, and concluded that, on balance, it was not worth writing off that technology because of the impact on the 10 million people who had bought DAB sets. We have said that all new technology should be DAB+ and future compatible so that further change is future-proofed and DAB+ is not excluded. As for the switchover date of 2015, the hon. Gentleman asked whether it was the only way we would get things moving. The Government believe that 2015 is an achievable date. The actual date that switchover happens will depend on the criteria for listenership and coverage being satisfied. We think it can be done by 2015, and that it is important to set a challenging target. The issue of £20 sets was raised. There are already some £30 sets. We have five years to go until 2015, so we remain confident that we will have £20 sets by then.*

Miss Anne Begg (Aberdeen, South) (Labour): *I am interested to hear what my hon. Friend says about the 2015 date. Can I take it from what he said this morning that 2015 is an aspiration to encourage the industry to move towards digital — to put their house in order and get things ready? However, if the coverage is not there in places such as the constituencies of the hon. Member for Orkney and Shetland (Mr. Carmichael) and the right hon. Member for Gordon (Malcolm Bruce) where there are a lot of hills, will the Government look at the date again? That date is not already fixed.*

Mr. Simon: *As I said, we believe it is an achievable date. If more than 50 per cent of listeners are not on digital by then, and if coverage is not similar to FM — 98.5 per cent — it will not happen on that date. If for any other unforeseen reason, we are not, as a nation, in good shape to do it by then, we will not do it. We will not switch over at an inappropriate time, but we believe that it can and should be done in 2015. As time ticks on, let me say that a relatively small and cheap piece of hardware will be available to convert in-car sets to something that works in the future as well as the present.*

[Sitting adjourned without Question put (Standing Order No. 10(11)).]

---

[410] http://www.publications.parliament.uk/pa/cm200910/cmhansrd/cm100112/halltext/100112h0011.htm

71.

19 January 2010

Labour MP says government's analogue radio switch-off "is absolutely potty"

House of Commons
18 January 2010
Oral Answers to Questions[411]

Rob Marris (Wolverhampton, South-West) (Labour): *My understanding is that the Government currently propose that analogue radio be switched off in 2013. If that is the case, it is absolutely potty. Will the Government reconsider?*

The Parliamentary Under-Secretary of State for Culture, Media and Sport (Mr. Siôn Simon): *My hon. Friend is, for once, slightly wrong on the detail. The policy is that we move to digital in 2015, but not that analogue radio be switched off. Most big radio stations will move to digital, but smaller commercial and community radio stations will stay on FM and will be, as I have said, on the same dial as the big digital stations.*

House of Commons
18 January 2010
Written Answers to Questions[412]

Theresa Villiers (Chipping Barnet) (Conservative): *To ask the Minister of State, Department for Transport with reference to the answer to the right hon. Member for East Yorkshire of 23 June 2009, Official Report, column 768W, on motorways, what assessment he has made of the effects on the level of motorway congestion of the DAB radio service Traffic Radio since its introduction.*

The Parliamentary Under-Secretary, Department for Transport (Chris Mole, Ipswich) (Labour): *Traffic Radio is one of a suite of Highways Agency information services designed to provide road users with access to the very latest traffic information.*
Research has shown that awareness and usage of information services can influence levels of motorway congestion. It is not possible to directly correlate the impact of Traffic Radio to motorway congestion due to the complexity of assessing one information service in isolation from the others. In addition, information is only one of a series of measures that can contribute towards congestion reduction.
The Highways Agency is undertaking a piece of research to evaluate whether the anticipated benefits of Traffic Radio, as outlined in its original specification, have been realised. This work is due to be completed by April 2010 and will be supplemented by information from the agency's annual Measuring Improvements in Network Information Services survey.

---

[411] http://www.publications.parliament.uk/pa/cm200910/cmhansrd/cm100118/debtext/100118-0003.htm
[412] http://www.publications.parliament.uk/pa/cm200910/cmhansrd/cm100118/text/100118w0019.htm

72.

23 January 2010

Canada: DAB radio "is no longer a replacement for analogue AM and FM services"

The Canadian government has published a consultation that proposes to re-allocate radio spectrum previously used for DAB radio to fixed and mobile wireless devices. The consultation document narrates the story of the failure of DAB radio in Canada.

"In 1996, the [Industry Canada] Department published an allotment plan to accommodate all existing and some new FM and AM radio stations by providing each with a DAB assignment. The dedication of the sub-band 1452-1492 MHz for DAB was justified on the expectation that DAB would replace analogue FM and AM stations, and that the associated spectrum would be released for new wireless services. Canada adopted the Eureka 147 standard for DAB, which was widely accepted by European countries and others. In response to the interest of broadcasters to offer some non-broadcasting services using DAB broadcasting facilities, the Canadian Radio-television Television Commission (CRTC, 1996) and the Department (1997) made provisions to permit a limited amount of non-programming services.

Starting in the late 1990s, the CRTC licensed 76 DAB stations in Toronto, Windsor, Montréal, Vancouver, Victoria and Ottawa. In addition, the CRTC approved a stand-alone ethnic commercial radio station. After a promising start, the roll-out of DAB has virtually come to a stop and some stations have ceased operation. The marginal development of DAB services in Canada can be attributed to several factors.

- *First, consumers have only had limited access to high-priced DAB receivers.*
- *Secondly, the United States, with its influential market, is implementing HD digital technologies on the shoulders of the analogue FM and AM channels.*
- *Most European countries have implemented DAB services in the VHF band III (174-230 MHz) instead of the anticipated L-band. Only a few countries have pursued DAB in the L-band. The success in such countries has been quite limited.*
- *Furthermore, Canada used a different channelling system that required the few receivers imported to Canada to be customized for this small market.*

An ongoing concern of Canadian broadcasters has been the inability to broadcast a significant level of new programming on DAB stations in order to attract subscribers during the transition phase. Moreover, as digital radio was implemented in only a few cities, without contiguous coverage over major transport corridors, car manufacturers are not installing DAB receivers in new vehicles for sale in Canada. Since then, with the availability of two subscription digital radio satellite services, the Canadian automobile manufacturers have proceeded to install satellite digital radio receivers in new vehicles.

In 2006, the CRTC launched a public review of its commercial radio policy, which included a review of the L-band DAB transition policy. It culminated in a new licensing model being adopted for digital radio broadcasting (Broadcasting Public Notice CRTC 2006-160). Some of the findings related to digital broadcasting and to the decision aspects with respect to this spectrum review are as follows:

- *the offering of new and innovative program content may raise consumer interest for DAB services, but technical quality alone will not drive demand;*
- *the FM and AM frequency bands will be needed in the future for radio broadcasting, and HD radio digital broadcasting could further enhance the importance of this spectrum for over-the-air radio broadcasting;*
- *the transition model of replacing analogue FM and AM stations with L-band DAB is no longer a tenable objective.*

In summary, several impediments will continue to affect the implementation of DAB services under the new licensing model, such as:

- *the lack of affordable L-band DAB receivers;*
- *the lack of factory-installed DAB receivers in new vehicles;*
- *the U.S. market influence of digital radio services using HD radio technology on existing analogue FM and AM channels; and*
- *the European market influence of having adopted DAB service in the VHF Band III (174-230 MHz) and their review of the L-band spectrum for a variety of technologies and service applications."[413]*

The Canadian government's policy that DAB radio *"is no longer a replacement for analogue AM and FM services"* follows on from US policy in debate that FM radio will be the universal radio platform to be included in all mobile phone handsets. There is an important lesson for the UK market in the Canadian conclusion that *"the offering of new and innovative program content may raise consumer interest for DAB services, but technical quality alone will not drive demand."*

DAB radio's death in Canada is demonstrated by World DMB's country page for Canada not having been updated since September 2008, while the last news entry on Canada's own digital radio website was posted in 2007.[414] The world map has lost the largest country to have implemented DAB and the only country in the Americas.[415]

---

[413] http://www.ic.gc.ca/eic/site/smt-gst.nsf/vwapj/dgtp010e-consultation.pdf/$FILE/dgtp010e-consultation.pdf
[414] http://www.worlddab.org/country_information/canada
http://www.cab-acr.ca/drri/news.shtm
[415] http://www.worlddab.org/country_information

73.

25 January 2010

Germany: "FM is and will continue to be the most important means of transmission for radio" say commercial broadcasters

Commercial radio broadcasters in Germany have published a policy paper emphasising that FM will continue to be the main broadcast platform for radio.[416] The VPRT, a trade association of 160 commercial broadcasters (70 of whom are active in radio), this week responded to the draft Work Programme for 2010 set out by the Radio Spectrum Group [RSPG] of the European Commission. One of its proposed work streams had been *"to discuss the pros and cons of indicating a target date for analogue radio broadcasting (FM) switch-off."*[417]

The formal response from VPRT sets out powerful arguments why current FM spectrum (referred to as 'Band II') will continue to be radio's most important platform for broadcasting:

"The future development in the different frequency bands (especially Band II) is of utmost importance to our radio service members. Therefore, we are seriously concerned about the fact that RSPG is considering a target date for analogue radio broadcasting (FM) switch-off.

Nevertheless, we see the need to think about future developments and possible usages of Band II which comply with the provisions of the GE84 [Geneva radio conference of 1984] agreement and ensure FM services which are able to operate free of any interference. At the same time, we would like to stress the necessity of adapting and developing those GE84 provisions to ensure the continuity of Frequency Modulation (FM) and the future usages in Band II. However, this needs to be achieved without further co-ordination at international level.

1. Band II with FM is and will continue to be the most important means of transmission for radio
Band II with FM is the most important means of transmission for VPRT's radio service members. Also, in the foreseeable future, Band II with FM will remain the basis of commercial activities for private radio stations. This is grounded by two reasons: firstly, the heavy usage by listeners and, secondly, the very high market penetration of FM receivers. Currently, more and more new Band II FM receivers are establishing in the market. As a consequence, the receiver basis is modernised constantly. Modern communication devices, such as mobile phones, smart phones and media players, integrate Band II FM receivers and ensure an even wider availability of FM. Switching off FM transmission is therefore neither realistic, nor can it be crowned with success.

2. FM in Band II is of utmost importance in the case of catastrophe

Due to the extremely high penetration of FM devices in Europe and its heavy usage, Band II is the only reliable way to inform the public in the case of catastrophe or need of contacting citizens in an emergency. This was recently proven when a blizzard hit Germany at the beginning of January. This is also valid in case of a regional power cut, as many devices are powered by batteries. For the time being, no digital receiver (DAB, etc.) has been developed for the operation with battery.

3. Band II is a small but efficiently used frequency range

The so-called Band II is the frequency range between 87.5 and 108 MHz and only represents 20.5 MHz. Nearly every single frequency is used in this bandwidth. Together with the broad receiver penetration and very high usage by the listeners, this small bandwidth is very efficiently used. In the last few years, receivers have been significantly developed which today results in an enormous improvement of their reception quality. Millions of listeners are convinced by the characteristics of FM. Even under very difficult circumstances for receiving a signal, a very good reception is possible. On the other hand, other (digital) systems are disconnected in a very early stage, which is rather disadvantageous. The usage of Band II is still "state of the art".

4. No migration or partial migration of the services in other frequency spectrum

VPRT rejects any proposals which include the shifting of the current usage from Band II to other frequencies. This would bring the intensive and effective use of Band II to an end. Due to the lack of digital receivers, as well as of the absence of consumer demand for change and migration, a restart of a digital system would mean inefficiency and un-sustainability for a very long period of time. In other bands, there is enough space to introduce new systems. Band III (174 to 230 MHz, channels 5 to 12) and therewith corresponds to 56 MHz – is available.

5. Consideration of future developments of FM transmission after GE84

Since the Geneva conference of 1984 (GE84), different parameters of the FM usage in Band II have changed. In the meanwhile, different and changed sources of signals are available (music), the signal processing was adapted and a compression of the signals was introduced. The processing of the FM signals in the receivers is completely digital. 25 years after G84, the provision from the Geneva plan should be adapted and developed according to recent technical developments.

6. There is a chance for new standards with unlimited parallel FM operation

In the medium term, there are different options to develop the use of Band II. The use of FM has to remain, due to the heavy use described in point 1.4. An unlimited parallel FM operation offers the opportunity of financing additional engagements from the remaining FM transmission. Further developments with new standards based, on additional unlimited parallel operation of FM, is a chance for economical efficiency. Therefore, it is necessary to adapt the ETSI spectrum mask ITU-BS.412-9, as well as other ITU-R recommendations, by keeping the guidelines for aeronautical services (VOR and ILS). In this way, the "envelope concept" which was already used in the GE06 plan could be kept. Therewith, new standards under an adapted ETSI spectrum mask would be possible without interfering with the existing FM transmission and conditions. In this case, a new planning conference would not be necessary.

We support a conversion to digital assignments if the FM transmission can be maintained without any limitation. In this case, a switch from single FM transmission into a digital transmission would be possible, without discriminating other FM transmissions.

We do not see a future for technologies which are linked to a switch-off of the FM transmission.

7. Interference with FM through new standards have to be avoided
As already mentioned in point 6, it must be avoided that future technology developments cause any interference to the existing FM transmissions. A reduction of the current coverage caused by future developments is not acceptable for VPRT members. Some aspects of the technical developments are promising but, due to a lack of information, a full evaluation is not possible.

8. No international re-planning of Band II or of parts of it
Due to the very intensive and effective use of Band II, we do not see any need for a long and very costly international re-planning. The GE84 plan should be supported in its principles and adapted as mentioned in point 6. Even a re-planning of certain parts of the Band II would not lead to any benefit, as the complete Band II is used and needed in the future.

9. Re-adjustment of Band II at national level is necessary
However, we always have been calling for a re-adjustment of Band II at national level in order to balance the relation between public broadcasters and private broadcasters. At the moment, we face an imbalance with regard to the amount of frequencies, as well as the frequency capacity, held by public broadcasters on the one hand and by private broadcasters on the other hand. We therefore ask for a readjustment which takes the actual demands into account. The introduction of new standards would carry forward the current imbalance.

10. Further research and economical comparison are necessary
Next to the research and comparison, with respect to the technical characteristics and parameters of the available standards, further research is needed to complete a substantial evaluation. For the time being, an evaluation of the economical and financial factors is still missing. We therefore ask to also take those aspects into consideration.

Summary of VPRT comments
- *Band II with FM is and will continue to be the most important means of transmission for radio*
- *FM in Band II is of utmost importance in the case of catastrophe*
- *Band II is a small but efficiently used frequency range*
- *No migration or partial migration of the services in other frequency spectrum*
- *Consideration of future developments of FM transmission after GE84*
- *There is a chance for new standards with unlimited parallel FM operation*
- *Interference with FM through new standards have to be avoided*
- *No international re-planning of Band II or of parts of it*
- *Re-adjustment of Band II at national level is necessary*
- *Further research and economical comparison are necessary*

Berlin, January 2010"

---

[416] http://www.vprt.de/index.html/de/positions/article/id/115/or/2
[417] http://rspg.groups.eu.int/consultations/consultation_workprogramme2010/rspg_draft_workprogramme2010.pdf

74.

26 January 2010

Government admits: no assessment of impact on small FM commercial radio stations of Digital Economy Bill

House of Commons
21 January 2010
Written Answers to Questions[418]

Digital Broadcasting: Radio

Mr. Alistair Carmichael (Orkney and Shetland): *To ask the Secretary of State for Culture, Media and Sport what assessment he has made of the likely effects of digital switchover under the provisions of the Digital Economy Bill on the ability of local commercial radio stations without a digital path to continue to broadcast on the analogue spectrum.*

The Parliamentary Under-Secretary of State for Culture, Media and Sport (Mr. Siôn Simon): *No specific assessment has been made of the impact of the radio provisions set out in the draft Digital Economy Bill on local commercial stations remaining on FM after the digital radio switchover. However, these provisions, and the proposals in the Digital Britain White Paper, were made following 18 months of consultation with the radio industry, which included representatives of small local commercial stations.*

I am continuing this dialogue with the industry with the specific purpose of ensuring that local radio can continue to thrive on FM after the digital radio switchover.

Digital Broadcasting: Scotland

Mr. Carmichael: *To ask the Secretary of State for Culture, Media and Sport what percentage of Scottish households are able to receive digital radio services; what plans his Department has to increase coverage for digital radio in Scotland before 2015; what the cost of implementing those plans will be; and who will pay for the implementation.*

Mr. Simon: *The Spectrum Planning Group, which formed part of the Digital Radio Working Group, reported in November 2008 that 77.8 per cent. of the population in Scotland had access to indoor Digital Audio Broadcasting (DAB).*

Coverage of digital radio broadcasting in the UK continues to increase and both the commercial and the BBC's national multiplexes now reach about 90 per cent. of the UK population. The Digital Radio Upgrade programme will require new investment in building and improving DAB coverage and reception. To this end we will be working with the BBC and commercial operators to ensure coverage of DAB is comparable to FM by the end of 2014.

The Digital Britain White Paper was clear that the investment need to increase coverage will need to come from both commercial operators and the BBC.

Mr. Carmichael: *To ask the Secretary of State for Culture, Media and Sport what percentage of households in Scotland receive (a) digital commercial radio and (b) BBC digital radio services.*

Mr. Simon: *The Spectrum Planning Group, which formed part of the Digital Radio Working Group, reported in November 2008 that 76.2 per cent. of the population in Scotland had access to digital commercial radio services, while 77.8 per cent. of the population in Scotland has access to BBC digital radio services. Figures are based on indoor coverage.*

---

[418] http://www.publications.parliament.uk/pa/cm200910/cmhansrd/cm100121/text/100121w0005.htm

75.

27 January 2010

France: "2010 will be the last chance" to launch digital radio

"2010 will be the last chance for digital terrestrial radio" to launch in France, said Alain Mear, vice chairman of the CSA [media regulator] digital radio working group. *"If digital radio does not start in 2010, there will be no digital radio."* He was speaking at a roundtable meeting held 15 January at the Senate to discuss the future of radio, according to RadioActu.[419] Mear argued that the government needed to set a deadline for the ending of radio broadcasting on FM and AM. *"This is the moment of truth"*, he said.

However, the view of some radio groups represented at the meeting was that radio in future would be delivered to listeners via a mix of platforms. Concern was also expressed about the financial cost of launching a new digital broadcast platform. Michel Cacouault, representing the commercial radio trade body Bureau de la Radio, predicted that there could be *"no development without an economic assessment"*. He stressed that the sector's declining advertising revenues in 2009, resulting from the global economic crisis, had required *"all the big groups to face re-organisation and downsizing."*

Arnaud Decker, director of corporate relations at media group Lagardere Active, said that *"the authorities must ask themselves how the competitiveness of the national music radio networks can be maintained"* when the launch of digital radio would increase broadcasters' costs.

Jacques Donat-Bouillud, director of radio at transmission provider TDF, said that the cost of providing digital terrestrial radio to a population of 40m would be around 500m Euros per annum. For a single station requiring national coverage in France, he said that digital transmission would cost 3m Euros per annum, compared to 6m Euros per annum for FM transmission.

Pierre Bellanger, CEO of commercial station Skyrock, pointed out that *"60% of mobile handsets are internet enabled"* and cautioned that *"there is no single correct answer to the question about the future of radio. The solution is a hybrid and, ultimately, the listener will be the winner."*

Bruno Patino, director of state radio station France Culture, agreed: *"[State] Radio France must participate in the digitalisation of broadcasting. Digital radio delivered by IP will be there, but that will not kill the broadcast platform."*

---

[419] http://www.radioactu.com/actualites-radio/122604/

76.

28 January 2010

Government: agreement to finance DAB radio upgrade "not expected until late 2010"

Recent government correspondence (see below) has confirmed that:
- DAB upgrade is *"unlikely to be an easy task"*
- DAB upgrade is unlikely *"to be resolved quickly"*
- DAB upgrade requires agreement about the current levels of FM coverage
- DAB upgrade requires agreement of a plan for building out DAB
- DAB upgrade still requires agreement on the level of investment required
- the government *"hopes to have a comprehensive plan by the end of 2010"*
- the DAB upgrade funding issue still has to be agreed between the BBC and the commercial radio sector, which is not expected until late 2010.

House of Lords
Delegated Powers & Regulatory Reform Committee
Second Report of Session 2009-10 [excerpt][420]
re: Digital Economy Bill
17 December 2009

"Clause 36

20. Section 58 of the Broadcasting Act 1996 makes provision for the duration and renewal of national and local radio multiplex licences, and in particular specifies the grounds on which an application may be refused. Section 58 is in Part 2 of the 1996 Act.

21. Clause 36 inserts a new section 58A into the 1996 Act enabling the Secretary of State by regulations subject to affirmative procedure to "amend section 58 and make further provision about the renewal of radio multiplex licences" and for that purpose to amend other provisions of Part 2 of the 1996 Act. There is a "sunset" provision preventing the power being exercised after 31 December 2015.

22. It is impossible to tell from the Bill whether the policy is that the licences should or should not be renewable at all, let alone for what period or on what grounds. Indeed, paragraph 56 of the memorandum candidly admits that the relevant policy decision has yet to be made. We draw attention to the skeletal nature of the power in clause 36, to enable the House to examine it further and determine whether it is justifiable in this context."

Attached is the Memorandum by the Department for Culture, Media and Sport and the Department for Business, Innovation and Skills which explained:

"Clause 36: Renewal of radio multiplex licences: Amendment of Broadcasting Act 1996
Powers conferred on: Secretary of State
Power exercised by: Regulations
Parliamentary procedure: Affirmative resolution

Clause 36 adds a new section 58A into the Broadcasting Act 1996. That provision contains a power to amend Part 2 of the Broadcasting Act 1996 (and in particular section 58) by regulations for the purpose of making further provision about the renewal of radio multiplex licences. In particular, regulations made under this power may make provision about the circumstances in which OFCOM may renew a licence, the period of such renewal, the information that OFCOM may require from an applicant, the requirements that an applicant must meet, the grounds for refusal of an application, payments to be made and further conditions that may be included in a renewed licence.

The reason for providing for a power to amend Part 2 of the Broadcasting Act 1996 by order in this way, rather than making amendments in the Bill, is that the decisions about whether or not to extend radio multiplex licences are dependent on an agreed industry wide plan for rolling-out DAB to match FM coverage. This planning process can only begin when (a) OFCOM have the power to allow multiplexes to merge, which requires the new powers to change the frequencies allocated to multiplexes set out elsewhere in the Digital Economy Bill, and (b) when funding issues between the BBC and the commercial sector are agreed; which is not expected until late next year. The power conferred on the Secretary of State will be subject to a sunset provision, so that it cannot be exercised after 31 December 2015.

Given that the power provides for amendment of primary legislation relating to the regime for renewals of licences, we consider it appropriate that any order made under this power should be subject to the affirmative procedure."

Letter from the Rt Hon Lord Mandelson, Secretary of State, Department for Business, Innovation and Skills to the Chairman of the Delegated Powers & Regulatory Reform Committee [excerpts][421]
January 2010

"1. I am writing in response to the Committee's Second Report of Session 2009-10 published on 17 December which addresses the Digital Economy Bill.......

Independent radio services
Report paragraph 22: clause 36: Renewal of radio multiplex licences

22. The Committee considers that it is impossible from the Bill to determine whether the policy is for licences to be renewable and, if so, for what period and on what grounds. The Committee draws attention to the "skeletal nature of the power in clause 36, to enable the House to examine it further and determine whether it is justifiable in this context".

23. The Department notes that the Committee does not recommend removal or amendment of this provision. The Government intends to further explain the rationale for the clause during the Committee stage of the Bill.

24. It is the Government's aim to work with broadcasters and multiplex operators to agree how to build out the Digital Audio Broadcasting (DAB) infrastructure to meet FM coverage levels, one of the criterium that needs to be met in setting a date for digital switchover for radio. This is unlikely to be an easy task, or indeed to be resolved quickly. Among other things, it will require agreement about the current levels of FM coverage, the plan for building out DAB and the level of investment required. The Government hopes to have a comprehensive plan by the end of 2010.

25. The Government believes that a key component of this planning will be the ability to alter the terms of multiplex licence renewals. The existing section 58 of the Broadcasting Act 1996 allows OFCOM to renew radio multiplex licences granted prior to 30 September 2006 for periods of 12 or 8 years (depending on when the licence was granted). However, the Government recognises the need to reduce, as much as possible, the impact of infrastructure build-out on digital stations.

26. One way this can be achieved is to allow multiplex operators to spread the cost of any new investment over a longer licence period. This is why the Government has proposed new powers in section 58A to amend the provisions about the renewal of multiplex licences. The reason that the power is not more specific is because it will not be clear exactly how it will be most appropriately applied until the plan for the build-out of DAB is developed.

27. Exercise of the power will, in any event, be subject to parliamentary scrutiny due to the fact that any regulations will require resolutions of both Houses before being made."

---

[420] http://www.publications.parliament.uk/pa/ld200910/ldselect/lddelreg/24/2403.htm
[421] http://www.parliament.the-stationery-office.co.uk/pa/ld200910/ldselect/lddelreg/41/4106.htm

77.

30 January 2010

Sweden: transmission company proposes DAB radio re-launch in 2010/11

DAB radio in Sweden could be re-launched to the public in 2010 or 2011, argues radio transmission company Teracom, interviewed in Radio World.[422] Although Swedish state radio has been broadcasting on the DAB platform since 1995, the signal still only reaches 35% of the population. No DAB licences for commercial radio have yet been issued. In 2005, the Swedish government halted any further public investment in DAB radio due to poor consumer response.

Despite these setbacks, Teracom is attempting to stimulate Swedish interest in an upgraded DAB+ transmission system. It started DAB+ trial broadcasts in May 2009 and is conducting market research with the 500 people to whom it has supplied DAB+ receivers. The trial stations on DAB+ comprise four from state radio, eight from commercial radio and three community broadcasters.

Teracom pilot project manager Per Werner said one of the aims of the pilot is "to demonstrate to decision makers that the Swedish radio industry is ready for digital radio and that there is demand for new regulations allowing the industry to enter the digital era." He explained: *"If a decision is made to build a DAB+ network on a larger scale, a natural consequence would be to migrate the current DAB network to a DAB+ network. This is a decision for Sveriges [state] Radio, which is currently using the DAB network in a limited coverage area."*

Werner advised: *"There is consensus in the industry that a wide range of programmes from public service and commercial radio, as well as community radio, will be necessary to provide a compelling offer to listeners."*

---

[422] http://www.rwonline.com/article/93472

78.

4 February 2010

Small local stations and the Digital Economy Bill: debated in the House of Lords

The issue of the potential impact on small, local radio stations of the government's proposed 'Digital Radio Upgrade' was first raised here in July 2009, when I had stated:

"... the potential losers from Digital Radio Upgrade would seem to be:
- *commercial stations presently carried on local DAB multiplexes who might have to be bumped because there is no longer the capacity after amalgamation*
- *local commercial stations presently carried on their local DAB multiplex who will have to quit DAB because they do not wish to serve the enlarged geographical area after amalgamation of multiplexes (for example, the cost of DAB carriage for Kent/Sussex/Surrey is likely to be considerably higher than Kent alone)*
- *new entrants*

In the rush to frame proposals in Digital Britain that respond to the circumstances of the large radio players with substantial investments in DAB infrastructure, it might appear that the voices of the smaller local commercial radio stations have got lost in the stampede of lobbying. These stations might be small in number but many of them remain standalone, so they will not benefit financially from the relaxation of co-location rules. Digital Britain is condemning many of them to remain on FM (or AM), leaving the large radio groups to dominate the DAB platform.

Although the proposals in Digital Britain have been framed to 'help' local commercial radio, overwhelmingly they will reduce the financial burden of group radio owners with local station operations in adjacent areas, and of group owners who have invested in DAB infrastructure. There is little in the way of financial benefits for independent local commercial stations, or for potential new entrants, both of whom face being crowded out of the DAB platform."

Last night, the House of Lords debated three amendments, amongst several, to the Digital Economy Bill which proposed that the voices of small, local FM radio stations and their listeners should be considered before the government commits the UK to digital radio switchover.

These amendments were eventually withdrawn after debate (see below) during which the government minister, Lord Young, made vague assurances that the views of listeners and others would be canvassed.

RadioCentre, the commercial radio trade body, subsequently commented: *"Clause 30 (which relates to the switchover powers) was debated in detail, with proposed amendments withdrawn following debate and assurances from Government."*

Amendments 236 & 237
Moved by Lord Howard of Rising[423]

236: Clause 30, page 33, line 17, after "to" insert " — (a)"
237: Page 33, line 19, at end insert "; and
( ) the needs of local and community radio stations; and
( ) the needs of analogue radio listeners."

Lord Howard of Rising: *My Lords, in moving this amendment, I will also speak to Amendment 237. The amendments are designed to ensure that attention is paid to the local and community radio sectors and the many millions of analogue radio listeners — to which I should add the providers of satellite systems about which the noble Lord, Lord Maxton, spoke. They should all be listened to before any decision is taken about switchover. We on these Benches have not hidden the fact that we remain unconvinced that the Government's plans to switchover in 2015 are realistic. We do not believe that audiences will be ready by then. The audience must remain at the forefront of all our considerations when we debate these parts of the Bill. As drafted, the Secretary of State will have to pay heed only to Ofcom and the BBC. Despite the BBC's dominance in the radio industry, there is a strong argument that it would be helpful for community and local radio stations to be consulted. Indeed, the whole commercial radio sector should be included. It does not seem unreasonable to suggest that the Secretary of State consult the other parts of the industry that will be affected. It would also seem both reasonable and important for the Secretary of State to consider the needs of those who listen to analogue radio. The Government stated in their final Digital Britain report that they would start the countdown to switchover once digital listening made up 50 per cent of radio listening. That seems far too low. It would still mean that there were millions of listeners not using digital. Our amendment would ensure that the needs of those listeners were taken into account before the Secretary of State could nominate a switchover date. The amendment is simply an attempt to ensure that all who will be affected by switchover are considered before the Secretary of State nominates a date. As I have said, it does not seem unreasonable to ensure that listeners are placed at the forefront of these considerations. I hope that the Committee will agree. I beg to move.*

Lord Gordon of Strathblane: *My Lords, I think that I can offer some reassurance to the noble Lord opposite. Unless all those targets were going to be met, virtually the entire commercial radio industry would not support the clause, which it does, with one minor exception. The feeling is that this is an empowering clause that does not oblige the Secretary of State to set a date. Indeed, he can set a date and then withdraw it if precisely those targets mentioned by the noble Lord are not met. The radio industry seems to feel that the Government have got it right. I hope that that reassures him.*

The Lord Bishop of Manchester: *My Lords, in speaking to the amendment moved by the noble Lord, Lord Howard of Rising, I recall that in an earlier debate this evening, the noble Lord, Lord De Mauley, expressed deep disappointment that I was not supporting an amendment that stood in the names of the noble Lords. I hope that they will now feel slightly happier, because I support this particular amendment. I do so because the Bill as it stands provides very little safeguard for those who are living in remote areas, some of them perhaps still relying on long wave, let alone FM, for their radio reception. I take the point that the noble Lord, Lord Howard, made about the Digital Britain report. If I recall correctly the point was made at Second Reading that, when 90 per cent has been reached, there will still be one in 10 people — some of whom would presumably lose access via their radio to all the national, BBC and commercial radio stations — for whom we really ought to have the* *greatest concern. Short of listening via the internet — which I know the noble Lord, Lord Maxton, though no longer in his place, would be urging us to do — or Freeview, there is nothing that the 10 per cent would be able to do until the DAB signal catches up with the FM one. Through this and other similar amendments, I hope the Government will come to recognise that there are some very serious reservations about giving the Secretary of State the power to set the switchover date without proper statutory consideration of the wider impact of that decision on those communities who are often disconnected from British society physically, and those small stations that serve them. I am much in sympathy with the amendment.*

Baroness Howe of Idlicote: *My Lords, we shall come to rather more detail about this aspect shortly. I, too, support the amendment and the basis on which it is being put forward. We spent this morning taking evidence from the commercial radio stations, both from those which disagreed with the main grouping and those which had done some amount of research over time. The more one looks at this whole area, it is quite clear that there is a big problem about when this is going to happen, stretching into the future, causing a considerable number of problems. At the very least, this amendment requires others — those concerned and those involved — to be consulted. So, like the right reverend Prelate, I certainly support the amendment.*

Lord Young of Norwood Green: *My Lords, Clause 30 states that before nominating a switchover date, the Secretary of State must have regard to any reports submitted by Ofcom or the BBC under the terms of Section 67(1)(b) of the Broadcasting Act 1996. The purpose of these reports is to review how long it would be appropriate for radio services to continue to be broadcast in analogue form. These reports should have regard to the provision of digital radio multiplexes, availability of digital radio services and the ownership of digital receivers. In order to produce these reports, Ofcom is required to consult multiplex licence holders and digital radio service providers. In addition — here I address the concerns of the noble Lord, Lord Howard, the noble Baroness, Lady Howe, and the right reverend Prelate the Bishop of Manchester — the Secretary of State must, on requiring these reports, consult such persons representing listeners and such other persons as he thinks fit, as provided in Section 67(4). The noble Lord, Lord Howard, talked about plans to switchover in 2015. That is a target that we have set, not a precise date, as I hope he will recognise. The experience that we had with the TV switchover, which in some ways was even more fraught with difficulty, has been an outstanding success so far. One of the largest switchovers, in the Manchester area, recently went over without a hitch. We had a lot of preparation, help and assistance. We want to adopt a similar approach. It is not just about the 50 per cent of listeners. We have also talked about DAB achieving the same coverage as FM, which is something like 95 per cent these days. We are well aware of the importance of that. I also point out another factor which I think is important. The prices of reasonable quality DAB radios have been coming further and further down. That is important for less advantaged parts of our population. We are aware of the concerns expressed. We believe that the clauses have got it right. We understand the concerns, which is why I have taken time to give some further assurance. Given the breadth of the requirements to consult already proposed in the draft Bill and our commitment to consult widely before setting a date, we believe that the amendment is unnecessary. With the explicit assurances I have given, I hope that the noble Lord will feel able to withdraw the amendment.*

Lord Howard of Rising: *I thank the Minister for his remarks. I am grateful to the right reverend Prelate the Bishop of Manchester and to the noble Baroness, Lady Howe, for their support for the amendment and to the noble Lord, Lord Gordon, for his reassurance. All I am* *doing is asking the Government to pay attention to and listen to the listeners before they take too drastic an action and leave a lot of people very unhappy. From his remarks this would appear to be the case, so I beg leave to withdraw the amendment.*

Amendment 236 withdrawn.
Amendment 237 not moved.

Amendment 239A
Moved by The Lord Bishop of Manchester

239A: Clause 30, page 33, line 21, after "services," insert "while retaining the use of the FM Band for those local and community radio services, including special interest services, for which digital transmission using DAB is not a suitable method due to —
(i) the size of local DAB multiplex areas, or
(ii) the unavailability of capacity on the local DAB multiplex,"

The Lord Bishop of Manchester: *This amendment pursues further the issue about retaining the use of the FM band for local and community radio services. I invite your Lordships to put yourselves in the shoes of someone running a small FM radio station, serving a population of 100,000 people in and around, let us say, King's Lynn. As it stands, that station's future seems to be to bid for a space on a local multiplex. That would mean that it would begin broadcasting to almost 600,000 people. Such a shift would significantly change the character of that station. Listeners from a much wider area would start to phone in to programmes and would, for instance, start to demand their own slice of news output. The station manager would be forgiven for wishing to stay on the FM band as long as possible, until a better solution was found for smaller stations to find a home on the digital spectrum. Consider then the prospect of the only way of going digital to be to join an even larger regional multiplex which covers the whole of East Anglia and serves well over 1 million listeners. That is the kind of situation that is faced by a number of small commercial radio stations if digital rollout continues as planned. Their pathway towards a digital future seems shrouded in a kind of fog, partly because of the large size of DAB multiplex areas and the lack of capacity on some multiplexes .Forcing those small-scale stations to broadcast to much larger areas than their current coverage would alter their feel and alter their connection with their audience. It would undermine the integrity that they hold as local broadcasters and potentially damage their ability to service platforms which stimulate and reflect local democracy and social action. There is also a danger that counter-intuitively such stations would struggle to attract advertising spend from local businesses which do not wish to market themselves to audiences up to 100 miles away. So, socially and commercially, beaming local stations to regional audiences is, frankly, not in anyone's interests. As for the lack of space on DAB in some areas of the country, take, for example, the multiplexes currently covering Humberside or, in my own area, of Manchester where there is virtually no space available. Even with just the larger stations currently on board, capacity is running out in some areas. I understand that some stations that gain access are broadcasting in mono rather than in stereo in order to preserve bandwidth. The limitations of DAB for local and community stations are well acknowledged by Ofcom. Indeed, it is already planning for small-scale commercial and community stations to stay on FM in the medium term as the most appropriate technology for those stations in terms of both coverage and cost. The vacation of FM band space by the removal of national and large local stations would free up more capacity for smaller stations. Ofcom sees this as a natural staging post in radio's digital evolution. Nowhere, however, is this halfway house given a firm legal footing. In case my amendment is misinterpreted as a* *luddite attempt to hold back the march of progress, let me make clear that it is intended as a temporary but crucial platform to support the transition to digital of small-scale local stations, which not only serve geographically defined areas, but also identity-defined and interest-defined groups. Your Lordships may remember the furore caused when in 1992, BBC Radio 4 proposed to transfer use f its long-wave frequency to a rolling news service. The BBC reckoned that FM reception was good enough and the vast majority of the country could pick it up without a problem. But those living in remote parts of the United Kingdom, and even in exceptionally hilly inland parts, knew different. The determination to push ahead with cutting Radio 4 from the long wave was met with purposeful if well-mannered resistance, as one might expect from Radio 4 listeners. In fact, I am told that the sight of 200 protestors in tweed and twinsets marching down Upper Regent Street was enough to help the BBC to see the error of its ways. Let us not make a similar mistake. After all, FM had at that point been around for almost 40 years. We have had DAB for less than a decade. The future of local radio — which is so crucial to forging community cohesion and identity, and promoting local social action and democracy — should not be left to chance. That must mean embracing a multi-platform ecology which creates a pathway towards digital broadcasting for local radio, retaining space for them on FM until such time as a digital platform offers them the right environment to continue what they do best. I beg to move.*

Lord Clement-Jones: *My Lords, the right reverend Prelate has said so much of huge value — an absolute tour de force on behalf of ultra-local radio. Of course from the remarks made in the previous amendment, I not only put my name to this amendment, but fully endorse what he said. It is that kind of certainty which is crucial, and this is a very elegant way of keeping it in the Bill. I hope consideration will be given to that, because — and this is not intended as a pun — a signal is needed in this area. We need a very strong signal — not just a digital or political signal — to FM radio, to ultra-local radio, that they have a future which is secure. That is exactly what the right reverend Prelate, whom nobody could accuse of being a luddite, has advocated.*

Baroness Howe of Idlicote: *My Lords, the right reverend Prelate has put it beautifully: "multi-platform ecology". I like that and it sets the pattern for the future rather well. Clearly, this amendment hits on the crucial area of what will happen in the mean time and what is to be done concerning FM. We need some reassurance on this point; I think the Minister said that it would be around indefinitely. At the moment, we know that Ofcom grants only short licences. There have been quite a number of complaints from the radio stations that not to have the certainty granted to them by, say, a 10-year licence means that their likelihood of failure is considerable. That side of things would be helped if the Minister could confirm that FM will definitely be there, and that licences can be given as people gradually go over to digital and more space on FM becomes available. We know that there are minimal alternative uses for the FM spectrum besides transmitting radio. It is not therefore likely that the Treasury will want to make vast sums out of it, as it clearly did when it realised its potential. It has already had its fair share from that — or unfair share, depending on how you look at it. Please can we therefore have two assurances from the Minister? We should all try to move as fast as possible and with all encouragement towards the digital switchover, because we can see the disadvantages in having that laggard time that many seem to envisage. Indeed, one person giving us evidence today said that, concerning radio, there was really no possibility of a digital switchover; he was as dispirited as anyone could have been about the process. Any form of encouragement that the Minister can give would be welcome, certainly on the future use of FM and on longer licences. Those two things were very much endorsed by the people who gave evidence this morning to the Select Committee on Communications.*

Lord De Mauley: *My Lords, it is late and I shall be brief, but when we finally switch over to digital transmission it is important to be sure that the Government stay true to their promise that the FM spectrum will remain available for use by local and community radio stations. The Digital Britain report said that that was the Government's plan, but it would give a great deal more reassurance if such a promise was contained in the legislation.*

Lord Young of Norwood Green: *I thank the noble Lord, Lord de Mauley, for his brevity. It is not as if I am not paying attention, or giving this less than its due, but we have already travelled over some of this terrain. However, I shall endeavour to reiterate the assurances. For small, local commercial and community stations, both the coverage area of a digital multiplex and the cost of its carriage are too great at present. That is one reason why we believe that those stations are best served by continuing to broadcast on FM. We have also committed to retaining FM for radio after the digital radio switchover. Of course, the Government will not stop any station that wants to and can move to digital, but we will reserve capacity on FM for those which have no obvious route there. I want to address a couple of the concerns that the right reverend Prelate expressed. How will we support those stations which remain on FM? In order to ensure that stations on FM can operate and compete with services on digital after switchover, the Government have already said that we are committed to establishing a combined electronic programme guide for radio. That will allow listeners to access stations via the station name, irrespective of the platform carrying the service. Listeners will therefore move seamlessly between bands, selecting stations simply by name; that is currently not the case when listening to FM and AM stations on an analogue radio receiver. We are working with the industry on that issue and encouraging its development.*

Lord Clement-Jones: *We have been given that assurance on a number of occasions. The Minister in the Commons, who recently announced his resignation, sadly, has given that assurance about the electronic programme, but no date was put on it. It is simply that they are working on it. This is a crucial aspect in retaining people's ability to tune in to FM.*

Lord Young of Norwood Green: *I will see whether I can find any further information. This is a genuine commitment. It is all part of the backdrop against which this debate is taking place. We have set 2015 as a target, but there is not a headlong rush to it. We are trying to ensure a number of things, and this is one of them. I can give a progress report on where we are: it is a clear commitment. The right reverend Prelate asked about making FM continue to be attractive to advertisers and listeners. The key to the switchover of radio will be establishing three distinct tiers of radio — local, regional and national — which will provide unique content and are sustainable in their markets. The services that will populate FM will have a distinct role in providing very local material and reflecting the communities they cover. Due to the very local nature of their content and the refocusing of the large regional stations, these services will benefit from less competition for local advertising funding. I hope that is of some help. The noble Baroness, Lady Howe, asked whether Ofcom will offer analogue licences for longer than five years. The duration of analogue licences is a matter for Ofcom. However, it has suggested that, subject to the outcome of the Bill, it will consult on this issue. We support this process as there is clearly a strong argument for allowing analogue licences over a longer licence period. I, too, rather like the elegant phrase "multi-platform ecology". I wish I had thought of it myself. We have not included in the legislation a commitment to retain FM because the Bill is not intended to set out all the details of the digital radio switchover but to enable a switchover to take place how and when that is appropriate. We agree with the right reverend Prelate's phrase "a multi-platform ecology" — imitation will* *soon be the sincerest form of flattery on this one. To do this, Clause 30 provides for changes in the licensing terms of those services for which it will no longer be appropriate to continue on analogue once the switchover date is nominated. For those licences where analogue broadcasting is the most appropriate or only means of broadcasting, these powers need not apply and their terms will be unaffected, including the right to broadcast on an analogue frequency. The continuation of FM is therefore already provided for in this legislation and should be read alongside the commitments in Digital Britain. In the interests of time, I shall not say any more. I have tried to give as many assurances as I can. We share noble Lords' concerns. We want this to be successful. We should take heart from how successfully we have handled the switchover to digital TV. That has been a success story. Lots of concerns were expressed at the beginning, and we had to work to ensure that people who had fears about handling the new technology were assisted. We got it right. I am not saying that that should be a blanket assurance for everything, but we should not forget how well we handled that. It gives us a good background of experience upon which we can build. I thank the right reverend Prelate for this part of the debate. I trust that the assurances that I have given will enable him to withdraw the amendment.*

The Lord Bishop of Manchester: *I thank the Minister for his response to this debate. I am also grateful to the noble Lords, Lord Clement-Jones and Lord De Mauley, and the noble Baroness, Lady Howe, who made very helpful comments. I welcome the Minister's assurances about the continued provision for local stations to use the FM band. He said that we had been over this ground several times, but we have done so partly because of the seriousness of the issue and partly, as the noble Lord, Lord Clement-Jones, said later in the debate, because of the need to pin the Government down to get the precise assurance that people need. I am sure that those who run and who listen to these media services will feel encouraged by the general direction in which all this is going. Again, the noble Lord, Lord Clement-Jones, was absolutely right to say — it may have been a parlance, but it is a very good way of saying it — that these people really do need a clear signal from the Government that is very much along the lines of what the Minister has said. Having said that, I suspect that this needs to be repeated and to be made even clearer. As those of us who have the privilege of sitting on the House of Lords Select Committee on Communications know, there is a huge difference between the switchover to digital television — this has indeed gone very smoothly, although it is not without its teething problems — and the digitalisation of radio. It has been made very clear to us in the evidence that we have received that the whole business of the digitalisation of radio is much more complex. While the Government and all the digital facilitators need to be congratulated on what they have done in the switchover to digital television, let us not think that because that went so easily it will be the same for radio. There are some very different and deeper issues that we must look at. That said, I utterly agree with the Minister and all noble Lords who have contributed to the debate that we want to keep up the momentum. We really do want to go along with what the Digital Britain report has said and get ourselves going in the technological direction in which we are set. Unnecessary delays are certainly not welcome. Finally — I think the Minister alluded to this in an earlier debate — there is the issue of manufacturers moving towards products that use a combined station guide, rather as Freeview and satellite television do for television, so that people can choose stations by name, whatever the band they are using. This kind of mixed economy of stations, both analogue and digital, will be the simplest way of getting through the many complexities that are on our path. I am most grateful to the Minister, and I beg leave to withdraw the amendment.*

Amendment 239A withdrawn.

---

[423] http://www.publications.parliament.uk/pa/ld200910/ldhansrd/text/100203-0011.htm

79.

5 February 2010

Criteria and a date for digital radio switchover: debated in the House of Lords

The news that the government's proposed 'digital radio switchover' will no longer involve 'FM switch-off' was first written about here in June 2009, when I stated:

"In the 13-page radio section of the Digital Britain Final Report published yesterday, there was not one mention of the word 'switchover' in the context of 'digital radio switchover'. Neither was there a single mention of the word 'switch-off', as in 'FM radio switch-off'. Throughout the document's radio section, the new buzz phrase is 'Digital Radio Upgrade', meaning a drive to make DAB radio better and improve its consumer take-up. In Digital Britain, the notion of switching off FM radio broadcasting, notably for local stations, has been buried for good."

However, between June 2009 and now, this issue – that the government will NOT switch off FM broadcasting – has been clouded by continuing statements from some radio lobbyists and in many media outlets which imply that FM radios will no longer work after switchover. If I were cynical, I might think the motivation is to panic citizens into buying DAB radios – a consumer response which would itself help the industry reach the criteria necessary for 'switchover' to happen.

As for those criteria, and the government's favoured switchover date of 2015, I have consistently predicted that the criteria stand absolutely no chance of being reached by 2015. For a long time, the government and large parts of the radio industry loyally stuck with these targets, imagining that somehow, if you say something frequently enough in public, it will come to pass of its own accord. However, in recent months, with 2015 rapidly approaching, there has been a sudden U-turn. Until then, the radio industry had been vociferous in demanding the government fix a 'switchover' date as early as possible. Now, suddenly, a date – any date – is no longer deemed desirable.

A chapter last month noted that neither the criteria, nor the 2015 date for 'switchover', are mentioned anywhere in the Digital Economy Bill:

"Figures. Numbers. Dates. Criteria. This kind of factual evidence or hard data might obstruct a future decision to force consumers to switch to DAB radio.

……… the criteria and the switchover date that had been agreed upon by stakeholders, over two years of deliberations, have now quietly been relegated to oblivion.

When would digital radio switchover have happened if the agreed criteria had been implemented in law? Probably never.

When will digital radio switchover happen now? Whenever those in power want it to."

A debate on Wednesday night in the House of Lords (see below) considered several amendments to the Digital Economy Bill which would have put specific criteria back onto the legislative table and would have ensured that all radio stations (large and small, BBC and commercial) are given an opportunity to migrate from analogue to DAB transmission. These amendments were eventually withdrawn, after the government Minister offered vague assurances that FM would not be switched off *"for the foreseeable future"*.

As a result, on paper, if the Digital Economy Bill is ever legislated, 'digital radio switchover' (which no longer involves FM switch-off) can be made to happen whenever and however the government wants it to.

In practice, digital radio switchover will inevitably never happen. Consumers appear increasingly disinterested in DAB radio. Besides, we have yet to hear one radio station owner promise they will turn off their analogue transmitters forever and, in that instant, cut their listenership by half just because the other half are thought to be listening via a combination of DAB, digital TV and the internet. You would have to be mad to do that. And, yes, the radio industry could be considered 'crazy' for continuing to go along with this bizarre notion that it is prepared to cut off its own nose to spite its face.

Eventually, it will end in tears.

Amendments 238, 239 & 241[424]
Moved by Lord Clement-Jones

238: Clause 30, page 33, line 19, at end insert —
"(2A) The Secretary of State may not nominate a date for switchover —
(a) unless it can be established that all local commercial radio stations will have the opportunity to move to digital audio broadcasting,
(b) until the proportion of homes in each of the four nations of the UK able to receive —
    (i) national BBC services,
    (ii) national commercial radio services,
    (iii) local BBC services, and
    (iv) local commercial radio services,
via digital audio broadcasting is equal to the proportion able to receive them via analogue broadcasting,
(c) until digital audio broadcasting accounts for at least 67 per cent of all radio listening, and
(d) until digital audio broadcasting receivers are installed in 50 per cent of private and commercial vehicles."

239: Page 33, leave out line 21 and insert —
"( ) must ensure that all commercial and BBC radio services broadcasting in the UK have the opportunity to switchover on the same date,"

241: Page 34, line 1, leave out "2" and insert "4"

Lord Clement-Jones: *In moving Amendment 238, I shall speak to Amendment 239 and 241. It is a wonderful thing to make one's debut at 8.30 pm. I will try to do it with some brio. This amendment is designed for two particular purposes: not only to explore some of the detail of this clause, which seeks to implement a very important part of the Bill, but to test the* *coherence of the Government's policies and objectives in this respect. On these Benches we recognise many of the benefits of moving from analogue to digital; of course we do. However, we believe the Government have been fundamentally poor in the way in which they have communicated their objectives and in the way in which they described both those and the processes involved in switching over. We are particularly concerned about the impact on local FM radio and what one might call ultra-local radio in that respect. A great deal of reassurance is needed about the future of FM. It is not just purely about freeing up space on the analogue spectrum for FM by migration from analogue to digital. The Government need to explain other aspects too. Why are we adopting DAB rather than DAB+, which would allow a much more local feel to our radio experience and enable many more ultra-local radio stations to migrate? The Government also need to explain — the Minister has made an attempt to do so — why the target of 2015 has been set. There are other details involved. One of the great forms of reassurance that the Government have tried to give to ultra-local radio FM stations is this notion of a common electronic programme guide. That is a great idea, but what is its practicality? What timescale is envisaged? The Government need to deliver much more clarity in this respect and other aspects. Will there be adequate supplies of radios that enable listeners to tune in to both FM and digital radio? It is fairly basic: will the domestic consumer be able to have access to those, or will most manufacturers switch over to pure digital? The availability of digital car radios is a particular concern. I shall illustrate where I believe there is some incoherence in this regard. It is almost as if two separate hands wrote Chapter 3b of the White Paper. That chapter is headed "Radio: Going Digital". Paragraph 20 of that chapter states: "However, it has always been our intention that the ultra-local services which remain on FM after the Digital Radio Upgrade should only do so temporarily". Later on, the paper — I leave the Minister to find his own copy — is written in very different terms. It talks about a much longer-term future for FM. Which is it? Will FM be a permanent or temporary part of the radio landscape? These things need sorting out as part of the Bill. Even if you talk to those intimately involved in the digital switchover, you will not get a precise answer about whether FM is here for the long term. As I understand it, technically, ultra-local radio is far better staying on FM almost in perpetuity, provided the two forms of spectrum can be afforded, and provided the receivers allow the receipt of both FM and digital for the foreseeable future. There is no particular reason why some of these ultra-local stations should indeed switch over. Why should they not be encouraged to stay where they are? That would seem to me to be a very coherent way of expressing government policy. However, of course, many operators want to have a broader canvas for their stations. That is entirely acceptable too. Where they wish to make that investment and have a broader scope for their radio services, that seems to be entirely right and proper. The Digital Britain White Paper talked about the triggers for the switchover to digital radio. In referring to those, the Minister said that it was not simply the case that half of all listening should be to digital and that the relevant date should not be set until national DAB coverage matches that of FM and 90 per cent of the population have access to local DAB. The clause we are discussing — this is what Amendment 238 is designed to do — does not refer to any of that. All it does is talk about the report submitted by Ofcom, to which the noble Lord, Lord Howard, referred. It does not give any direction to Ofcom. One assumes that Ofcom will refer to the government policy as set out in the White Paper, but there seems to be no duty on it to do that. That seems to constitute a gap in the provisions. We on these Benches believe that the Government need at least to give reasons why some conditions are not appropriate. We do not necessarily believe that all of these additional conditions are the right ones, but we do believe that certain of those should be additional to those stated, so that all local stations have digital migration pathways. Currently, more than 120 stations lack viable digital migration pathways. All categories of commercial and BBC stations should be treated equally, creating a level playing field going forward. The switchover date — 2015 — is based on an assessment of DAB listening rather than digital listening, which includes digital television and the internet, as this will replace FM and AM as radio's broadcast backbone. People are listening more and more to radio over the internet. As regards the whole issue of vehicle reception, I think that less than 1 per cent of cars can receive digital signal at the moment. I think the Government estimate that only 10 per cent of vehicles will be able to receive it in 2015. We are trying to get a more coherent policy and a switchover that accords more to the reality of what listeners are doing. There is absolutely no point in the Government being so far ahead of what listeners want to do that they create a reaction against their own plans. Amendment 239 is another measure designed to test the earnest of the Government. The Government have rather arbitrarily decided that switchover should take place two years after notice of it is given. Amendment 241 seeks to change that. There are concerns about the speed of switchover. Amending the notice period would allow more time to solve the issues around consumer take-up, such as those involving vehicles, and would allow digital solutions to be provided for all stations which want them. A four-year period would be far more realistic in terms of recognising the considerable challenges involved in the digital switchover. Amendment 239 states that the Secretary of State "must ensure that all commercial and BBC radio services broadcasting in the UK have the opportunity to switchover on the same date". That is another way of testing the Government's proposals in this respect. Are they saying that, at the end of the day, all these stations need to switch to digital, or are they saying that FM has a future in the long term? What is the purpose behind the switchover? It seems to me that the Government have not expressed their intentions very clearly. I hope that this is an opportunity for them to do so. I beg to move.*

The Lord Bishop of Manchester: *My Lords, Amendment 238 has been put eloquently by the noble Lord, Lord Clement-Jones, and with all the promised brio. However, he has set in that amendment probably rather an ambitious goal. Some may say that it is too ambitious. It may, of course, have the impact of delaying the switchover date from the much bandied-about year of 2015. However, as we heard from the Minister earlier, that is not a precise date. The amendment may also effectively force something of a reappraisal of how digital radio is rolled out between now and switchover. But, that said, I believe that neither of those consequences would be to the detriment of the well-being of the radio industry. The noble Lord referred to Ofcom. In order to make switchover manageable and help local stations remain viable in these challenging economic times, Ofcom has proposed a series of defined areas within which stations in the same commercial family will have more flexibility to share resources and locations. As your Lordships will know, that has been welcomed as a helpful step by many of the large and small commercial stations. Nevertheless, it throws up anomalies, especially where radio groups would be barred from broadcasting to some of their network from locations that are actually far closer to their listeners than those from which they are allowed to broadcast. I say this not to criticise Ofcom's sterling efforts in coming up with solutions to try to make life easier for these vital local stations. The regulator is trying to be flexible and responsive in a rapidly changing arena. Yet that is precisely the point. It is almost as if, at the moment, we are rushing towards a finishing line when the course has not even yet been chalked out. This amendment helpfully acknowledges that the route to an appropriate digital platform is not yet entirely clear, especially for many smaller stations. I have concerns over the specific proposal that sets a target of two-thirds of listening on DAB before switchover simply because it does not quite recognise the mixed radio economy that we are moving towards, to which the noble Lord, Lord Clement-Jones, referred. If a growing proportion of us are listening to radio via our computers, mobile phones and digital TV sets, with only a small remnant still listening on analogue, it would be unfortunate if the wording of primary legislation made it such that we could not set a* *switchover date just because DAB was specified as the only non-analogue platform. But, with those reservations, on the whole I support Amendment 238.*

Baroness Howe of Idlicote: *My Lords, I will briefly add my support for these amendments. There clearly has to be a rethink on this whole area. There are so many different interests and people who have been hoping and planning. At Second Reading, I remember my enthusiasm for immediate digital switchover as far as radio was concerned. It would be a splendid way for me to be able to enjoy my journeys up and down to Warwickshire without having to fiddle around and change the frequency on the radio, which is presumably a fairly dangerous practice anyhow. So, although one may not agree with every word, this points out to the Government that there is going to have to be a more up-to-date assessment of where we are and how quickly, if at all, we are going to reach the desired goal of switchover and meet the needs of local radio stations, particularly the commercial ones.*

Lord Young of Norwood Green: *My Lords, the proposed amendment inserts specific switchover criteria into the Bill which must be satisfied before the Secretary of State can nominate a date for switchover. We agree that it is important for the Government to consider a wide range of issues and views before setting a date for digital switchover. We seem to be referring to 2015 as though we have set a date. I reiterate that that is a target. As with TV switchover, the announcement of a target date is essential in uniting the industry in creating the impetus to ensure that progress towards switchover conditions is made. For example, within six months of 2015 being published in the White Paper, progress with car manufacturers had advanced further than in the previous six years. We still believe that 2015 is an achievable deadline and the certainty of a timeframe is in the best interests of listeners and the industry as a whole. However, I stress that it is still a target. We expect broadcasters and network operators to work towards that deadline but a final date will only be set when progress measured against consumer-led criteria is clearly on target. We do not believe it is appropriate to insert the level of detail proposed by this amendment into legislation as it would fail to grant sufficient flexibility to address the inevitable changes in the market over the next five years. As this amendment raises a number of themes, I will take each of the issues raised in turn. It was an omnibus address delivered with brilliant brio. First, the proposed amendment requires a DAB network large enough to accommodate all commercial radio stations before a switchover date is nominated. We believe that this is the right ambition, but impractical in the short term. With over 50 BBC services, nearly 350 commercial stations, and 200 licensed community stations, the current DAB infrastructure cannot support a wholesale move to digital. We believe a mixed landscape of FM and DAB, as set out in the Digital Britain White Paper, is a better solution. It will allow small commercial stations and community radio stations to remain on FM, which was a point of concern expressed by the noble Lord, Lord Clement-Jones. In the case of community radio stations, it will allow for more services to be launched. On the question of whether FM is temporary or permanent, FM is the right technology for community and small local radio stations for the foreseeable future, following switchover.*

Lord Clement-Jones: *My Lords, it is permanent.*

Lord Young of Norwood Green: *I feel I am being lured into saying "never say never" or "never say ever". We certainly see it for the foreseeable future following switchover. That is pretty generous. We are trying to reassure the noble Lord. It is the right technology for community and small local radio stations. We cannot predict exactly how technology will develop for ever and ever, but we are saying "for the foreseeable future, following switchover". That is a pretty good guarantee. Secondly, this amendment would require DAB* *networks in the UK to reach the same number of households as analogue. Again, we agree with the principle and we are working with the industry to secure this outcome as quickly as is feasible. However, the definition of analogue coverage is too broad, encompassing long wave, medium wave and FM. It also provides no clarity as to whether it is a high-quality stereo signal or a low-quality mono signal. This amendment would result in multiplex operators and broadcasters bearing the full cost of a near universal DAB network for the longest possible time. Thirdly, the amendment proposes to require that 67 per cent of all radio listening be to DAB before a switchover date can be set. We agree that the digital radio switchover should be market-led and dependent on the take-up and usage of digital radio. However, this criterion places too great an emphasis on DAB over other digital technologies. The issue is not whether listeners choose to consume radio via DAB, the internet or digital TV, but the extent to which they have access to digital radio technology and are using it. Also required in this amendment is that DAB receivers are installed in 50 per cent of private and commercial vehicles. We believe there would be significant challenges in measuring success against such a criterion. In addition, such a legal requirement would fail to take account of the adoption of in-car radio converters or future technological developments. Just as we had the development of the Digibox in relation to enabling analogue TVs to carry on working, so it is obvious that we will see the price of in-car converters come down and that technology developed. All DAB radios include FM: that was another point that was raised. We are working with car manufacturers with the aim that, by 2013, all radios sold with vehicles will be digitally enabled. There are already devices on the market that will convert an FM car receiver so that it can receive DAB, and the market in this area is likely to grow considerably. It is huge: there are millions of analogue car radios. We are trying to cover all the areas of concern. Amendment 239 would also require that all commercial and BBC stations have the opportunity to switch over on the same date. It removes the flexibility granted to the Secretary of State to nominate different switchover dates for different services. Although it remains the Government's intention that all services specified by the Secretary of State should switch over on the same date, flexibility is needed to ensure that the switchover can be delivered most appropriately for listeners. With TV, the gradual approach is working well, so we are making a plea here for flexibility. Amendments 241 would increase the minimum notice period from two to four years. We believe that this would unnecessarily extend the costs of dual transmission for broadcasters, and slow down the rollout of the DAB network. The noble Lord, Lord Clement-Jones, asked about our motivation. Part of it is the most effective use of the spectrum; another part is to ensure that people are given the best possible service. The noble Baroness, Lady Howe, said she had to fiddle with the knob on her car radio. It is time that she updated it to one with RDS, which will automatically change the frequency for her as she drives along. I cannot guarantee her a government handout, but it is well worth it. The right reverend Prelate was concerned about local radio and defined areas, and made a number of interesting points. We will look carefully at his remarks in Hansard and write to him on the issue. We will make copies available to all noble Lords who have contributed to the debate, because these are points of common interest. I apologise for going on at length, but this is an important debate that encompasses a wide range of issues. I have tried to address all the points raised and understand the concerns, and I hope that, given my response, the noble Lord will withdraw his amendment.*

Lord Clement-Jones: *My Lords, I thank the Minister for that comprehensive reply, particularly since it was on the run: it was not obvious from the amendments how broad the debate was going to be. He has addressed almost all the issues, except the one that I raised about DAB+. I forgive him for that, although if he does have an answer —*

Lord Young of Norwood Green: *I do have an answer, but forgot to make a note of the question. The advantage of DAB+ over DAB is the greater capacity that it provides for services on the same frequency, allowing a digital multiplex to broadcast around 20 rather than 10 services, thereby giving more stations the opportunity to move to digital. However, we believe that the benefits of DAB+ are more than outweighed by the negative impact on existing DAB listeners. Only a very small fraction of the 10 million digital receivers that have been sold are capable of receiving DAB+, and any technology change would make these receivers effectively redundant. That would significantly delay the switchover timetable and increase the time during which broadcasters would bear the burden of dual transmission costs. Nevertheless, the Government have been clear about their intention to promote receivers that are capable of receiving DAB+ and other digital technologies that are used across Europe. This will protect receivers against any change of technology in the future. We are trying to take into account the concern in that area. I apologise if I put that over quickly, but it will be on the record.*

Lord Clement-Jones: *My Lords, I thank the Minister for that suite of answers to the various points raised. I also thank those who took part in the debate; the right reverend Prelate and the noble Baroness, Lady Howe. Not every jot and tittle of the amendment necessarily carries all the weight that it should. We are looking for pointers from the Government. Although some of the White Paper sets out the conditions and criteria, they are not in the Bill and there is uncertainty, albeit among a minority of radio operators. The future of FM has not been made clear. I take the words "foreseeable future" in these circumstances to mean semi-permanent. There is a fear that FM stations "left behind" on analogue will be second-class citizens who will not be there for long — they will be hustled across into digital — when actually, in terms of cost and reception, ultra-local radio is probably best left on FM. It may be that the Government wish to release that spectrum, have a fire sale and flog off FM to the highest bidder — I know not. The Government need to be crystal clear about those issues. The use of the "mixed landscape" wording by the Minister was very healthy; no wholesale move in the circumstances is healthy, as it has been described to us that digital will be a great advantage to the larger national radio stations but not to the smaller ones. I understand what the Minister was saying about the vehicle aspect. I hope that he is correct in saying that the ability to convert to digital by various gizmos, which will not cost the public an arm and a leg, will be increasingly possible. I think he mentioned the 2015 gate but I do not know what percentage of in-car use was associated with that. Clearly, there is some considerable optimism about the switch taking place within the next three or four years. That is ambitious. On two years versus four years, I cannot help reflecting that the two-year period is more, not for the listeners' or the radio operators' convenience, but for the Government's convenience because they want a decision made and they want to be able to consider the future of FM after that. That is the impression given to date. In a sense, two years is a very aggressive period, once the notice has been given, in which to expect that migration to take place. That is the fear behind that short period. Basically it is for the Government's convenience rather than for allowing things to take place in an orderly fashion. This has been a very useful debate. I thank the Minister for what he has said. Obviously, some reflection is required. We have all had an enormous amount of correspondence from radio operators and no doubt we shall reflect with them between now and Report about whether further tweaks are required to this part of the Bill. In the mean time, I beg leave to withdraw the amendment.*

Amendment 238 withdrawn, Amendment 239 not moved, Amendment 241 not moved.

---

[424] http://services.parliament.uk/hansard/Lords/ByDate/20100203/mainchamberdebates/part016.html

80.

8 February 2010

Costs/benefits of digital radio switchover: why the government buried the evidence

Digital radio switchover was first mooted in the 1980s and started to gather momentum following the first UK public demonstration of the DAB digital transmission system at the Radio Festival in Birmingham in 1991. New Scientist magazine reported then that DAB radio could be up and running in the UK *"by the mid-1990s"*.[425] However, it was not until 1999 that DAB radio was launched publicly and DAB radio receivers were made available commercially.

Over this period of decades, it would have been sensible to commission some kind of cost-benefit analysis to assess if there were a potential net benefit to radio listeners, to the radio industry, and to the UK generally of embarking on a plan to convert the whole nation to digital terrestrial radio. If such an analysis was ever published, I must have missed it.

We are now in 2010, an incredible 29 years after research and development first started on DAB radio technology. The issue of digital radio switchover has been the subject of a succession of government initiatives since the end of 2007 (the Digital Radio Working Group, Digital Britain, the Digital Economy Bill). Has the government shared with the public a cost-benefit analysis which demonstrates that the public policy on digital radio pursued over the last 20+ years is somehow worthwhile? No. Does such a cost-benefit analysis exist? Yes. Can we see it? No. Where is it? Apparently, gathering dust on a government or Ofcom shelf.

How do we know this? A parliamentary committee recently delved into these facts during its current investigation into the issues surrounding digital switchover in the television and radio markets (see transcript below). Ofcom has remained remarkably silent on the issue of digital radio switchover in recent years. The regulator's director of radio, Peter Davies, was last seen speaking publicly about DAB in November 2008 when he admitted that new legislation would be necessary to salvage the DAB platform. A little earlier, in April 2007, Davies had prematurely declared that *"we are potentially at a Freeview moment with digital radio."*[426] Three years later, radio's 'Freeview moment' seems as far over the horizon as ever.

The first we knew that the government had commissioned some kind of cost/benefit analysis [CBA] for its proposed digital radio switchover was in November 2009 when the Digital Economy Bill was published. The government's accompanying Impact Assessment document stated:

"The partial Cost Benefit Analysis conducted by Price Waterhouse Cooper (PWC) for the Digital Radio Working Group, which is available on the DCMS website, suggests the Digital Radio Upgrade could reduce the total transmission costs for the radio industry from £87.9 million to £64 million...."

"First, by supporting greater investment in DAB infrastructure a greater number of consumers will have access to DAB and the quality of reception will improve. Secondly,

consumers will benefit from access to a wider range of services, specifically new national stations and functionality, such as pausing and rewinding live radio. Finally, the released analogue spectrum will allow for a greater range of community radio stations, as well as possible non-radio services. The PWC partial CBA for the Digital Radio Working Group suggests the value of these benefits could be in the region of £1.1 billion, over a period from 2009 to 2030....."

"The significant consumer costs of the Digital Radio Upgrade in the non-voluntary conversion of analogue sets to digital, including the cost of in-car conversion. The PWC report suggested the cost of such conversion to be in the region of £800 million, again over the period from 2009 to 2003."[427] [typo – "2003" should be "2030"]

Although this document stated that the PWC report was available from the government's web site, I have searched for three months and still never found it there. Nevertheless, the 91-page report entitled 'Cost Benefit Analysis of Digital Radio Migration', prepared for Ofcom by PWC on 6 February 2009, contains a number of very serious reservations that there will be ANY benefit from digital radio switchover, and it states:

*"The results suggest that there are relatively few up-sides to the estimates, and several significant downside risks. ... To a significant extent, the positive Net Present Value [NPV] of the Cost Benefit Analysis relies on two crucial parameters. The first is the Digital Radio Working Group [DRWG] recommendation that an enlarged regional [DAB] multiplex network should be implemented. Failure to implement would result in a substantial negative NPV. The second critical parameter is the time horizon. The results suggest that there is a very long pay-back from the DRWG policy 'investment' – the NPV turns positive after 2026. This result assumes that the existing multiplex licences are extended to 2030, as per the DRWG recommendations. Without the licence extension or any other policy instruments that provide clarity on the long term future of commercial radio, **the industry and consumers may fail to see the benefits of digital radio over the longer term.** Our analysis suggests that the NPV is negative should either of these two proposals not be implemented."* [emphasis added]

The PWC report explicitly noted for Ofcom the limitations of its analysis, as a result of the lack of consideration it had assigned to external factors. These paragraphs probably explain why the government has been so keen to keep the report away from public scrutiny:

"The scope of this study is limited to the assessment of the DRWG policy. The overall digital radio policy appraisal process would need to take into account other policy options and 'states of the world'. With this in mind, we highlight three issues in particular:

*1.   The impact of recession: **We have assumed no change in commercial radio sector structure and health** beyond a consensus view of advertising forecasts. As this CBA is conducted for the time period to 2030, short term recessionary impacts may have only a limited impact on the longer term outcome for the industry. On the other hand, the current economic downturn could still affect the short and medium term investments required for marketing or coverage extension, which in turn could delay the desired DRWG policy outcome.*

*2.   Other policy options: We recognise that to reach a view on this question of how to drive digital radio penetration and listening (which in turn delivers consumers' and citizens' objectives) requires a full assessment of the costs and benefits of a number of policy options; **this study has examined one, the DRWG policy. This is the only policy assessed in this study*** *and the policy is at an early stage of its development; Government and Ofcom could give consideration to other possible policy options. In addition, we recommend modelling a number of other 'business-as-usual' scenarios taking into account different assumptions, and assessing how they affect the CBA of the DRWG policy.*

3.   Other digital platforms: **This CBA assumes that DAB listening will continue to be the leading platform for digital radio listening.** *The DRWG has reinforced the view that 'a radio-specific broadcast platform is an essential part of radio's future', and that DAB is the 'most effective and financially viable way of delivering digital radio' for the medium to long term. A long term view needs to account for the possibility of technology obsolescence or replacement. At present, there is no consensus view that suggests otherwise. However, there are signs that internet listening may begin to take off if internet radios are more actively promoted and technologies such as WiFi or mobile broadband mature and become universally available. A number of the cost and benefit categories assume an impact from increasing the coverage of DAB (for example, consumer benefits from increased coverage is assumed based upon the incremental benefits to consumers who could not receive digital radio stations). Should these trends continue, or a more structural shift to internet to occur, there would be a smaller benefit from increasing the coverage of DAB; consumers either have alternative access to digital radio even within out-of-coverage areas, or would prefer a non-DAB solution when they receive DAB coverage."* [emphasis added]

In the 12 months since the PWC report was prepared, all three of these assumptions have been undermined by subsequent events:

1.   The health and structure of the commercial radio industry have changed considerably during 2009:

- Commercial radio's financial health has been impacted severely by the recession. The sector's revenues were down 19.5%, 10.8% and 12.5% year-on-year in the first three quarters of 2009. Revenues from national advertisers were down 28.8%, 16.1% and 16.5% respectively

- In January 2009, an analysis commissioned by RadioCentre found that "the [commercial radio] industry as a whole is now loss making". Hours listened, revenues and profitability have all fallen further since then

- At the beginning of 2009, Global Radio was the biggest owner of DAB multiplex infrastructure. Since then, it has disposed of its entire stake in the national DAB multiplex and most of its stakes in local DAB multiplexes to transmission provider Arqiva, demonstrating the radio sector's inability to generate profits from the DAB platform after 10 years

2.   The policy recommendations for digital radio switchover made by the Digital Radio Working Group have since been amended by the Digital Britain report and the Digital Economy Bill. The recommendations of the Working Group's Final Report published in December 2008 had included:

- *"Government should agree a set of criteria and timetable for migration to digital",* whereas no criteria or timetable are specified in the Bill

- *"A long term plan should be developed to move all services to digital",* whereas the Bill acknowledges that some local radio stations will never have the opportunity to migrate to digital

- *"The BBC should build out its national [DAB] multiplex across the UK to reach FM comparable levels [of coverage]",* whereas the BBC has acknowledged that such expenditure is constrained by the Licence Fee settlement

- *"The government should consider funding options to enable this important investment [in DAB infrastructure]"*, whereas the government has made no financial commitment to the build-out of DAB multiplexes
- *"The government must consider the case for a [import] duty exemption for digital radios"*, a proposal that is not mentioned in the Bill
- *"Consumer groups believe that, once an announcement [of digital switchover] is made, no equipment should be sold that does not deliver both DAB and FM"*, a proposal that has been dropped

3.   The DAB platform has failed to grow in 2009, as had been forecast by the government, Ofcom and the Digital Radio Development Bureau [DRDB]:
- Volume sales of DAB radio receivers were down 10%, down 1% and down 6% year-on-year in the three most recent quarters for which data have been released by the DRDB
- Listening to radio via digital platforms accounted for 20.9% of total radio listening at year-end 2009, compared to the 26% forecast by the government's Digital Britain report in June 2009 (and compared to the 42% forecast by Ofcom in November 2006)
- Listening to commercial radio via digital platforms accounted for 19.7% of commercial radio listening at year-end 2009, compared to the target 30% announced by RadioCentre in January 2007
- Total hours listened to digital-only radio stations at year-end 2009 were at their lowest level since 2007, demonstrating that digital radio content is failing to drive consumer take-up of digital radio
- Unused capacity on the DAB platform has increasingly been filled during 2009 by non-commercial, government-funded, listener-funded, religious or ethnic radio services, rather than by mainstream, mass appeal stations
- The commercial radio sector launched no completely new broadcast digital radio stations in 2009 (Absolute Xtreme was replaced by Absolute 80s), and the BBC is expected to announce cuts to its digital radio stations at the end of this month

As a result of these developments during 2009, the minimal, long term benefits from digital radio switchover identified by the PWC report a year ago are likely to have been diminished to the point where there may no longer be any benefit evident at all, even as far into the future as 2030. So how can the government still justify pursuing its policy of digital radio migration? It cannot, which is why it remains so reluctant to engage in an analysis of the facts, the numbers, the data and the evidence, all of which clearly show that this misguided, poorly executed, top-down attempt to switch radio broadcasting in the UK to the DAB platform is likely to become a 'white elephant' that has already cost the radio industry getting on for £1 billion.

House of Lords
The Select Committee on Communications
"Digital Switchover Of Television And Radio In The UK"
27 January 2010 [excerpts][428]

Witnesses:
Mr Stewart Purvis, Partner for Content and Standards, Ofcom
Mr Peter Davies, Director of Radio Policy & Broadcast Licensing, Ofcom
Mr Greg Bensberg, Senior Adviser, Digital Switchover, Ofcom.

Baroness McIntosh of Hudnall: *I feel we could get back on to slightly safer territory and the notion of cost and benefit. We understand that you commissioned a report from PWC last year into the costs and benefits of digital switchover in radio, but you didn't publish it. We know, therefore, what we have learned from the DCMS about what it said. It appears that it found, for example, that the benefits could – and I emphasise the word "could" – outweigh the costs by £437 million after 2026, but that conclusion is hedged about with quite a lot of caveats to do with what would have to happen in order for that good outcome to eventuate, and that if those things didn't happen, then quite quickly you would get into a position where the costs would outweigh the benefits. Can you tell us a bit about that report? In particular, can you tell us why you haven't published it? Do you think that, given what it appears to say – I choose my words carefully – about the constraints on potential for benefit, that it should have been available to inform the Government's digital policy? Can you also tell us about your own impact assessment on radio digital migration, which I believe you have been asked to undertake? Will this include a full cost-benefit analysis? When are you intending to publish it? ….. [edited]*

Mr Purvis: *There are a lot of questions there. Peter commissioned the piece, so I am going to ask him to talk to them, but let me say that you have talked about informing the Government's decision and one of the main points of doing this was to help inform the Government's decision. It was a government decision as to whether this information should be published or not. But, we felt, as part of the ….*

Lord Gordon of Strathblane: *Sorry, it was your document, though, wasn't it?*

Mr Purvis: *No, it was actually a PWC document.*

Lord Gordon of Strathblane: *It was commissioned by you.*

Mr Purvis: *Commissioned by us, yes.*

Lord Gordon of Strathblane: *Surely, it would be your decision to publish.*

Mr Davies: *We were asked to commission it by the Government. We then commissioned it from PWC with a lot of input from various government departments and then submitted it to the Secretary of State.*

Chairman: *So you decided not to publish it.*

Baroness McIntosh of Hudnall: *Who owns it?*

Mr Purvis: *Whenever you commission a document from an outside source, in a sense the ownership of the detail must lie with the people who actually did the work, but, in a sense, when you commission it, obviously you commission it with a purpose and the purpose was to give it to the Government.*

Baroness McIntosh of Hudnall: *With respect, that is not necessarily true.*

Mr Purvis: *No, there are options.*

Baroness McIntosh of Hudnall: *I have work commissioned from me and it may be, and often is, on the understanding that the ownership of what I produce falls to the person who commissioned it from me.*

Mr Purvis: *Yes, that's true, but in terms of the ownership. But in the sense of the responsibility for the detail of the commission, the source of the commission must inevitably take its full share of that. But there are a number of options which apply when these pieces of work are done. On this particular occasion, it was decided in conjunction with the Department that work would be sent to the Department. Perhaps the most important thing is for Peter to respond to your characterisation of the work, but, in a sense, we have not hidden the piece of work. Indeed, I think it is now available to you. Is that right?*

Baroness McIntosh of Hudnall: *In, as they say, a redacted form.*

Chairman: *Just to be absolutely clear, the Department asked you to commission the work from PWC. Is that what you are saying?*

Mr Purvis: *They asked us to commission the work. Did they ask us specifically from PWC?*

Mr Davies: *Not specifically from PWC.*

Chairman: *The Department said to Ofcom, "Ofcom, you go and commission this particular work." Is that the position?*

Mr Davies: *Yes.*

Chairman: *You then got the work which then came back to you and then you sent it to the Government and the Government said, "We're not going to publish this in full."*

Mr Davies: *I think they have certainly made it available to various groups. I think consumer groups have had it for some time.*

Chairman: *Fine. There will be no problem, therefore, in this Committee having the full report.*

Mr Davies: *I think they have made available the redacted version rather than the full report. The reason for that is some of the numbers in there are commercially sensitive, but there is no reason why the Committee should not have the full report.*

Mr Purvis: *You certainly have seen the conclusions.*

Baroness Howe of Idlicote: *I just wonder who has paid for it. Has it come out of your budget?*

Mr Davies: *Yes.*

Baroness Howe of Idlicote: *Even more indication of ownership.*

Baroness McIntosh of Hudnall: *Shall we go back to the questions. We now know why you didn't publish it. Am I right in thinking that, notwithstanding the fact that you did not publish it, it did influence the Government or is in the process of influencing the Government as far as their policy on digital migration goes?*

Mr Davies: *I think it is one of the inputs to government thinking, certainly. We were very careful when we sent it to the Secretary of State to make clear what all the caveats were. You are absolutely right, there are a lot of caveats around it. This is a piece of work which is at a very early stage of the process. We were very clear to government that they should not use this as the means of making a decision, but it might help to inform the decision.*

Baroness McIntosh of Hudnall: *The thing that is slightly troubling – perhaps only to me, but a bit – is that when you see what appears to be evidence that the costs and benefits are, let's say, finely balanced, or could be, that the drive towards digital migration, one might think, was driven more by the technology than by the needs either of the broadcasters or the consumers. That's the question that seems to me still to hang in the air. Is this technology-led or is it consumer-led, if we wrap into 'consumers' both the people who are the end-users and the people who are using the technology to deliver a service?*

Mr Davies: *I think that is why there are so many caveats around it, because it needs to be, as you say, consumer-led. So, some of the conditions that would need to be met for the figure to come out positive are that coverage needs to be built out, that the content proposition needs to be right, that a lot of the benefit in there is from additional choice for consumers. That is obviously down to industry to provide. That is not something that either government or Ofcom can do. One of the main caveats was the need to roll out the regional layer [of DAB multiplexes] that we were talking about earlier, to become a new national layer, so providing more choice of mass market stations, if you like. So it is absolutely consumer-driven, but where that leads you, I think it is probably too early to say, and, as you say, it is very finely balanced.*

Baroness McIntosh of Hudnall: *What about your own impact assessment?*

Mr Davies: *We haven't done an impact assessment yet.*

Baroness McIntosh of Hudnall: *But you have been asked to – correct?*

Mr Davies: *At some point in the future. I think the Digital Britain report said that we would be asked to do one, but we haven't been asked to do one yet. Obviously we would need to do that and we would need a much fuller cost-benefit analysis before any final decision was taken.*

Baroness McIntosh of Hudnall: *So that's a future thing.*

[425] http://www.newscientist.com/article/mg13117784.200-radio-sans-frontieres-by-the-mid-1990s-people-driving-across-europe-should-be-able-to-tune-into-their-favourite-radio-programmes-in-hifi-wherever-they-are.html

[426] http://pqasb.pqarchiver.com/smgpubs/access/1259491541.html?dids=1259491541:1259491541&FMT=ABS&FMTS=ABS:FT&type=current&date=Apr+22%2C+2007&author=Steven+Vass&pub=Sunday+Herald&desc=Analogue+radio+could+be+history+by+2020+.+.+if+Ofcom+wins+the+day+RADIO%3A+SWITCHOVER+RADIO%3A+SWITCHOVER+Optimistic+projections+on+digital+listening+suggest+end+of+MW+and+FM&pqatl=google

[427] http://www.ialibrary.berr.gov.uk/uploaded/Digital%20Economy%20Bill%20IAs%2020104.DOC

[428] http://www.publications.parliament.uk/pa/ld200910/ldselect/ldcomuni/100/100.pdf

81.

9 February 2010

Renewal of national commercial radio licences: debated in the House of Lords

House of Lords
8 February 2010 @ 1723
Digital Economy Bill
Committee (7th Day)[429]

Clause 31 : Renewal of national radio licences
Debate on whether Clause 31 should stand part of the Bill.

Lord Clement-Jones: *My Lords, before I propose that the clause not stand part, I must apologise. As a result of the way in which the business of the House has been organised today, I shall not be able to be here for about two hours of the Committee's proceedings. I very much regret that, as many important matters remain to be debated. However, since the business was switched at extremely short notice — I hope that the Whips are whipped for it in some future incarnation —*

Lord Davies of Oldham: *Oh!*

Lord Clement-Jones: *I am of course not referring to the noble Lord, Lord Davies. Moving this business from Tuesday to Monday at very short notice is not a happy situation. I therefore hope that Ministers will give full and frank responses as if I were present. I am very grateful to my noble friend Lord Addington, who has kindly agreed to step into the breach when I am not able to put the arguments. I propose that Clause 31 should not stand part. Under this clause, the national analogue radio stations talkSPORT, Classic FM and Absolute Radio are receiving valuable seven-year extensions to their licences. In exchange, the existing licensees have been asked to give their support to an early switchover, with the proposed 2015 date coming much earlier than that recommended by the Government's 2008 Digital Radio Working Group. However, there is a view among some operators that extensions to these licences are not worth the damage to radio of a digital switchover policy which assumes an unrealistic timetable for digital switchover and which fails to provide solutions that allow all local radio stations to move to digital. They do not accept that as a reasonable quid pro quo for an early switchover. They believe, on the contrary, that the industry's engagement with the digital radio switchover proposal has been distorted by its interest in licence extensions which are essentially to do with the attractiveness of the current analogue model for radio rather than the proposed digital model. Their view is that Clause 31 will deprive the Government of revenue due from re-auctioning the licences for these national analogue stations. However, the Government have failed to publish an assessment of how much revenue will be lost to the Treasury under this approach. The Government need to justify the advantage of the clause against the background of the following factors: that the sums lost to the Treasury will clearly amount to tens of millions of pounds over the lifetime of the extended licences; and the lack of evidence about whether digital investment by the holders*

247

of these licences will continue without the extensions. On the face of it, many are already, contractually or otherwise, committed to digital even without this.

Lord Howard of Rising: *My Lords, although I share a number of the noble Lord's concerns, I do not think that removing the clause would be helpful. It is a facilitating clause that enables the move to switchover at a later date, and it does not set in stone when the switchover will take place or indeed that it must happen. It is more important that the Secretary of State considers a range of issues before nominating a switchover date than that the process in its entirety is stopped. I believe that the level of digital radio listening should be much higher than the Government have suggested. It would also be very much better if the fact that the FM spectrum will remain in use for local and community radio stations was on the face of the Bill. More progress should be made in creating a help scheme and a recycling scheme. We should be focusing on these issues rather than on an attempt to derail the digital switchover process completely.*

The Lord Bishop of Manchester: *My Lords, I recall that last week the noble Lord, Lord Clement-Jones, and I supported each other's amendments, but sadly that relationship is about to be broken albeit, I hope, temporarily. To allow the Bill to pass without this clause would pose a real problem for the entire digital radio project. The three commercial stations currently granted national analogue licences cater for a broad range of tastes, from Beethoven and Brahms to Bon Jovi, via the latest soccer score from Bolton Wanderers. Their collective appeal has been vital to encouraging digital take-up by listeners, with around a fifth of their current audiences now listening via a digital platform. To disrupt that migration would be rather unwise. Re-advertising these national licences with just a few years to run before we expect to switch off the service seems to be sending the wrong signal to both the industry and to listeners. It seems to suggest that we are not fully committed to digital as the future, that we doubt whether we will be in a position to switch over the bulk of national stations in seven years, and that we can expend less energy on the steps that are undoubtedly still needed to get listeners to switch to digital, especially through pushing down the cost of DAB radio sets and through getting DAB into more cars as standard. I do not think that any of those things are the right course. If, as I understand it, the message from the legislature to the private sector is to be, "We want you to invest in this new technology, market it to your listeners and encourage them to adopt the new listening platforms", surely we cannot keep expecting these companies to keep on writing blank cheques. We all appreciate that digital platforms are still in their relatively early days. It has to be remembered that not one digital radio station has yet posted a profit. For their pioneering endeavours, they deserve the stability that this reprieve offers them. One does not often hear pleas for breaks for business from these Benches, but this is a case of tidying up the licensing regime to make it serve the purposes of the digital age.*

Lord Eatwell: *My Lords, I declare an interest as chair of the consumer panel of Classic FM. This panel is entirely independent of the company. It is devoted to maintaining the standards of Classic FM and the widespread broadcasting of classical music by the independent sector. If this clause does not stand part of the Bill, your Lordships should be aware that the future of Classic FM will be severely compromised because it is a requirement of existing law that the analogue licences are auctioned. As at present conceived, analogue licences do not have a clear format specification. There is not a licence for classical music. There is simply a licence for non-speech, which is the licence held by Classic FM. If these national stations were to be auctioned in the near future, I would be willing to bet the noble Lord who is opposing that Clause 31 shall stand part of the Bill at least a bottle of claret that this licence would be* *secured by a pop music station, and that Classic FM would disappear. I wonder whether the noble Lord has taken into account that possibility in his proposal.*

Lord Young of Norwood Green: *My Lords, key to supporting the drive to digital is to encourage and to allow broadcasters to invest in their digital futures. Experience shows that licence renewals, which are linked to the provision of a digital service, are a key incentive. At a time when the Government are asking the industry to contribute to a focused and intense drive towards digital, we believe that it would be wrong to remove this incentive. Clause 31, alongside Clause 32, would allow Ofcom to grant a further renewal period of up to seven years to analogue licence holders who also provide a digital service. Clause 31 relates specifically to the national analogue licences, although the rationale for the decision for extending the renewal is identical for both national and local licences. I do not want to take up too much time because noble Lords who have contributed to this debate have put many of the arguments excellently. The noble Lord, Lord Howard, talked about the necessity to maintain the clause. The right reverend Prelate displayed a very catholic — I hope he does not mind me using the word — taste in music from Beethoven to Bon Jovi, which I liked. In his analysis of the need for Clause 31, he is absolutely right. As he said, we cannot expect companies to carry on writing blank cheques. We need to give them an incentive. My noble friend Lord Eatwell's analysis of Classic FM was exceedingly apposite. We believe that this clause is essential for the reasons stated by a number of noble Lords. In those circumstances, I support the Motion that this clause stands part of the Bill.*

Lord Clement-Jones: *My Lords, I thank the Minister for that reply. I also thank other noble Lords for contributing to the debate with some fairly bloodcurdling prospects. However, I do not think that the Minister has answered the question about why these extensions are required. I put this proposal somewhat as a devil's advocate. By and large, I believe that the majority of the radio industry is behind the scheme as put forward by the Government, but there is a significant minority of interest which is not. That is why I put forward the clause stand part debate. But if I was in their shoes, listening to what the Minister had to say, I would consider that his arguments were entirely circular and that the Government have done this because they needed to and that this was the best way forward. I do not think that any real forensic argument has been put forward by the Minister. I could probably put forward rather better arguments than the Minister has. I certainly could have put my finger on areas where investment is needed, since I have been briefed by some of the major radio players. The Minister has been extremely half-hearted in responding. This is the one bit of this Bill which is the Government's opportunity to set out their stall in terms of their digital radio policy, other than the amendments we have already dealt with. We had quite a useful debate on our last Committee day, but the Minister has not really answered the questions in a robust way. Certainly, he has not set out the stall for the Government's policy in terms of the extensions of these national analogue radio stations. We are talking about digital radio switchover. What is it about these extensions that will make those radio stations invest more when they migrate to digital? That is what it is all about. The Minister did not even attempt to talk about the amount of money that the Treasury would forgo. Some estimates have put that as high as £73 million, which is a large amount of money. I do not think that the Minister dealt with that either. The Minister has been extremely disappointing. I do not think that that minority of radio stations will be particularly happy to hear the Minister's lack of engagement with their arguments. It is almost as if he has taken a view that only a minority of radio stations is concerned, that the bulk of the radio industry is quite happy and that therefore that minority will be overridden without so much as a buy your leave. That is an unfortunate position to be in. This House, above all, is about rational debate and about putting forward the arguments. To be frank, in previous amendments to this clause, the*

Minister put forward some useful points — he certainly did in response to some of mine — but when I have tried to elicit an overarching policy, he has been lacking and I have been somewhat disappointed.

Clause 31 agreed.
Clause 32 agreed.
Amendment 241B not moved.
Clauses 33 and 34 agreed.

Clause 35 : Local radio multiplex services: frequency and licensed area

Amendment 241C
Moved by Baroness Howe of Idlicote
241C: Clause 35, page 39, line 3, leave out "local"

Baroness Howe of Idlicote: *My Lords, this amendment, which relates to the provisions for digital radio, seeks to allow for the efficient use of the radio spectrum and for a potential increase in radio listening choice for the people of Northern Ireland. Although national BBC services are available via digital radio in all four parts of the United Kingdom, the national commercial multiplex is unavailable in Northern Ireland. The reasons for that are historical and technical, and relate to how the same frequencies were used in the Republic of Ireland. The result is that stations, including Absolute Radio, Planet Rock, BFBS radio and Premier Christian Radio, cannot be heard digitally in Northern Ireland. There is some hope that the spectrum position will change. However, as currently worded, even if that spectrum were to become available, Ofcom would not have the powers to allow it to be used by the national commercial multiplex. That is clearly an anomaly and, I suspect, an oversight. It would result in the inefficient use of spectrum and an artificial restriction on the radio-listening choice for some citizens. This amendment seeks to correct the situation and, without obliging, would enable Ofcom to increase the coverage of the national commercial multiplex. Were this to become technically possible, Ofcom would follow the process already proposed for similar expansion of local digital radio or multiplexes using the framework already in the Bill. This amendment, while modest and not contentious, will have benefits for the people of Northern Ireland and clearly will be welcomed by the radio industry, so I hope that the Government will be prepared to accept it. I beg to move.*

Lord Young of Norwood Green: *My Lords, this amendment would allow Ofcom to vary the frequency or licensed area of national, as well as local, radio multiplex licences. On the face of it, this is not an unreasonable change and would potentially enable the national commercial radio multiplex to extend its coverage to Northern Ireland. However, Clause 35 was structured specifically with reference to local radio multiplexes so as to allow them to merge or be extended in order to close the gaps in local radio multiplex coverage in the UK not currently served by DAB. Simply removing the word "local" from the text may not be the best way to achieve the desired result. Consideration needs to be given to what variation powers Ofcom should have with regard to national multiplex licences and to the basis on which such powers should be exercised. We have some sympathy with what the noble Baroness is trying to achieve and the Government will consider this issue before Report. With that assurance, I hope that the noble Baroness will feel able to withdraw the amendment.*

Baroness Howe of Idlicote: *My Lords, I am pleased to hear that, even if this amendment is not entirely appropriate according to the Minister, serious consideration is going to be given to how this can be made possible. Under those circumstances, I beg leave to withdraw.*

Amendment 241C withdrawn.
Amendments 241D to 241F not moved.
Clause 35 agreed.

Clause 36 : Renewal of radio multiplex licences
Debate on whether Clause 36 should stand part of the Bill.

Lord Clement-Jones: *My Lords, Clause 36 deals with the renewal of radio multiplex licences and it inserts a new Section 58A after Section 58 of the Broadcasting Act 1996. The House of Lords Delegated Powers and Regulatory Reform Committee, which we always listen to with some respect, had some interesting words to say about this clause: "It is impossible to tell from the Bill whether the policy is that the licences should or should not be renewable at all, let alone for what period or on what grounds. Indeed, paragraph 56 of the memorandum candidly admits that the relevant policy decision has yet to be made. We draw attention to the skeletal nature of the power in clause 36, to enable the House to examine it further and determine whether it is justifiable in this context". I am merely a humble hand maiden of this House in tabling this clause stand part debate, and I hope that the Minister can give us further enlightenment.*

Lord Young of Norwood Green: *I have never had to respond to a hand maiden before in this House. I am still wrestling with that analogy. The Government stated in the Digital Britain White Paper that we would work with the industry to agree a plan to build out the DAB infrastructure to current FM coverage. We recognise the need to limit as much as possible the impact of such build-out on radio stations. One way this can be achieved is to allow multiplex operators to spread the cost of the investment in the new infrastructure by extending the period of their licence. We have suggested that licences could be extended up to 2030. The renewal of multiplex licences as a means to support digital radio was first introduced in the Broadcasting Act 1996. However, these renewal powers only apply to licences which were granted within 10 years of the 1996 Act coming into force. Therefore, there are a number of multiplex licences which are currently not eligible for a renewal. If renewals are to provide a real support to the build-out of DAB coverage to FM levels, they need the flexibility to achieve three objectives: first, to allow the extension of the licence period for those licences which are already eligible for, and in some cases have already been awarded, a renewal under the existing terms; secondly, to allow the renewal to apply to all multiplex licences, including those not currently eligible within the existing provisions; and thirdly, to ensure that any further renewals are awarded with conditions which link them to the progress to digital radio switchover, and more specifically to an agreed build-out plan and timetable. The link to a DAB coverage plan for switchover, which is likely to take a year to agree, is why we believe these powers are most appropriately applied via an affirmative order. I note concerns about the breadth of the order-making powers and I hope that I have satisfied noble Lords that they are justified because of the range of changes needed to implement this policy.*

Lord Clement-Jones: *I thank the Minister for that brief but — I hope to discover on reading Hansard — informative statement. As somebody who is not fully conversant with the radio multiplex licence variations, that was not the clearest possible answer I could have asked for.*

I hope that it will make sense on further consideration. It seemed to tell me that the Government need the maximum possible flexibility without having determined exactly which licences require extension. I am not sure that takes us a great deal further than what the House of Lords Delegated Powers and Regulatory Reform Committee said, but perhaps, as I say, on reading Hansard it will all become blindingly obvious.

Clause 36 agreed.
Clause 37 agreed.

---

[429] http://www.publications.parliament.uk/pa/ld200910/ldhansrd/text/100208-0007.htm

82.

10 February 2010

Digital radio stations: listeners abandon ship

The latest RAJAR radio audience data demonstrated one thing clearly: the UK radio industry's strategy for its digital stations is in tatters. Audiences for digital radio fell off a cliff during the last quarter of 2009. This did not appear to be the result of any specific strategy shift (no station closures, only one minor format change) but more the result of increasing public malaise about the whole DAB platform and the radio content that is presently being offered on it (plus a little Q4 seasonality). The figures speak for themselves.

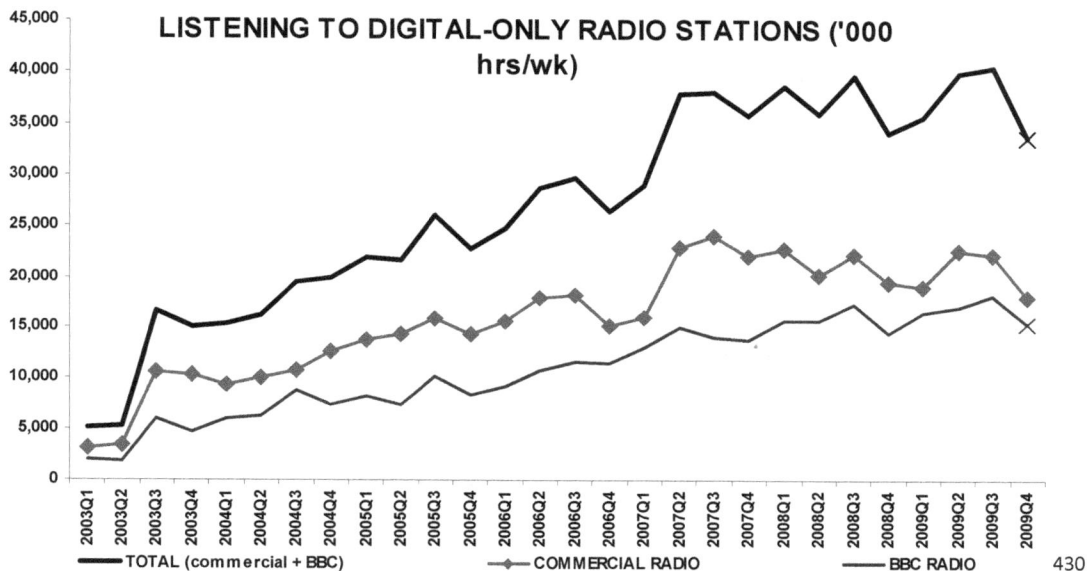

LISTENING TO DIGITAL-ONLY RADIO STATIONS ('000 hrs/wk)

TOTAL (commercial + BBC)   COMMERCIAL RADIO   BBC RADIO   430

Total listening to digital radio stations is back down to the level it achieved in 2007, following a period of sustained growth between 2000 and 2007. Far from moving towards some kind of exponential growth spurt as the industry had expected, total listening now seems to have plateau-ed. It appears that market saturation has already been reached for much of the content presently available on digital radio platforms, considerably earlier than had been anticipated, and at a level of listening that cannot justify these stations' existences for their commercial or BBC owners.

In the commercial sector, only Planet Rock has maintained its momentum, probably a reflection of its commitment to offering its listeners genuinely unique content. Elsewhere, the jukebox music stations have suffered massive falls in listening, possibly a result of their ease of substitution by online offerings such as Spotify and Last.fm, and of owner Bauer's policy to curb investment in digital radio broadcast platforms and content.

253

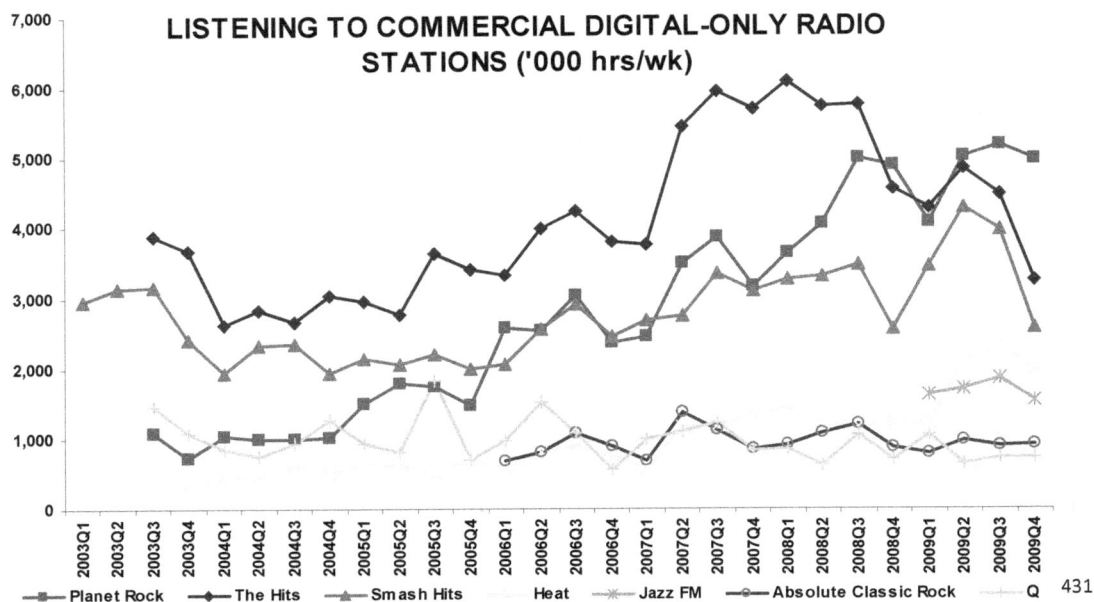

LISTENING TO COMMERCIAL DIGITAL-ONLY RADIO STATIONS ('000 hrs/wk)

Planet Rock · The Hits · Smash Hits · Heat · Jazz FM · Absolute Classic Rock · Q

431

Commercial radio has talked the digital talk for years about striving to make DAB a successful platform, vaguely promising new digital radio 'content' that it has still not delivered. Instead, it has spent the last few years cutting costs, consolidating, lobbying the government, complaining about the BBC, closing its digital stations and contracting out its DAB capacity to marginalised broadcasters (religious, ethnic, government-funded and listener-supported stations) that will never attract mainstream audiences to the platform (and whose listening is not even measured in the RAJAR audience survey).

From the listener's perspective, the only thing that has happened to the DAB platform in recent years is the disappearance of commercial digital stations such as OneWord, TheJazz, Core, Capital Life and Virgin Radio Groove. For the average consumer, the arrival of Traffic Radio, Premier Christian Radio or British Forces Broadcasting Service are hardly replacements.

A report commissioned by RadioCentre from Ingenious Consulting in January 2009 concluded:

*"Commercial radio is now at a crossroads with respect to DAB. It needs **either** to accept that the commercial challenges of DAB are insuperable and retreat from it – such a retreat, because of contractual and regulatory commitments, would be slow and painful; or strongly drive to digital."*

In the year since this report was prepared, commercial radio has done neither. Instead, it has spent a small fortune on parliamentary lobbying, not one iota of which has had a direct impact on 10 million increasingly baffled DAB radio receiver owners. These latest RAJAR data convey their clear message that content is their only concern.

For the BBC, the problem is somewhat similar. With the exception of Radio 7, listening to its digital radio stations remains unimpressive, despite them benefiting from massive BBC cross-promotion over many years. Some stations are outright disasters – Asian Network is listened to less now than it was almost seven years ago, when only 158,000 DAB radios had been sold. Some stations are simply not suited to the DAB platform – 1Xtra targets a youth audience who listen to a lot of radio online and via digital TV, but who have little interest in

DAB (particularly as DAB is not available in mobile phones). Some stations will become redundant in an increasingly on-demand world – Radio 7 would eventually be little more than a shopfront for the huge pick'n'mix BBC radio archive to be made available to consumers online.

LISTENING TO BBC DIGITAL-ONLY RADIO STATIONS ('000 hrs/wk)

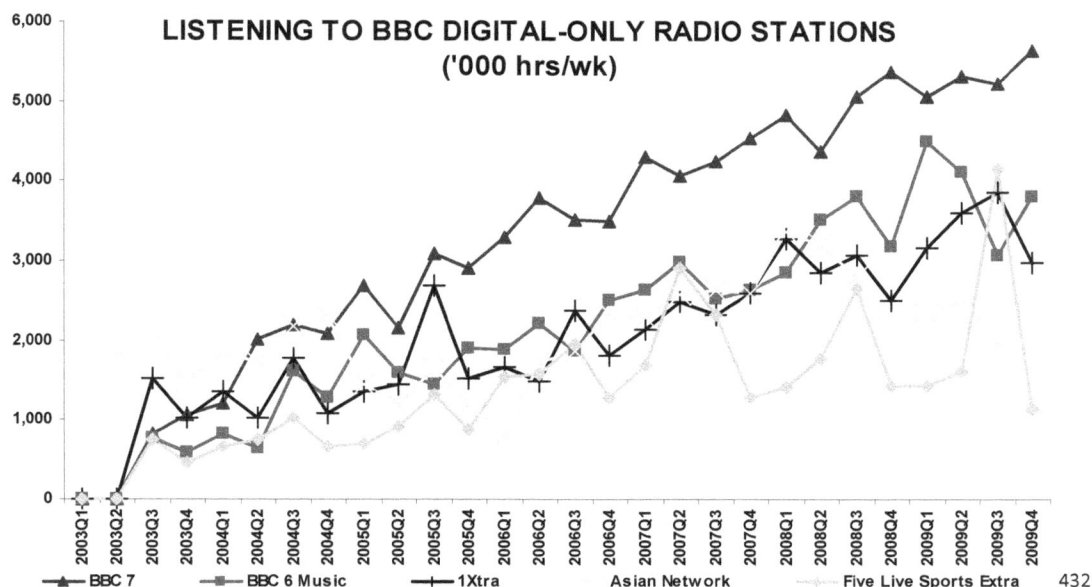

For the BBC, it is becoming increasingly hard to justify spending, for example, £12.1m per annum on the Asian Network when its peak audience nationally is only 31,000 adults. Broadcast platforms such as FM attract huge audiences for a fixed cost, making them the most efficient distribution system for mass market live content. As a result, Radio 1 costs us only 0.6p per listener hour. By comparison, the Asian Network is costing 6.9p per listener hour, probably making it more expensive to 'broadcast' than to send each listener a weekly e-mail attaching the five hours of Asian Network shows they enjoy.

The BBC should still be congratulated for creating new digital radio services in 2002 that attempted to fill very specific gaps in the market which commercial radio was unlikely to ever find commercially attractive. This is precisely why we value a public broadcaster in the UK. However, the BBC digital radio strategy over the last decade has suffered from:

- The BBC's evident inability to successfully execute the launch of genuinely creative, innovative radio channels that connect with listeners (GLR, the 'new' Radio 1, the original Radio 5)
- The BBC pre-occupation with constantly creating new 'broadcast channels' when most niche content is more suited to narrowcasting and delivery to its audience via IP (live, on-demand or downloaded).

For the UK radio industry, its digital 'moment of truth' has belatedly arrived. A new strategy now has to be adopted which does not continue to raise the DAB platform to the level of a 'god' that has to be worshipped above all others. The future of radio is inevitably multiple-platform and the industry's focus has to be returned to producing content, rather than trying to control the platforms on which that content is carried.

I suspect that Tim Davie, director of BBC Audio & Music, will eventually lead these winds of change, following in the wake of director general Mark Thompson's pronouncements at the end of this month as to where the internal financial axe will fall. Where the BBC leads, commercial radio will inevitably (have to) follow.

The future digital radio strategy is likely to be 'horses for courses'. Rather than all radio content being delivered via all available platforms, it will in future be delivered only where, how and when it is most demanded by listeners. Our economic times make this mandatory. The DAB platform's mass market failure will make it necessary.

---

[430] source: RAJAR
[431] source: RAJAR
[432] source: RAJAR

83.

12 February 2010

Norway: every fifth radio sold is an internet radio, every eleventh is DAB

In Norway, sales of internet radio receivers are booming. During the final quarter of 2009, 22% of radios sold were internet radios, up from only 1% a year earlier, according to Norwegian website Sandnesavisen.[433] By comparison, only 66,000 DAB radios were sold in 2009, 9% of the total 729,000 radios sold during the period, according to data from the Electronics Industry Foundation.[434] Additionally, 200,000 mobile phones were sold in Norway during 2009, none with DAB capability but many with integrated FM radios.

Erik Andersen, information officer for the Electronics Industry Foundation, said: *"We estimate that there are somewhere between 12 and 15 million FM radios that are in regular use around the country while, in comparison, only approximately 290,000 DAB radios have been sold in recent years."*[435]

Andersen is not surprised by the rather low DAB radio sales, and finds it problematic that a date for FM band switch-off has not been announced which could be referred to. *"We have no desire to mislead customers so, as long as politicians do not give us a switch-off plan, we advise enquiring customers to buy an FM radio,"* said Andersen.[436]

Jarle Ruud, acting general manager of Digitalradio Norge, the organisation charged with ensuring a speedy and smooth transition from analogue to digital radio in Norway, said: *"This is a classic 'chicken and egg' situation, both in terms of sales volumes versus a switch-off date, and the channel selection on DAB."*[437] Just as consumers are hesitating to buy a DAB radio before the government announces a switch-off date, Ruud thinks there are many radio stations that refuse to invest in digital transmission equipment because of the potentially lengthy and costly period of dual distribution.

The sales trend came as no surprise to Øyvind Vasaasen, media manager at NRK (state broadcasting) and chairman of Digitalradio Norge, who said: *"This is what we predicted. We have remarked to the authorities that sales appears to have stabilised at around 60,000 DAB radios sold per year, and this is also the result for 2009. It shows that the sales trend is far too slow."*[438]

Geir Friberg is managing director of Norway's largest local radio group, Jærradiogruppen, which owns more than 20 of the approximately 150 local stations in Norway. He said: *"There is really no great difference between FM and DAB. There are too few radio stations that broadcast on DAB today, and listeners cannot hear the difference between FM and DAB."*[439] Friberg believes that adding more FM stations may even be a better policy than DAB.

Per Morten Hoff, general secretary of IKT Norge (Norway's IT industry interest group), queried the accuracy of the published sales figures: *"More and more consumers have eyes for internet radios, and the Norwegian company TT-Micro has alone sold 20,000 web radios.*

Since these radios can also receive FM, DAB and 13,000 internet stations, they are strangely classified as 'DAB radios' in the statistics. This is apparently to show that DAB is not a total failure. The true DAB sales are probably no higher than 40,000 units, and that figure is getting smaller. There is also the riddle that sales of mini-TV's are included in the sales statistics of DAB radios."[440]

Jørn Jensen, president of World DMB Forum, responded: *"Hoff does not know what he's talking about. Radio broadcasting on DAB and DAB+, as well as mini-TV's in DMB, all use DAB technology."*[441]

At the beginning of February 2010, the Norwegian media regulator, Medietilsynet, submitted a 145-page document to the government on the digitisation of radio broadcasting. This is in preparation for a White Paper on digital radio that will be published by the government during 2010. The report was commissioned to map the Norwegian local radio industry's opinions on the migration to digital radio and how it could be achieved. It collated responses from 55 parties and it concluded:

"The results of the survey reveal that virtually all local radio licensees believe that the digitisation of local radio will have economic consequences for them as businesses. The players fear that a transition will be so costly that only a few of the current licensees will survive digitisation. The consequence may be that ownership diversity disappears and that the market is left with big group operators. Besides mentioning competence-building, information and predictability, the majority of licensees believe that the most important measure the government can contribute to the digitising process is financial support in one form or another.

Local radio licensees who participated in the survey are roughly evenly split down the middle on their view of whether local radio should be digitised or not. Those who are negative to digitisation weigh the economic aspect heavily. When asked what distribution platform they see for themselves as the primary platform for local radio in the future, 52 percent responded analogue broadcasting via the FM band, 33 percent broadcasting via a digital platform such as DAB, DRM, etc and 9 percent broadcasting via the internet......"

"Amongst local radio licensees, 78 percent are negative about a switch-off date for analogue local radio broadcasting. 18 percent are positive about a switch-off date."[442]

Vigleik Brekke, chairman of the Norwegian Association of Local Radio (NLR) said that there was no economic reason for many of his 150 member stations to migrate to digital transmission. *"Closing FM radio will create problems for stations,"* he said. *"They will not be able to survive."*[443]

Professor Lars Nyre of the University of Bergen's Institute of Media Studies said he was sceptical about DAB, especially as FM radio seemed to be perfect for listeners.[444]

Asked whether the government's White Paper will include a switchover date, Norwegian Culture Minister Roger Solheim said: *"This is an important issue, and we aim to present the facts in 2010. One of the key questions we want to work with in the White Paper is whether we should fix a switch-off date."* Asked if his predecessor Trond Giske's policy still held sway that half of Norway's households should own a digital radio before a switch-off date is announced, Solheim said: *"It is certain that digital radio will come, we see that too, in the developments. The question of the timing, and the state's role in it, we must come back to in the White Paper."*[445]

According to industry data, one in three households in Norway buys a new radio each year. Each household owns between 6 and 7 radios.[446]

[433] http://www.sandnesavisen.no/index.php?option=com_content&task=view&id=1856&Itemid=149
[434] http://www.aftenposten.no/forbruker/digital/article3513446.ece
[435] http://www.aftenposten.no/forbruker/digital/article3513446.ece
[436] http://www.aftenposten.no/forbruker/digital/article3513446.ece
[437] http://www.aftenposten.no/forbruker/digital/article3513446.ece
[438] http://www.nrk.no/nyheter/kultur/1.6974994
[439] http://www.nrk.no/nyheter/kultur/1.6974994
[440] http://www.tu.no/it/article236046.ece
[441] http://www.tu.no/it/article236098.ece
[442] http://www.medietilsynet.no/Documents/Rapporter/100205_Digitalisering_lokalradio_min.pdf
[443] http://www.journalisten.no/story/60379
[444] http://www.journalisten.no/story/60379
[445] http://www.aftenposten.no/forbruker/digital/article3513446.ece
[446] http://www.tu.no/forbruker/article235674.ece

84.

19 February 2010

European commercial radio trade group says 'no' to universal FM switch-off date

At the start of its annual conference held in Brussels on 11/12 February 2010, the Association of European Radios [AER], the trade group representing 4,500+ European commercial radio stations (including RadioCentre members in the UK), issued the following press statement:

"AER considers that setting a date for the switch-off of analogue radio services is currently impossible. Indeed, the question of which kind of technology will be used should be solved first. Hence, broadcasting in FM and AM shall remain the primary means of transmission available for radios in all countries, with the possibility to simulcast in digital technology, until market developments enable a potential time-frame for general digitisation of radio. Transition to any digital broadcasting system should benefit from a long time-frame, unless there is industry agreement at national level to move at a faster rate."[447]

This statement followed on from a policy paper the Association had published a few days earlier, responding to the European Commission's Radio Spectrum Policy Group plans for its draft Work Programme. The paper said, in part:

"It should be underlined that, in most of Europe, currently and for the foreseeable future, there is only one viable business model: free-to-air FM broadcasting on Band II. Thus, Band II is the frequency range between 87.5-108 MHz and only represents 20.5 MHz. Across Europe, nearly every single frequency is used in this bandwidth. Thanks to the broad receiver penetration and the very high usage by the listeners, this small bandwidth is very efficiently used. On-air or internet-based commercially-funded digital radio has indeed not yet achieved widespread take up across European territories. These two means of transmission will be part of the patchwork of transmission techniques for commercially-funded radios in the future, but it is hard to foresee when.

So no universal switch-off date for analogue broadcasting services can currently be envisaged and decision on standards to be used for digital radio broadcasting should be left to the industry on a country-by-country basis.

Radio's audience is first and foremost local or regional. Moreover, spectrum is currently efficiently managed by European states and this should remain the case: national radio frequency landscapes and national radio broadcasting markets are different, with divergent plans for digitization, diverse social, cultural and historical characteristics and with distinct market structures and needs….."

"Finally, AER would like to recall that European radios can only broadcast programmes free of charge to millions of European citizens, thanks to the revenues they collect by means of advertising. These revenues are decreasing all through Europe due to two factors: the shift

towards internet-based advertising, and the recent financial crisis. For 2009, radio advertising market shares were forecast to decrease by 3 to 20% all across Europe, compared to 2008.

In some countries (e.g. France and the UK), a part of the revenues derived from the TV digital switchover was supposed to be allocated to the support of digitisation schemes for radio. This is no longer the case. In most countries, it is still unclear who will bear the costs of the digitisation process.

However, any shift towards digital radio broadcasting entails very long-lasting and burdensome investments. Nevertheless, some individual nations may wish to proceed with a move to greater digital broadcasting at a faster rate, as there will be no 'one-size-fits-all' approach.

So, any shift towards digital radio broadcasting will most likely require a very long process. Decision on the adequate time-frame should be left to each national industry: as a matter of principle, transition to any improved digital broadcasting system should benefit from a long time-frame, unless there is industry agreement to move at a faster rate.

It should also be recalled that commercially-funded radios are SMEs, and are in no position to compete for access to spectrum with other market players. So, market-based approaches to spectrum (such as service neutrality or secondary trading) should not be enforced in bands where commercially-funded radios broadcast or may broadcast.

To end up with, AER would like to recall that, in most of Europe, currently and for the foreseeable future, there is only one viable business model: free-to-air FM broadcasting on Band II; hence:

• no universal switch-off date for analogue broadcasting services can currently be envisaged
• now and for a foreseeable future, commercially-funded radios need guaranteed access to spectrum, in all bands described above. Besides, no further change to the GE 84 plan [the 1984 Geneva FM radio broadcast frequencies agreement] should be suggested, but the plan should be applied with consideration to the technological development (and its enlarged scope of possibilities) throughout the past 25 years
• any shift towards digital radio broadcasting should benefit from a long and adequate time-frame. "[448]

---

http://www.followthemedia.com/newsfromyou/newsfromyou20022010.htm?PHPSESSID=f36fb0ea4e0e430f1ff5f141c79025f2
[448] http://www.aereurope.org/content/view/693/43/lang,en_GB

85.

23 February 2010

Denmark: government contemplates DAB radio's future

Politicians in Denmark are presently considering what to do about digital terrestrial radio. The next funding period for state radio runs from the beginning of 2011 to the end of 2014 and will have to take into consideration a political decision taken in June 2009 to forge ahead with the migration from analogue radio to the DAB platform. Presently, only state radio stations are available on DAB.

The governing Liberal Party says it *"wants to create real competition in the radio market"* because state broadcaster Danmarks Radio has *"a de facto monopoly over radio in Denmark and so the diversity and choice in radio remains limited."* The party decided in 2009 that it would support FM switch-off as a means to create a *"new start"* for radio in Denmark. To ensure that DAB will prove an attractive investment proposition for commercial radio, FM would be switched off between 2016 and 2018. The Liberal policy document states:

"An expansion of DAB radio allows for a potentially very large number of different radio channels. This expansion is in full swing in all major European countries, several of which already offer between 60 and 100 digital radio channels, while Denmark currently only offers 20 channels in total. The technological opportunities need to be exploited for the benefit of listeners. And development of DAB also provides the opportunity to create a balanced radio market, where there are real opportunities for development of commercial radio, which has proven not to be possible on the FM band. We should therefore concentrate on the offensive deployment of DAB. England has the full support of all radio operators in the country who decided to work for the switch to DAB in 2015. Recently, France decided that DAB radios must be in all French cars from 2014 and, in March 2009, the German Broadcasting Commission decided to roll out DAB+ in Germany. Although progress has been slow, and in some places has become completely stagnant, there is no longer the same uncertainty that DAB (Eureka147 standard) is the future of radio."[449]

The Liberal Party document goes on to cite the 'success' of DAB in the UK and mistakenly believes that UK government policy is to switch off FM (a fallacy shared by many):

"Today, around 8% of total radio listening [in Denmark] is on DAB. In England, that figure is 12.7% and soaring. To ensure that DAB is an attractive platform for both listeners and investors, it is important that DAB radios are made attractive by adding more unique content. This is the same set of conditions which must be present in other countries that anticipate the same trends. For example, Norway has estimated that at least half of the population must have a DAB radio before [analogue] switch-off can be announced. England will switch to DAB in 2015. Moreover, it is recommended that all cars sold in England from 2013 must be equipped with DAB radios. This decision is backed by both the public service and commercial radio stations. The assumption is that DAB listening will reach 50% in 2015 and coverage will reach 90% of the country and all main roads. Lord Carter's report 'Digital

Britain' in January 2009 also defined a set of criteria that must be met in order to close [analogue radio] permanently."[450]

Ellen Trane Nørby, media spokesperson for the Liberals, said: *"Danmarks Radio is proposing that we switch off analogue radio in 2015, but they offer no viable way to get Danes to listen more to DAB."* Nørby believes that the biggest driver for DAB take-up will be when state radio channels are progressively removed from FM: *"The people who listen to DAB are heavy listeners to the channels they cannot find anywhere else. If the offerings on DAB are not unique, why would you buy a DAB radio?"*[451]

Nørby said that the transition to DAB needs to adopt a realistic timescale: *"Today, we are in a situation where most listening takes place on FM and each of us has about five FM radio receivers, compared to only 1.5 million DAB radios [sold in total]. Therefore, it is important that we have a real debate over the future of digital radio, and how we progress safely from FM to DAB. The honest discussion we propose is in contrast to others who say simply that we must switch off [analogue radio] but do not explain how."*[452]

Although the Liberal policy document is resolute about digital radio switchover, Nørby appears more circumspect in interviews. In December 2009, she said: *"It is not enough simply to say that FM will be switched off in 2015 without considering a discussion on content. …. Driving the take-up of digital television was the Danish desire to have a flat-screen TV. I have seen no Christmas boom – and we have not seen this for many years – in the sales of DAB radios. That's why we Liberals have been so critical of the migration plan, because DAB has not won as much grassroots support as you could ask for."*[453]

The government proposals drew a critical response from Michael Christiansen, chairman of Danmarks Radio. He said: *"The Liberal Party proposal is a barely disguised attempt to destroy Danmarks Radio's stations. … The proposal totally disregards modern Danish music by moving the P3 channel over to DAB at a time when it its coverage is not enough. … Replacing P3 [on FM] with a more or less indifferent foreign [commercial] radio channel, I cannot take seriously. We want to help the drive towards DAB, just as we have driven the migration to digital TV. But no one thought of closing down DR1 or DR2 [TV stations] on analogue until digital TV signals were available [to all]. The notion is so far out that I find it hard to relate to."*[454]

---

[449] http://www.venstre.dk/fileadmin/venstre.dk/main/files/oplaeg/Radiosiderne_medieoplaeg.pdf
[450] http://www.venstre.dk/fileadmin/venstre.dk/main/files/oplaeg/Radiosiderne_medieoplaeg.pdf
[451] http://www.business.dk/medier-reklamer/venstre-vil-slukke-p3-paa-fm
[452] http://mediawatch.dk/artikel/v-dab-overgang-skal-vare-ni-aar
[453] http://politiken.dk/kultur/tvogradio/article866331.ece
[454] http://www.berlingske.dk/kultur/venstre-vil-oedelaegge-drs-radiokanaler

86.

25 February 2010

Digital radio switchover: a Broadcasting Minister's last day: "there is nothing that I am trying not to say"

Apologies: this is an extremely long chapter. I suggest it is worth persevering with because, more than any other ministerial statement, the transcript below illustrates perfectly the government's mistaken determination to press ahead with its policy to adopt DAB radio. During the weeks it has been sitting, the House of Lords Communications Committee has become increasingly adept at understanding the flaws and contradictions in the Digital Economy Bill's radio clauses, which is why their dialogue here with the Minister crackles with suspicion.

Siôn Simon handles his government script with the aplomb of a dodgy used car salesman, combined with the smug self-satisfaction of a schoolboy who can proffer a verbal comeback to any question asked of him. In December 2009, Simon had to repay £20,000 in parliamentary expenses for a second home he was renting from his sister between 2003 and 2007. Embarrassingly, the Broadcasting Minister reveals here his mistaken belief that the government's proposed 2015 digital radio switchover date is written into the Digital Economy Bill. It isn't.

I could itemise the many flaws in the Minister's replies to the Committee's questions, but it would spoil your reading fun. Your starter for ten – Halfords does not stock DAB radios.

House of Lords
Select Committee on Communications
"Digital Switchover of Television and Radio In The UK"
10 February 2010 [excerpt][455]

Witnesses:
Mr Siôn Simon, a Member of the House of Commons, Minister of State for Creative Industries
Mr Keith Smith, Deputy Director, Media in the UK, Department for Culture, Media and Sport

Chairman: *Good morning. Thank you very much for coming, you are very welcome to the Committee. Now, Mr Simon, you are the Broadcasting Minister.*

Mr Simon: *For the moment, until the end of today.*

Chairman: *Well, that is what I was going to raise with you. How long have you been in this post?*

Mr Simon: *I have been in this post since, I think, the beginning of June last year and today is my last day, so this will be my final appearance before a Select Committee and I am looking forward to it tremendously.*

Chairman: *Well, before you look forward to it too immensely, let me ask you a question, and I ask the question because, on my reckoning, there have now been five broadcasting ministers since May 2006. That is a very high turnover. Do you think that these frequent changes help in making policy?*

Mr Simon: *I think I should have been appointed in 2005, but ----*

Chairman: *But Number Ten actually omitted to do so?*

Mr Simon: *They omitted to do so and, as you know, you will have to ask the Prime Minister about the appointment of ministers because that is his decision, not mine.*

Chairman: *Well, the resignation of ministers, on the other hand, is very much in the hands of ministers, is it not? You are not even going to take the Digital Economy Bill through?*

Mr Simon: *I have had great enjoyment and satisfaction dealing with it as it has been through all the various stages of consultation with stakeholders across industry and Parliament since I inherited the White Paper. The White Paper was published the week I was appointed, the Digital Britain White Paper.*

Chairman: *By your predecessor who also only lasted a year.*

Mr Simon: *I think he did an outstanding job actually, I must say. Having come in and inherited his output, I think Lord Carter did a very, very sophisticated, subtle and impressive job across a whole range of industry sectors.*

Chairman: *At any rate, you enjoyed it so much that you are not going to stay for a couple of months?*

Mr Simon: *It has been an absolutely tremendous pleasure and privilege. I think it is one of the best jobs in Government. I have lots of reasons that I am stepping down as I am deciding to do other things and I am also stepping down from Parliament.*

Chairman: *What are the other things you are going to do?*

Mr Simon: *I am going to run, if I can, to be the first elected Mayor of Birmingham, which will probably be a couple of years, but is a job worth doing and worth planning for.*

Chairman: *Well, as a Birmingham MP for 27 years, I am probably the person round this Committee who has most sympathy with that ambition, although I do not think the job actually exists, let alone you are going to win it. Quite seriously, do you really think that these swift changeovers of ministers actually do the policy process any advantage?*

Mr Simon: *I do seriously have to tell you that I am a junior minister and junior ministers, as you know with all your experience, are not responsible for the appointment of ministers, the policy towards the appointment of ministers or the length of tenure of ministers. You need to take that up with the Prime Minister; it is not a matter for me.*

Chairman: *Well, I doubt very much whether the Prime Minister is going to agree to come to this Committee, given all the other things he has got on his mind just at this period in Government, but never mind.*

Lord Maxton: *Do we know who your successor is going to be?*

Mr Simon: *I do not know.*

Lord Maxton: *Presumably, somebody will have to be appointed this afternoon if you are going this afternoon?*

Mr Simon: *Again, it is a matter for Downing Street, not for me or officials.*

Chairman: *So you are going this evening and we do not know who is going to take your place?*

Mr Simon: *I think it is fairly common procedure that it is after the one minister has resigned that the next one is appointed, so I assume there will be an announcement from the Prime Minister in due course about what arrangements will need to be made.*

Chairman: *Okay, I do not think we are going to get much further on this particular path. Let us ask you about the digital switchover. This is a short inquiry that we have conducted and most complaint, I think, has been about radio. There seems to be enormous uncertainty amongst the public about what is happening and indeed what the case for radio switchover is. We have just been talking to consumer groups and, just to give you some flavour of what they said initially, one said he could not see any advantage, another said it was difficult to see what the benefits are and a third said that they were concerned about the expense of throwing away sets, and I think that is one of the issues that has come through from some of our correspondence as well. What would you say are the advantages of digital switchover as far as radio is concerned?*

Mr Simon: *I think there are several advantages. Firstly, digital has practical and technical advantages of benefit to the consumer, so there is a whole range of extra functionality and interactivity that you get from a digital radio set that you will not from analogue. It is also the case that the FM infrastructure, the transmission infrastructure of FM, is ageing, FM is an old analogue technology, and the likelihood is that in the medium term the question will arise anyway of whether this infrastructure can economically be renewed and the likelihood is that it probably would not be economic to renew this infrastructure. What you would be faced with in that case would be a piecemeal disintegration of the FM infrastructure in a disorderly way and an inevitable move by the market towards digital. What the Government is, therefore, doing is trying to help manage this move in an orderly and efficient way.*

Chairman: *So what would you say to the member of the public who has written a not untypical letter to us in which he said, "We have acquired a large number of FM radios over the years, all of which work perfectly. Five of these are used regularly in different rooms. Why should we ditch these for no good reason?"*

Mr Simon: *Well, firstly, they will not necessarily have to ditch them; it will depend what services they are using them to listen to, and it may be that there are different members of the family using different sets in different rooms at different times to listen to different services. It is reasonable to assume, therefore, that in a typical family some of those sets in* *some of those rooms will be used to listen to the kinds of local commercial services or community radio services which will remain on FM, indeed which will be expanded and have their presence secured on FM and remain available. Now, clearly it will be the case that, in order to listen to major national or large regional broadcasters after 2015, consumers will need some kind of upgrade to listen to digital. It is very likely that a relatively cheap, small add-on which converts an analogue set to a digital set will become available before the switchover date. We are talking to manufacturers, and I cannot guarantee what manufacturers will manufacture, but we think it very likely that a small, cheap converter will be available and people will purchase over time new sets.*

Chairman: *Well, we will come to that point. Can I just ask one broader point though than that. Ofcom commissioned a cost-benefit analysis of digital radio migration from Price Waterhouse. That report found that the benefits might outweigh the costs only after 2026. Is that the basis upon which you are planning as well?*

Mr Simon: *I think that the report that Ofcom commissioned was into the recommendations of the Digital Radio Working Group, which were different from those which eventually made it into the Digital Britain White Paper and the Digital Economy Bill, so, to be honest, it is not really a straight comparison because the Price Waterhouse report was not into what is going to happen, it was into a different set of recommendations by a different group.*

Chairman: *I hear what you say, but why has the Price Waterhouse report not been made public?*

Mr Simon: *Clearly, there has been a little bit of difficulty about this. There is no sense at all in which it was intended not to be made public. It is now on the DCMS website. It should have been on the DCMS ----*

Chairman: *In a redacted, blacked-out form.*

Mr Simon: *I believe that the form that it is in on the website is redacted to remove commercially sensitive information, which is the usual practice with commercially sensitive information, but I am told that the Committee has been supplied with an unredacted version by Ofcom, and any stakeholders who have asked for a copy have been sent a copy. We should have put it on the website more quickly. There were technical difficulties to do with the report not having been written internally and being supplied in the wrong format and so on which meant that it did not get on the website. There has been no intent whatsoever, and there is no intention, to keep private the report.*

Chairman: *Well, we, I gather, have received the report this very morning, so we obviously have not looked through it yet.*

Lord St John of Bletso: *I noticed in the written evidence from DCMS that one of the challenges of the digital radio switchover will be converting the occasional radio listener rather than the avid listener who has already invested in DAB sets. There has been also a commitment to a further cost-benefit analysis of the digital radio upgrade. What is the timetable of this proposed new cost-benefit analysis, and will you wait for the outcome of this analysis before taking further decisions to go forward?*

Mr Simon: *I think the commitment with the digital radio upgrade is for a full impact assessment and a cost-benefit analysis particularly of the need for, the case for and the* *design of a possible digital radio help scheme. As to the sense in which there is a cost-benefit analysis of the whole programme, rather than a kind of discrete piece of commissioned work, like the report we were just talking about, this would be a constant, ongoing process of review which is starting this year and will be constantly reviewed, measured and updated as the programme unfolds.*

Lord St John of Bletso: *We have, as I understand it, 90 per cent still on analogue and ten per cent on digital, but have you estimated the effects of the costs and revenues to the different radio stations and the bodies arising out of digital migration, and how will profitability change and who will benefit? That is really what we are trying to get at.*

Mr Simon: *I think the percentage already listening on digital is higher than that, I think it is more like 20 per cent, and we are committed to not switching over until listenership on digital is at least 50 per cent and coverage is at least comparable to FM, which would be 98.5 per cent. Can you just tell me a bit more as I was not quite sure exactly what you wanted in your subsequent question?*

Lord St John of Bletso: *It was just the phasing of the transfer as far as not just the timing of the cost-benefit analysis, but what the effects would be and the costs and revenues for the different radio stations and others.*

Mr Simon: *The intention is that the radio market be much more distinctly than it currently is organised into three distinct tiers, so a national tier at the top, which would be digital, a large regional tier, which would also be digital, and then at the lowest end a local tier, which would be small, local commercial broadcasters and community radio stations who would remain on FM. The effect would be that some currently medium-small commercial broadcasters would probably grow their broadcast areas and migrate on to digital. Some of that size could potentially shrink slightly and stay on FM, the underlying dynamic being that it is much cheaper to stay on FM than go on to digital, and there would be small commercial broadcasters for whom it would not be economic to migrate to digital which has inherently a much bigger footprint.*

Lord Maxton: *I am not quite clear where the BBC would fit into that because they provide national and local services.*

Mr Simon: *If I may say so, my Lord, that is a very good question and the answer is that in the crucial matter of building out extra transmitter infrastructure so that the coverage of digital matches by 2015 the current coverage of FM, which is about 98.5 per cent of the country, the assumption is that the commercial sector and commercial operators would fund that build-out as far as it was commercially and economically viable, and then the assumption is that the BBC, with its obligation to provide a universal service, would fund the probably seven or eight per cent of the build-out which was not commercially viable.*

Lord Maxton: *But they were only talking about 90 per cent, the BBC were last week.*

Chairman: *Is it going from 90 to nearly 100 per cent?*

Mr Simon: *Very roughly. It is between 90 and 98.5 which, it is assumed, would not be commercially viable.*

Chairman: *But you are assuming that the cost of that will be taken from the licence fee?*

Mr Simon: *I am assuming that the BBC would be building those transmitters. We are talking about a cost of probably between £10-20 million a year.*

Chairman: *So it would be taken from the licence fee?*

Mr Simon: *I think the assumption is that the BBC would be able to absorb that within its current budgets.*

Chairman: *Well, the BBC said to us that actually they would like to talk to you about the licence fee, so there may be a constructive dialogue to be had there!*

Lord Maxton: *Obviously, what you are saying does imply a fairly major restructuring of the radio industry. I assume you are going to be consulting on this, are you?*

Mr Simon: *We are, and have been, consulting continuously. I have had two summits with small commercial operators who, at the beginning of the process of talking to them, were a bit concerned.*

Lord Maxton: *Can I link this to one of the Government's other quite right policies, which is extending broadband to everybody. Where does that fit into radio because it seems to me that people are already selling internet-available radios so that you can pick up radio stations from wherever?*

Mr Simon: *It is all part of the same, I think, pretty relentless drive towards digital, but they are different strands of the same broad movement rather than being directly interlinked or dependent, so internet radio listenership forms part of the digital radio listenership, but it actually forms a very small part of the digital radio listenership and there is still a case and a clear need for an explicitly transmitted-over-the-airwaves radio digital output as well; you cannot do it all on the internet.*

Baroness McIntosh of Hudnall: *I just wanted to go back to two things that you said in your last couple of answers, Minister, and the first one was about the 50 per cent target for digital listenership before making the decision to migrate to digital. We had evidence from the witnesses who were here immediately before you that there is a strong likelihood that the 50 per cent of people who, at that point, would not have taken up the digital option may include a high proportion of vulnerable listeners. Do you have any observations, or indeed has your Department done any research, to demonstrate whether or not that is likely to be the case? Also, you made the point about the BBC becoming responsible for making up the shortfall between 90 and 98.5 per cent, and the evidence that they gave us last week demonstrated, or they believe it demonstrates, that the level of investment to produce that last eight to ten per cent's worth of coverage is enormously much greater than the level of investment that has been necessary, or will be necessary, to arrive at 90 per cent. I think, from memory, they were talking about having 90 transmitters currently and needing to invest in a further 140 in order to meet that remaining ten per cent. That is a pretty big ask, and I just wonder whether you would, in the light of that, be prepared to reconsider your answer to Lord Fowler about where the money is going to come from.*

Mr Simon: *If I could take those two, in the first case we have not commissioned research about that yet. Intuitively, I suspect that you are probably right and that there will be a disproportionately high number of, for instance, older, disabled or other vulnerable people in* *the cohort that is not digital by the time we switch over, so, for that reason, we will be commissioning a full impact assessment and cost-benefit analysis, looking into exactly those issues and using that information to determine, and what would be the details of, some kind of digital switchover help scheme in just the same way that we commissioned research which informed the digital TV switchover Help Scheme that we have ultimately put into place and which is, I think, widely held to have been pretty successful. On the second question, we have been talking to the BBC very closely and very recently about this and, no, I am pretty clear that what I said initially was right, that we are talking about costs of somewhere between £10-£20 million.*

Baroness McIntosh of Hudnall: *Sorry, £10-20 million per annum, you said earlier?*

Mr Simon: *Per annum, yes.*

Baroness McIntosh of Hudnall: *Over what period? An additional net £10-20 million spend for the BBC into an indefinite future or over a specified period of time?*

Mr Simon: *No, over the period of the switchover, after which the BBC's costs would decline by up to £40 million a year because they will only be transmitting on one platform and the cost of analogue will be deleted. The basic principle is that the market will pay the cost as far as it is viable. For those people who live in areas, presumably almost always more rural areas, where it is not commercially viable to build out digital transmitters, then the BBC, because it has in its Charter an obligation to provide a universal service, will likely be bound to build out those transmitters. In our conversations with them, they have recently seemed to recognise that and I certainly have not had a sense from them that they believe that the cost would be prohibitive.*

Chairman: *I am bound to say, it is not the flavour of the evidence that they gave to us last week. I think Lady McIntosh makes an extremely good point here. Rather than continuing this, we had better just recheck with the BBC what exactly it is that they want and require here, but, I have to say, your evidence is slightly at variance with what the BBC were telling us.*

Baroness Howe of Idlicote: *I think quite a number of the questions I was going to ask have been answered, but I am still not clear about this £10-20 million that you are talking about with the BBC. Presumably, this will be extra money over and above what they are getting at the moment that you are negotiating with them?*

Mr Simon: *This is not money that we anticipate giving to them. It is a cost which we anticipate will accrue to them when they are building out new transmitters in order to meet their obligation in respect of universality.*

Baroness Howe of Idlicote: *So there will be a difference in that they are prepared to go up to 90, but not beyond that?*

Mr Simon: *The commercial sector will build out to 90 per cent and for the rural areas, the seven or eight per cent beyond that which the commercial sector would not build to because it would not be commercially viable, the assumption is that the BBC will build out that step further and that that will cost them about £10-20 million a year during the period of switchover, which they will later recoup as they make savings from not transmitting anymore on analogue.*

Chairman: *So you are cutting the budget of the BBC by £10-20 million in that period?*

Mr Simon: *Well, I do not have any control over the budget of the BBC.*

Chairman: *You are telling them what to do!*

Mr Simon: *I am not telling them. I am just telling you what the assumptions are about where the likely build-out will come from.*

Baroness Howe of Idlicote: *I also want to go back to this whole business of just when the radio switchover is actually going to happen as there are huge question marks over this. I know the target is 2015, but, as you probably also heard, a lot of people do not even want it to happen. If we are looking at FM and the extension of time that may well be required, is the Government prepared to give a guarantee that FM will remain right the way through whatever period it is, even if it takes another ten years or even longer for the total switchover to take place?*

Mr Simon: *I think the Government can guarantee that FM will remain for the foreseeable future. The Government cannot give guarantees indefinitely, but for the foreseeable future the part of the FM spectrum which will be used for local commercial and community radio will continue to be available for that.*

Baroness Howe of Idlicote: *Well, could you please define for me the 'foreseeable future'? How far beyond 2015 would that be?*

Mr Simon: *I cannot put a number on it, but I would have said it would be well beyond 2015, well beyond.*

Baroness Howe of Idlicote: *Well, 2020, say?*

Mr Simon: *My personal guess, for what it is worth, is probably well beyond that.*

Baroness Howe of Idlicote: *You are probably aware that there will be, and there are already, efforts to get on the face of the Bill a guarantee that it will stay for as long as that.*

Mr Simon: *A point I would make is that, as far as we are aware, we cannot find any evidence anywhere that the FM spectrum will be, going forward, of any particular economic value to anybody. We are not aware of anybody being likely to want it and certainly to want to pay anything in order to use it for anything else. As long as it remains viable for people to use it for radio, as long as people want to continue to use it for radio and as long as the infrastructure still works, then there is no reason at all why it should not continue. It could be another 50 years.*

Baroness Howe of Idlicote: *But you did in fact say that it was an ageing infrastructure. Have you got any guess about how long it would remain a viable infrastructure to be used?*

Mr Simon: *I honestly do not know. There is no sense whatsoever in which I am prevaricating or equivocating, there is nothing that I am trying not to say, but I simply cannot give you any certainty because there is no certainty.*

Baroness Howe of Idlicote: *I wonder whether Mr Smith might have an answer.*

Mr Smith: *I have nothing really to add to what the Minister has said. Again, and it is a personal view, but I would expect FM to be available well into the future and I think I would share the Minister's view of beyond 2020, but we do not know precisely when. The infrastructure is an ageing infrastructure, but who knows how long it might last.*

Lord Inglewood: *I would just like briefly to turn back to the question of extending the DAB national network because it has been measured in the discussions so far by reference to the percentage of the population, but, since the population is scattered relatively randomly across the country, of the final ten per cent of the population, how much in terms of the total number of transmitters and transmission stations does that represent? In fact, in terms of the total cost of rolling this out, that last ten per cent is going to cost a great deal more than the first ten per cent, is it not?*

Mr Simon: *In answer to the first question, nobody knows how many transmitters ----*

Lord Inglewood: *No, but just an order of magnitude; I am not interested in specifics.*

Mr Simon: *I do not know.*

Lord Inglewood: *But it is much more than ten per cent of the total cost of the thing, is it not?*

Mr Simon: *Certainly, you would expect for the non-commercially viable, rural final percentage that the unit cost would be higher than doing the middle of big cities obviously, yes.*

Baroness McIntosh of Hudnall: *Just on the question of who is listening, there appears to be some evidence, including from Ofcom's research, that most people who listen to the radio at the moment are pretty happy with what they have got now and that actually this notion of extensive choice and interactivity, which is the key selling point really of digital radio, as you said yourself at the beginning of your evidence, actually does not weigh that heavily with most consumers. We also have heard evidence from John Myers, who was commissioned by the DCMS to report on this, that there is a considerable oversupply of radio services in the sense that there are too many stations out there fulfilling need which is perhaps not as great as they need it to be to be commercially viable, so are you convinced that there really is the consumer-led demand for a digital switchover, or is this really being driven by the fact that the technology exists and, because we can do it, we will do it?*

Mr Simon: *In the first instance, I think the clearest evidence that there is a demand for digital radio is that in the last ten years ten million digital radio receivers have been bought, and many of them in the earlier years quite expensively and many of them repeat purchases. For a new technology from a standing start where the existing technology remains in place and is dominant, I think that is clear evidence of significant demand and it is a market that continues to grow.*

Baroness McIntosh of Hudnall: *Before you go on from that, could you just tell us what the percentage of the absolute number of radio sets is, as far as it can be estimated, in the country?*

Mr Simon: *It is impossible to say because there are wildly different estimations of how many radio sets ----*

Baroness McIntosh of Hudnall: *Well, on what do you place reliance?*

Mr Simon: *The estimations of the number of analogue sets in existence, which I can think of having heard, varies between about 40 million and 100 million plus, and it is a technology that is 100 years old, so that is a massive accumulation of equipment and it is, therefore, very difficult to compare with a ten-year-old technology. Out of maybe 50 million meaningful analogue sets, if there have been ten million digital sets sold in the last ten years, I think that shows significant consumer demand for a new technology. On the second question about whether the market is oversupplied, firstly and fundamentally, the size of the market is a matter for the market and for the consumer and it is not ultimately my job to manage the size of the market. However, one thing that we are trying to do with this move, as I have said, is to stratify the market into large national, large regional and very local services, which should mean that, rather than everybody in their rather disorderly fashion competing perhaps on unequal terms in one chaotic market, people can do business more efficiently in a more discrete market, and local commercial radio stations, for instance, can do local commercial advertising that will not overlap and be in competition with big national chains and so on.*

Baroness McIntosh of Hudnall: *So, leaving aside just the generality of choice as a benefit in itself, if you had to say what the benefits are to consumers of the particular way in which the Government is driving the digital switchover in radio, what would you put as your two or three most significant benefits? Would it be, for example, an extension of nationally available commercial broadcasters on digital platforms, or what would it be?*

Mr Simon: *I think it would be, firstly, that the technology offers additional benefits, so it offers for most people more stations more clearly defined and a better mix of the different tiers, additional functionality, interactivity; digital radios do things that analogue radios do not. Secondly, I think it offers the consumer an orderly, consistent and coherent pathway to a digital future which is inevitable. The FM technology is ageing and the economics suggest that it would not be economic for major broadcasters to replace that infrastructure nationally and, rather than let it disintegrate in a piecemeal way and let it be replaced in a piecemeal way and have consumers all over the country who no longer can get the services that they have been used to, I think it is appropriate for the Government to try and manage this transition in a coherent way.*

Chairman: *What I do not quite understand is that you talk about new services being provided, but they are going to be provided in an industry, or you are relying on them being provided in an industry, a particularly commercial industry, which is going to be competing with the BBC, which obviously one wants to see, but that commercial industry itself is struggling, not to survive, but certainly struggling in very, very difficult economic circumstances at the moment. I do not quite see where this expansion is going to come from.*

Mr Simon: *I am not saying that the aim of the policy is to expand the market for commercial radio. The aim of the policy is to manage the transition in a way that works both for the businesses of the broadcasters and for the consumers and to manage the experience in a way that works for the consumer. As you say, the commercial radio market in terms of revenue has shrunk by a third over the last ten years as advertising revenues have shrunk for all broadcasters, print media and so on. It is a business in which people are under real* *pressure and that is why, among the overwhelming majority of commercial broadcasters, there is a strong support and an appetite for a clear, managed, relatively swift, but not rushed, pathway to digital where, although in the first instance they will have some additional cost, they can see a future in which they can do more business and make more money.*

Lord Inglewood: *Just really arising out of that and something you said earlier, it is not part of your case, is it, that, if the analogue network were not ageing and hence degrading, that would not be the reason for moving to digital? The reason for moving to digital is not because the analogue network is simply breaking down, is it?*

Mr Simon: *The fact that the FM infrastructure will need to be replaced in the short to medium term and that it looks like that replacement would not be economic is certainly one of the underpinning factors.*

Lord Inglewood: *But that goes back to the question of Lady Howe's, does it not, that the foreseeable future is actually defined by the capability of the FM infrastructure to deliver the services it currently does?*

Mr Simon: *I am not sure I understand the question. I understood her question.*

Lord Inglewood: *Yes, but she said, "How long is it going to go on for?" and you said, "For the foreseeable future", which was fair, but I think in fact that the way it is defined is that it will go on, according to the evidence you have subsequently given us, for as long as the FM transmission infrastructure is capable of delivering it. When it collapses, that is ----*

Mr Simon: *They are two different points though. There is the question of how long it will be possible for some people locally to continue to broadcast on FM, which is a different question from at what point do businesses need certainty about the capital investment decisions they might need to take in the future to renew the major national infrastructures.*

Lord Inglewood: *Sorry, but I may have misunderstood your evidence, in which case I apologise, but I thought what you said was that you did not believe that the industry was prepared to invest for the further life that the FM infrastructure would require.*

Mr Simon: *That is correct.*

Lord Inglewood: *If that is the case, why are they prepared to invest in the infrastructure for digital, which is part of the proposal that you are describing to us?*

Mr Simon: *Because renewing an old technology with its limited functionality does not make economic sense to them, whereas investing in a new technology with additional functionality does. When it comes to making the investment decisions, they are not going to buy the old kit again, they are going to buy the new kit.*

Lord Inglewood: *I understand the argument, thank you. The other thing I wanted to ask you about was right at the outset when there was a bit of verbal sparring going on between yourself and our Chairman, reference was made to things, such as cheap sets and set-top boxtype devices that would enable analogue radios to pick up digital and so on, and you used the subjunctive, that they might well become available, it was very likely. Is there any assurance that you have been given by any of the manufacturers about the actual provision* *of these products and, if there is not, as events move on, if it becomes apparent that a lot of this hardware may not be available, will that delay and/or postpone any digital switchover?*

Mr Simon: *Their adaptor-type technologies are already available for cars, so you can already buy a little thing in Halfords that you can put on your analogue car radio that will enable it to receive the digital signal. I do not think that you can easily get hold of such a product to adapt a domestic transistor radio, but the technology already exists. I am led to believe that they would not be difficult to manufacture and there is no reason to believe that they would be. We are talking to manufacturers who say that such products are on the way.*

Lord Inglewood: Do you know in what sort of price range these things might be?

Mr Simon: *I honestly do not off the top of my head.*

Lord Maxton: *On Amazon, £65.*

Mr Simon: *Currently?*

Lord Maxton: *For the car one, and you have to buy an aerial as well.*

Mr Simon: *We expect all of these prices to come down a lot.*

Chairman: *They have already come down quite a bit, have they not?*

Mr Simon: *Yes, they have come down tremendously. Digital radios themselves, the cheaper sets, you can now get a set for less than £25, which is very, very greatly less than the cheapest even a couple of years ago. As the market expands, the costs will come down, and it is expanding all over Europe. We are also working with, and encouraging, manufacturers to use, without being too technical, what is called the 'World DMB Profile One chip', which is compatible all across Europe, and that will give them great European economies of scale which again should make it easier for them to produce even cheaper sets and adaptors. In answer to the question, if nobody produces these things at affordable cost, would that delay switchover, I think it would depend what effect that had on the market. If that meant that nobody bought them and nobody bought any digital radios either and we did not get over the listenership threshold, then yes, it would delay. If it simply meant that people did not bother with an adaptor, but bought new digital radios, then it would not.*

Lord Inglewood: *I sense you sense that this is not actually going to be a problem.*

Mr Simon: *That is my belief.*

Baroness Eccles of Moulton: *I think, Minister, it is quite clear to us that there is going to be plenty of equipment around for converting, multi-chip, et cetera, et cetera, but why are we continuing to base the whole of our digital radio switchover on what is becoming an outdated system, which is DAB, as opposed to what other countries have done, which is either to scrap DAB or start off with DAB+? Why have we allowed ourselves to get caught in being committed in an early stage of the development of the switchover to what is now not the most up-to-date and modern form of broadcasting?*

Mr Simon: *That is a very good question which lots of people ask. The answer is that we were in this country by far the earliest adopters as a market of digital radio and we took it up far* *more quickly and far earlier than anybody else, as a result of which the amount of stock in the market that you would have to write off in this country is vastly greater than European comparators. There are ten million DAB sets, many of which have been bought at early adopter prices in the market in the UK and, if we abandoned DAB, it would mean that all those early digital adopters had their digital investment written off after a very few years, which just seems really counterintuitive in terms of how to drive the market towards digital and hardly also would seem likely to inspire much confidence in those people to make them very likely to buy another digital radio of the new standard. What we are saying is that the new sets henceforth should have the DMB World Profile One chip in, which is DAB, DAB+ and DMB compatible, and it would mean that you could use it anywhere in Europe and receive all that mix of signals. It does not preclude us in the future from moving to a technology which is greater than DAB, but it is a cost-benefit decision that has had to be taken and the clear consensus, not just in the Government, but of the regulators and stakeholders, has been that writing off the ten million DAB sets that have already been bought would be counterproductive.*

Baroness Eccles of Moulton: *Is including the new chip in any radio that comes on to the market from hereon going to be mandatory in the same way as fitting seatbelts in cars became mandatory?*

Mr Simon: *Not in the same way in the sense that, I think, seatbelts in cars was a piece of primary legislation, but mandatory in the sense that we will work with manufacturers in order to draw up the technical specifications for the new generations of sets, and they will be the formal technical specifications which will be adopted, hopefully, European-wide, so in that sense mandatory.*

Baroness Eccles of Moulton: *So you would not be able to sell the set if it did not meet the technical specifications?*

Mr Simon: *You would not be able to sell an approved set. It would not be a crime to sell a set, but it would look like a dodgy set.*

Lord Maxton: *France is setting the law which says that by 2012, I think, all radios sold must be DAB+ and that would include all car radios, which is the market area which we really have not looked at, but it is a very important area. Is it right that 20 per cent of all radio listening is in cars? That is a very large percentage, but it is a very difficult area to switch from FM to digital, even with the device which we have talked about already, is it not?*

Mr Simon: *Our targets are switchover by 2015 and new cars all being digitally radio-ed by the end of 2013. That is our clear target. Now, obviously, even if all new cars are digital by 2013, that still leaves the majority of the car stock analogue for a while beyond 2015, at which point one hopes that the price of the adapters, which you can already buy in Halfords, has come down considerably from the price that you mentioned on Amazon. I did not think they were that expensive in Halfords.*

Lord Maxton: *Well, they are £79 in the shop and, to be honest, with the reviews they have had of them, you have to almost buy an external aerial.*

Chairman: *I would not take him on on this!*

Mr Simon: *I am not going to, not for a minute am I going to! He is obviously speaking with some authority.*

Lord Maxton: *It will cost you another £12 or £13 on top.*

Mr Simon: *All I can say is that manufacturers assure us that this technology will be getting cheaper, better and more widely available over the three years before cars go all digital and over the five years until digital switchover. That is a long time for them to make big improvements.*

Lord Maxton: *I have not got one yet, by the way!*

Mr Simon: *I guessed!*

Chairman: *But you said that the target date is 2015?*

Mr Simon: *Yes.*

Chairman: *That remains the target date, does it?*

Mr Simon: *That is the firm target.*

Chairman: *That is a firm target date and, therefore, will that go into the Bill as a date, the Digital Economy Bill going through at the moment?*

Mr Simon: *Is it not on the face of the Bill?*

Mr Smith: *The Bill sets out the conditions which would have to be met for the target to be set.*

Chairman: *But I think I am right, am I not, that 2015 is not there as a target?*

Mr Smith: *It is not on the face of the Bill, you are absolutely right.*

Chairman: *Why not?*

Mr Simon: *Because it is a target.*

Chairman: *An aspiration?*

Mr Simon: *No, I am happy with the word "target". It is a firm target, but it is not an inflexible or dogmatic target. It relies on our having met the listenership test, met the coverage test and, if we do not meet those tests, then we will not hit that date. We believe that we can, and should, hit that date.*

Lord Inglewood: *It will not be brought forward?*

Mr Simon: *That is very optimistic.*

Lord Inglewood: *I am just asking, not suggesting.*

Mr Simon: *I think it is unlikely.*

Chairman: *So you may miss the target. That is basically what you are saying.*

Mr Simon: *I am saying that, if we do not satisfy the criteria, then we will not do it if the nation is not ready.*

---

[455] http://www.publications.parliament.uk/pa/ld200910/ldselect/ldcomuni/100/100.pdf

87.

28 February 2010

Car industry: "gaps in digital coverage are a major deterrent to [the] introduction of digital radios"

The Society of Motor Manufacturers and Traders [SMMT], representing 500+ companies in the UK car industry, has submitted written evidence to the House of Lords Communications Committee inquiry into digital radio switchover. Its members have itemised a number of concerns about the practicalities of the government proposal that all new cars be offered for sale with DAB radios by 2013:

- *"the apparent perception that the markets for in-vehicle radios and domestic radios are similar, if not identical, and that any assumptions about the speed of take up can be applied to both markets*
- *the timeline for adapting the existing vehicle parc [cars already on the road]*
- *the continued availability of traffic information after 2015 to those driving vehicles which are not digitally-enabled*
- *the extent of radio transmitter coverage*
- *the need for broadcasters to promote the advantages of digital radio to consumers to create demand*
- *safety and security issues arising from the use of digital convertors*
- *the need for pan-European approaches to the introduction of digital radios in vehicles."*[456]

The key issue raised by SMMT concerning the necessary robustness of DAB in-car reception across the whole of the UK would require a massive investment from the radio industry to rectify:

"SMMT members are clear that the gaps in digital coverage are a major deterrent to their introduction of digital radios as standard equipment. As outlined [below], any vehicle manufacturer bears the reputational risk if a radio in one of its products appears not to work properly. Drivers have become accustomed to the gradual deterioration in FM reception which occurs throughout parts of the UK and recognise this is not the fault of their radios. At the present stage of digital roll-out, shortcomings tend to be blamed on the vehicle manufacturer.

SMMT members therefore welcome the statements in the [Digital Britain] report that:
- *one of the criteria for deciding the date of the Digital Radio Upgrade will be whether national DAB coverage is comparable to FM coverage and that local DAB radio reaches 90% of the population and all major roads*
- *the BBC should begin an aggressive roll-out of the national multiplex to ensure that its national digital radio services achieve coverage equivalent to FM by 2014.*

However, there is also a need for a plan to enable reception on those stretches of road, primarily tunnels and long underpasses, where reception goes 'dead' for a short period. At present, for instance, FM coverage in the Dartford Tunnel is addressed by special measures. In shorter tunnels, the FM signal tends to deteriorate but not disappear, whereas the digital signal disappears entirely."

SMMT noted that:

"There appears to be an assumption that the market for in-vehicle radios and that for domestic radios have similar, if not identical, features. In fact, they differ in five main ways:
- *in the automotive market, the vehicle itself, not the radio, is the reason for the purchase*
- *vehicles are required to undergo an approval process which is far lengthier than any applying to consumer goods*
- *the sizes of the two markets and their dynamics are vastly different, where customers purchase new radios more frequently than they do vehicles*
- *if a radio in a vehicle fails, or even only appears not to work properly, blame is attached to the vehicle manufacturer, whereas the reputational risk if a domestic radio fails is borne by the radio manufacturer*
- *in automotive applications, the radio is not static. It moves between transmitters and, therefore, complete and national coverage of the digital radio network will be required."*

SMMT's concern for new cars is that:

"meeting a deadline of 2013 will be a challenge for vehicle manufacturers who began product development in 2009, but we expect it to be achievable. A bigger challenge is represented by those models already on the market or most of their way through the development cycle, where the manufacturers will have to decide whether to divert engineering resources to the task of digitally-enabling them or provide new vehicles with digital convertors."

SMMT's concerns for the cars already on UK roads are:

- *"The [Digital Britain] report suggests that the majority of the vehicle parc should be converted to digital by 2015, with low-cost convertors for the remainder.*
- *Vehicle manufacturers are certain that retrofitting of digital radios on a large scale is impractical. Vehicles' electronic systems have become increasingly integrated; often, the radio is part of this integration and cannot easily or economically be replaced. A radio has to operate in the vicinity of sensitive electronic components, and poor integration has a detrimental effect on other systems.*
- *Retrofit also affects the perceived quality of the vehicle:*
    - *antennae have to be chosen very carefully – reception from an internal antenna may be poor if a vehicle is fitted with infra-red reflection glass, or if a magnetic antenna base is fitted to an aluminium body*
    - *poor refitting of trim items removed to permit a retrofit will cause rattles.*
- *Drivers will, therefore, be reliant on the use of digital convertors to enable continued use of their analogue radios after 2015. As vehicles have very long lives, most of the vehicles first registered since 2006, if not earlier, will still be in use in 2015. It is likely that over 20 million vehicles will have to be so fitted, and very likely that most of the necessary sales will be made in the few months before the date for digital migration. The commitment for a cost:benefit study to be conducted before any digital migration date is announced is therefore welcomed by vehicle manufacturers because it should firmly identify the progress made towards digitally-enabling the car parc."*

The message from the car industry seems clear – why should they risk their reputations by installing DAB radios that will suffer poor reception due to lack of a robust DAB radio transmission system in the UK?

The bigger question is – why would consumers pay extra for a DAB car radio that offers increasingly little additional mainstream content over a standard FM radio?

---

[456] http://www.parliament.uk/documents/upload/SMMT.pdf

88.

3 March 2010

DAB radio receiver sales: never let facts get in the way of a big number

A newsletter arrived in my in-box today from Digital Radio UK, the new organisation charged with making DAB radio a success. It told me some startling news:

"By the end of 2009, when buying a radio, more than three quarters of people chose a digital one."[457]

And, just in case I did not believe this fact, immediately beneath, it told me the same thing again:

"New sales figures reveal that, when buying a radio, more than 75% of people choose a digital one."

I did not believe it. All the previous data from the radio industry had shown that DAB radios are around 22% of total radio sales, as demonstrated in the graph below.

DAB RADIO RECEIVER SALES (% total receiver sales)

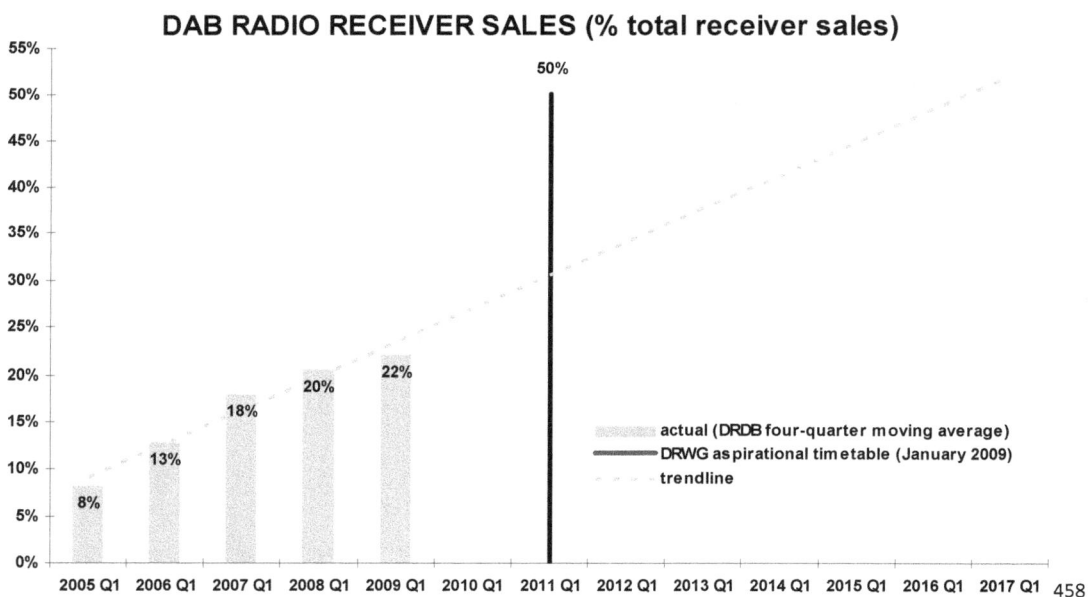

A year ago, the government's Digital Radio Working Group had set an 'aspirational' target for DAB radios to be 50% of total radios sold by the beginning of 2011. As this graph clearly shows, the odds of successfully coming anywhere close to that target are zero.

Maybe something revolutionary had happened in the consumer market for the proportion of DAB radios sold to have suddenly surged from 22% in Q1 of 2009 to 75% by year-end. It was extremely puzzling.

Then I read an extraordinary letter that Ford Ennals, chief executive of Digital Radio UK, had written to the House of Lords Select Committee on Communications on 15 February 2010. It said in part:

"I thought [...] that it might be useful if I wrote with the very latest radio sales data. Encouragingly, it shows that, during 2009, consumers increasingly chose digital sets over analogue ones.

*I thought it clearest to present the data in a simple table, which is attached, but it may be useful if I explain a couple of the terms used. **Where the data refers to 'kitchen radios' it means the kind of sets that you and I would call 'a radio'** i.e. a set whose sole function is to listen to the radio.*

Where it refers to 'all radios', these figures include those pieces of electrical equipment which happen to have a radio chip in them (e.g. a hi-fi where the main reason for purchase may be to listen to CDs or an MP3 player where listening to downloaded music is the primary function).

As you can see, by Christmas 2009, 76% of people buying 'a radio' chose a digital one…… "[459] [emphasis added]

Aha! Now I think I understand. The only way in which it is possible to contrive that more than three quarters of radios sold are digital radios is to arbitrarily create a completely new definition of 'radio'. In this brave new world, only a 'kitchen radio' will now be called a 'radio'. (The truth is: 76% of people who purchased a kitchen radio during December 2009 bought a digital radio, though the proportion for the whole of 2009 was 63%.) Every other type of radio is no longer defined as a radio. This new definition of 'radio' would completely exclude:

- Micro systems
- Clock radios
- Tuner separates
- Handhelds
- Boomboxes
- In-car radios
- Audiovisual systems
- Home cinemas
- Docking stations
- Dect phones [?]
- Mobile phones
- LCD TVs
- Record players

This seems like a long list of products which, if they also happen to include a radio, will no longer be defined as having a 'radio'. How can a 'clock radio' not be a radio? How can a 'tuner' not be a radio? I know this long list to be a comprehensive definition of 'radio' because it was the very definition of 'radio' used by the Digital Radio Development Bureau, the forerunner to Digital Radio UK, in its published data. Of course, that was last year. In 2010, 'radio' seems now to have a whole new definition.

What can I say? However desperate you might be to try and make DAB radio a success, how is it justifiable to deliberately mis-state data so outrageously in print? And to Parliament?

---

[457] http://radiocentre.cmail1.com/T/ViewEmail/y/F4AD8FBE8ACCAFAD/5F04C4F224D87E97C67FD2F38AC4859C
[458] source: Digital Radio Development Bureau
[459] http://www.parliament.uk/documents/upload/DigitalRadioUKsuppukdso.pdf

89.

6 March 2010

Lord Fowler: "There will be a public outcry if we get the radio digital switchover wrong. There could be a very big row indeed about this."

The Digital Economy Bill was debated in the House of Lords this week in its 'Report Stage'. Once again, amendments that had been proposed specifically to take into account the views of listeners and small radio stations were rejected by the government. As it stands, the Bill only requires Ofcom and the BBC to be consulted before the government can take a decision about switchover from analogue to DAB radio.

Parallel with the progress of the Bill through the House, Lord Fowler has been chairing a separate Select Committee on Communications inquiry into digital switchover. Through its weekly meetings, where it has collected copious evidence from witnesses, it must be becoming increasingly obvious to the Committee that the government plan for digital radio switchover is an undignified mess. Lord Fowler's growing displeasure with this situation surfaced during Debate of the Bill:

"I do not intend to pre-empt our [Committee] report, but I must say that it is generally a very important issue with the public and that there will be a public outcry if we get the radio digital switchover wrong. There could be a very big row indeed about this. …. I think I probably speak for the committee when I say that there is public confusion at the moment about what exactly the plans mean to the individual consumer, and I cannot believe that that is a sensible way of proceeding."[460]

The government's frosty response, delivered by Lord Davies, conveys everything:

"[Lord Fowler] told me, as if I did not know, that there could be the most enormous row if this switchover went wrong."

The government simply refuses to listen to commonsense on this issue, even from the chairman of the Lords Communications Committee. As a result, a *"very big row"* about digital radio switchover is indeed inevitable, probably at Easter, and more so following publication of the BBC Strategy Review.

Here is the 'radio' part of the debate in full:

The House of Lords
Parliamentary Debate
3 March 2010

Digital Economy Bill
Clause 30 : Digital [radio] switchover[461]

Amendment 137
Clause 30, page 36, line 33, after "to" insert —
"(a) "

Amendment 138
Page 36, line 35, at end insert —
"( ) the needs of local and community radio stations; and
( ) the needs of analogue radio listeners"

Lord Howard of Rising: *My Lords, I tabled Amendments 137 and 138 again simply to get more detail from the Minister, as his assurances about these points were not wholly convincing. The amendments would give the Government an explicit requirement to take into account the views of radio listeners and local and community stations. The Minister argued that this was unnecessary because of the breadth of the requirements to consult that are already proposed and the commitment to consult widely. The problem with such vague assurances is that they can be quickly forgotten. The [Bill] currently states that the views of the BBC and Ofcom should be given due regard before the Secretary of State nominates a date for the digital switchover. It does not say too much about consulting widely or taking into account in any way those who are most affected by the switchover — the listeners. I hope the Minister can give more encouragement that the listener will not be forgotten in this whole process. I beg to move.*

Lord Clement-Jones: *My Lords, I commend the amendments, which are a very constructive way of seeking further assurance from the Minister. Indeed, they very much reflect the concerns that I expressed from these Benches in the Clause 30 stand part debate in Committee. Assurances about the future of analogue radio in particular are so important. The noble Lord, Lord Young, and I engaged in a slightly semantic conversation about whether FM's existence would be perpetual or whether it would simply be there for the long term. I think the assurances were that it would be there for the long term, which did provide some reassurance. However, the interests of the ultra-local stations and the consumers of the product of those stations are extremely important, and I very much hope that the Minister can cast more light on the future of analogue in the face of the digital switchover.*

Lord Fowler: *My Lords, I congratulate my noble friend on the amendment — I am now rowing back frantically — on a very important issue. It is so important, in fact, that the Select Committee on Communications is currently engaged in an inquiry on precisely this — the digital switchover — although a number of reasons have been adduced as to why it should be called not a switchover but various other names. I do not intend to pre-empt our report, but I must say that it is generally a very important issue with the public and that there will be a public outcry if we get the radio digital switchover wrong. There could be a very big row indeed about this. My only reservation about the amendments is that I can think of quite a number of other issues on which I would like the Government's assurance. There are, for example, 20 million car radios out there. What will happen to those? How will they be converted? What are the plans? There are so many issues here that either we will have a totally comprehensive list or we will simply have to ask the Minister at this stage for his current views. I think I probably speak for the committee when I say that there is public confusion at the moment about what exactly the plans mean to the individual consumer, and I cannot believe that that is a sensible way of proceeding. My noble friends on the Front Bench have raised a crucial issue to which we will have to return again, and very soon.*

Baroness Howe of Idlicote: *My Lords, I absolutely agree with what has been said so far. This is one of the greatest eye-openers. As we have proceeded with this Bill — particularly as it has run parallel to the deliberations of the Select Committee which the noble Lord, Lord Fowler, is chairing so ably — we have come to realise just how important radio is to so many people, whether to the disabled or to all of us, listening as we do for a vast amount of our time to the radio. However, this is clearly one of the areas in which there is still a need to reassure people locally. The idea was that analogue transmission could be switched off once 50 per cent of listening is to digital radio. Then there was the business of how long FM would be available once it is more or less accepted that there will be a change. As regards the production and selling of cars, the issue is when there will be sufficient technology to convert radios already in cars and to convert some DAB radios to the right level. No one is trying to argue for a moment that the quality of digital radio will not be valued. But getting to that point will need a lot of reassurance to citizens. I would be grateful, as would I am sure other noble Lords, for further reassurance from the Minister that FM will be available ad infinitum, but certainly well beyond the point of switchover. That would do a great deal to reassure noble Lords who have looked into all this. But much more importantly, the citizens and the consumers — I come back to them because I am looking at this issue from both viewpoints — are crucial. I hope that the Minister will be able to give that reassurance.*

Lord Gordon of Strathblane: *My Lords, I do not have any problem with the sentiments behind the amendments. The only problem is that if those points are listed, it would look as though that is what the Government or Ofcom should give priority to, but they are only three of a myriad number of conditions to which they must give attention. Specifying that is almost counterproductive.*

The Lord Bishop of Manchester: *My Lords, I rather echo that point. In Committee, I expressed, as did many noble Lords, concerns about local and community radio stations and about the extension of FM. These are very important matters, but as other noble Lords have indicated in this short debate, there are other areas as well. In all this, I hope that we will continue to recognise that, while it has often been said that the switchover for television has gone very smoothly, the complexities in relation to radio are far greater. While supporting so much that lies behind these amendments, it would be a great shame, in a sense, to wreck it by omitting rather than being inclusive.*

Lord Davies of Oldham: *My Lords, I am grateful to all noble Lords who have spoken in this short debate, particularly the right reverend Prelate the Bishop of Manchester and my noble friend Lord Gordon, for identifying the weaknesses of the amendment and the nature of the issue on which the Government need to take care. Perhaps I might say that if I was not going to take care after the Opposition Front Bench and the noble Lord, Lord Clement-Jones, had spoken in support of the amendment, I certainly was after listening to the noble Lord, Lord Fowler. First, he told me, as if I did not know, that there could be the most enormous row if this switchover went wrong. I could not agree with him more and I accept entirely what the right reverend Prelate has said. The switchover from analogue to digital for television is much easier than this exercise because of the diversity of radio opportunities and provision. But the noble Lord, Lord Fowler, produced an even greater anxiety for me when he mentioned car owners. He is right that we would not dare to get that wrong. I know that we are not far from a general election, but the idea that the Government are about to alienate 20 million car owners by telling them that their radios are defunct, out of date and will not work is somewhat unrealistic. The conversion of car radios is an important point that has to be established before a digital switchover could conceivably be considered a success. We have* been clear that an affordable in-car converter is key to the success of digital radio switchover. There are already devices on the market which will convert an FM car receiver to receive DAB. One would predict that this market will expand very rapidly. Very few markets move quite as quickly as the car accessories market, which helps to guarantee the sale of cars. That point therefore will be taken into account, as will the other points about the importance for the Government of effective consultation before such a switchover could take place. We have made clear that, for the foreseeable future, the Government will consider FM radios to be part of the broadcasting firmament. Radio stations will want to combine to broadcast on FM to take account of the points that the right reverend Prelate drew to the attention of the House. What date will all this be effected? That is a pointed and precise, but nevertheless very difficult, question. We have indicated that 2015 is ambitious, although it is achievable. If we do not set a target, there is no stimulus to all those who can make a contribution to effecting this successfully to get to work and do so. So we want a date and have identified 2015, but we recognise that it is a challenge. However, we accept the concept behind the amendments; namely, that the fullest consultation will be necessary. Otherwise, the almighty row anticipated by the noble Lord, Lord Fowler, will descend upon the Government who get it wrong. Why do I resist the amendments, as we did in Committee? It is simply because consultation is written into the Bill already. We could not dream of going forward or of proposing that the Government could go forward with an issue of such significance to our people without the fullest consultation in order to guarantee that we do not fall into those dreadful traps to which noble Lords have called attention. Again, I am grateful to the Opposition Front Bench for drawing our attention to the necessity for care and consultation. That is part of the Bill and the amendments are unnecessary. Having stimulated a further debate, after the extensive one we had in Committee, I hope that the noble Lord will withdraw his amendment.

Baroness Howe of Idlicote: *My Lords, can the Minister clarify the point about which a lot of people are concerned; namely, that whenever the point of switchover occurs, FM will continue beyond that point? A lot of small operators are very concerned about that.*

Lord Davies of Oldham: *My Lords, I wanted to indicate that. If I did not make it clear enough in my reply, we see FM continuing, but we also see the kind of criteria that will be necessary before we begin the process of significant switchover. As I have indicated, the Government will move with the greatest care with regard to this issue, as we have with television switchover. Noble Lords will know of the care that we have taken to make sure that groups who might not be able to make that switchover effectively because of limited resources are given support. Radio is much more complex and difficult, as the right reverend Prelate made clear. The Government are fully seized of that, which is why consultation is written into the Bill on this issue.*

Lord Howard of Rising: *I thank the Minister for his comments, and I thank my noble friend Lord Fowler for his support. I was delighted to hear some support from the noble Lord, Lord Clement-Jones, after the sandbagging that I received from the noble Baroness, Lady Bonham-Carter. Having raised the issue and heard how sympathetic the Minister is to the potential problems — even though he dodged with his customary skill committing himself specifically to consulting listeners — I beg leave to withdraw the amendment.*

Amendment 137 withdrawn. Amendment 138 not moved.

---

[460] http://www.publications.parliament.uk/pa/ld200910/ldhansrd/text/100303-0011.htm
[461] http://www.publications.parliament.uk/pa/ld200910/ldhansrd/text/100303-0011.htm

90.

13 March 2010

DAB converters for portable analogue radios? It's a "no no"

All of us would like to invent a 'killer application' that could captivate consumers with its usefulness, change the future direction of technology, and make millions. But there is a big difference between inventing one in our heads and turning it into a technical reality in the marketplace.

The converter/adapter that is able to magically transform a portable analogue radio into a DAB radio is one such invention. It exists in the heads of the DAB radio lobby as a means to persuade politicians that mass consumer conversion to DAB is a possibility rather than a pipedream. Unfortunately, it does not exist in reality.

When the notion of such a converter was mentioned last year, I examined the analogue portable radios scattered in almost every room of our home. The only access to their internal electronics that some of them allow is via a headphone socket – and when you insert anything into that, the loudspeaker cuts out. So how exactly could any kind of gizmo be 'added' to such radios to transform them into DAB?

My doubts were confirmed when Intellect, the trade organisation that represents UK radio receiver manufacturers, wrote to Parliament in February 2010 and stated: *"Whilst it is technically feasible, there are currently no products on the market that can adapt an analogue radio to receive DAB signals."*[462]

Subsequently, Laurence Harrison of Intellect presented evidence in person on this issue to the Lords' Communications Committee: *"A converter would have to include within it pretty much all the components, bar the speakers, of a standard digital radio anyway. Therefore, the cost differential for a converter will be minimal between that and just buying a new digital radio."*[463]

The converter is a prime example of the radio industry's current pre-occupation with technology being the answer to its problems. Last week, Steve Orchard (former group programme director of GWR, former operations director of GCap) wrote an opinion piece which proclaimed: *"DAB is vital to commercial radio's future."*[464] What?? Sorry?? Surely, it is 'content' which is vital to the future of commercial radio, just as it always has been, and just as it always will be. Content = listening = advertising = revenues = profit. Whereas: DAB = platform = infrastructure = investment = risk.

The radio industry desperately needs a strategy that focuses on producing content, rather than focusing on DAB. We already have platform businesses such as Arqiva whose function is transmission infrastructure such as DAB and FM; and we already have consumer electronics companies that produce radio receiver hardware. I don't see Arqiva or Roberts trying to produce radio shows, so why does the radio industry so desperately want to control platforms and invent hardware?

As ever, the challenge for the radio industry is to create content that is sufficiently compelling, regardless of the platform. Consumers gravitate to content, whatever platform that content is on. The history of radio has demonstrated this time and time again. For example:

- 90% of the population listen to analogue radio for around 20 hours per week (on FM and AM platforms that the radio industry has lobbied to have shut down)
- BBC Five Live and TalkSport attract 5% and 2% shares respectively of all radio listening, despite being broadcast on AM (a platform that commercial radio lobbied the regulator in the 2000s to write off for mainstream formats)
- Pirate radio with poor FM reception continues to attract significant audiences in cities (stations which the radio industry has long lobbied to be shut down, despite itself not offering consumers any comparable content)
- Atlantic 252 attracted a 4% share of all UK radio listening in 1994, despite broadcasting from Ireland on Long Wave (a platform the BBC tried to shut down in 1992)
- Ricky Gervais' radio show remains the most downloaded podcast ever, despite never having been broadcast and only ever having been made available as an online download (a platform largely ignored by commercial radio).

Sometimes, it seems that parts of the radio industry have stumbled so far away from their core product, content, that the eventual outcome might even be (to adapt Steve Orchard's comment): 'DAB is a vital part of commercial radio's death'. The sector's profitability is already zero. This is no time for distractions that will not directly put bums on seats.

The quotes below offer more detail on recent dialogue concerning the mythical DAB adapter.

....

"For customers who don't want to buy a new radio set, it will be possible to convert existing sets to digital instead. An adaptor device will come onto the market soon that will cost around £50 and, in time, conversion may cost less than a new radio set."[465]
Digital Radio UK
2 December 2009

....

House of Lords
Select Committee on Communications
20 January 2010[466]

Witnesses:
Ford Ennals, Chief Executive, Digital Radio UK
Barry Cox, Chairman, Digital Radio Working Group

Lord Gordon of Strathblane: *There is, I might suggest, a vital difference. It is comparatively easy and cheap to convert a television set to digital with a set-top box that you can buy from Tesco for £20. Can you do that to an analogue radio set?*

Mr Ennals: *I fully expect that there will be low-cost converters available. We were talking to companies which were making these last week, and they are talking about DAB adaptors for* *about £20 or £25. When the DTT Freeview development started, those products were costing over £100. The market will become more competitive, prices will come down. You can replace your radio for £25 with a digital radio. There will be a burden of cost on the consumer, but it is significantly more affordable than it would have been in the past.*

Lord Gordon of Strathblane: *If it is as cheap to buy a new digital set as it is to buy a converter, there is a fair disposal problem involved in 50 to 100 million radio sets that are good to go to the rubbish dump.*

Mr Cox: *There is undoubtedly a difference with television because you can keep your old set and put the adaptor on it. I heard what Ford was saying, and it would be useful if some adaptors come on the market, but the likelihood is that many of those analogue sets will have to be disposed of.*

....

Intellect [UK trade association for the electronics industries]
Written evidence to the House of Lords Communications Committee
1 February 2010[467]

"Converting analogue radios to digital:
Whilst it is technically feasible, there are currently no products on the market that can adapt an analogue radio to receive DAB signals. Our members would undoubtedly produce such devices should a clear market demand ensue following the passing of the Digital Economy Bill. However, simply adapting an analogue product will not allow listeners to enjoy the full range of benefits that DAB can offer. With some entry level digital radio receivers costing as little as £25, adapter devices are likely to cost more than digital receivers at the start."

....

House of Lords
Select Committee on Communications
24 February[468]

Witness:
Laurence Harrison, Director, Consumer Electronics, Intellect

Lord Gordon of Strathblane: *What about converters for what are known as 'kitchen' sets?*
[....]

Mr Harrison: *Converters – if you like, a set-top box for an analogue radio – are technically possible. I think we need to look at just how appealing that would be for the listener. A converter would have to include within it pretty much all the components, bar the speakers, of a standard digital radio anyway. Therefore, the cost differential for a converter will be minimal between that and just buying a new digital radio.*

Lord Gordon of Strathblane: *So they are not going to fly off the shelves?*

Mr Harrison: *It will depend on just how much the individual values their analogue set. Of course, converters would also come into play if you are talking about, for example, a large expensive hi-fi system; they would work for that, and if you like the sound quality of that hi-fi then a converter may be an option, but I do think we need to be careful, purely because we* *know that the price differential, for example, will not be that great between a converter and a standard digital set.*

Lord Gordon of Strathblane: *So for big, stand-alone hi-fi sets with colossal speakers and everything else it might make sense but for the small 'kitchen' portable a no no?*

Mr Harrison: *We know that some manufacturers are looking at the possibility of introducing a converter, so it may well be that some of those do come to market. I just think for the context we need to be aware of what that converter will look like, and how appealing it may be. I think your assessment is correct.*

[...]

Lord Maxton: *There is a major difference; with your existing television all you need is a box.*

Mr Harrison: *Indeed.*

Lord Maxton: *A converter, basically. With radios that is not the case.*

Mr Harrison: *That is true.*

Lord Maxton: *You do not have to get rid of the televisions but you do have to get rid of the radios.*

Mr Harrison: *That is absolutely true. All I would say on TVs – you are absolutely right and I do not want to downplay the situation at all ......*

---

[462] http://www.parliament.uk/documents/upload/Intellect.pdf
[463] http://www.publications.parliament.uk/pa/ld/lduncorr/uncorrcomm240210ev5.pdf
[464] http://www.thedrum.co.uk/news/2010/03/05/13011-radio-boss-commercial-sector-will-gain-nothing-from-6-music-closure
[465] http://www.charleswest.org/?p=431
[466] http://www.publications.parliament.uk/pa/ld200910/ldselect/ldcomuni/100/10012007.htm
[467] http://www.parliament.uk/documents/upload/Intellect.pdf
[468] http://www.publications.parliament.uk/pa/ld/lduncorr/uncorrcomm240210ev5.pdf

91.

17 March 2010

France: digital radio "is not progressing one inch"

A meeting of radio sector stakeholders on Monday 15 March 2010 at France's media regulator, the CSA, failed to progress the plan to launch digital terrestrial radio this year. According to Le Point, the commercial broadcasters – RTL, Europe 1, NRJ and RMC – demanded a moratorium.[469] State broadcaster Radio France is one of the few continuing to support the CSA's plan to launch digital radio, delayed from 2009 to mid-2010, using the T-DMB transmission standard.

A member of the Bureau de la Radio trade organisation commented: *"There is no economic model [for digital radio]. The choice of the [T-DMB] broadcast standard adopted in Bercy is very expensive. The upside for listeners is not sufficient for us to fund a third broadcast platform to add to the existing Long Wave and FM [platforms]. … We are disappointed because this meeting has not enabled anything to progress. What happens next?"*

According to Le Point, the regulator has responded only with *"radio silence".* Its headline read: *"Digital terrestrial radio is not progressing one inch."*

---

[469] http://www.lepoint.fr/actualites-medias/2010-03-16/technologie-la-radio-numerique-terrestre-n-avance-pas-d-un-pouce/1253/0/434215

92.

22 March 2010

The DAB challenge: most radios stay tuned to one station most of the time

A 'thought piece' by Ipsos MediaCT, entitled 'The Future of Radio', identified the many challenges for the government's proposed digital radio switchover to be successfully implemented by the 2015 target date:

- *digital listening share has to more than double in just four years*
- *the UK's DAB coverage [..] is currently around 90%, but there is now the need to extend it across all the UK population*
- *a requirement to improve the quality of [DAB] reception and sound*
- *the issue of people having to replace their analogue radio sets*
- *less than 1 in 10 of these [existing radio] sets is DAB, so a very significant number of replacements need to be sold*
- *all manufacturers are committed to producing sub-£20 sets in the next two years*
- *more digital radios need to be fitted in new cars and more digital converters need to be sold for existing cars*
- *take-up of digital platforms has been steady, but not remarkable*
- *digital listening has a long way to go to meet the Government's targets*
- *there are a number of barriers to overcome to meet the demands of the Digital Britain Report, which require investment – in a recession – and co-operation between manufacturers and broadcasters*
- *DAB will have to be marketed properly and quickly*[470]

A Capibus study conducted by Ipsos found that a high proportion of radio receivers were tuned to the same station most of the time:
- 86% of kitchen sets
- 79% of bedroom sets
- 74% of living room sets
- 70% of car radios

Ipsos asked:

"What happens when the switchover occurs and the station now only broadcasts on DAB? Do listeners go out and buy a new DAB set for each room in the house or switch their listening to another station or stop listening? This will be a major issue for stations and their audiences. It will be the listener who will be in control of radio's digital destiny."

---

[470] http://mediatel.co.uk/newsline/2010/01/14/is-radio-ready-for-the-digital-switchover/

93.

24 March 2010

Government: digital radio switchover in 2015 "still on track"

Tony Lloyd MP Manchester Central
Chair of the Parliamentary Labour Party
23 March 2010 @ Imperial War Museum North, Salford[471]

"One of the commitments that the government has already made is the switchover to digital [radio]. That will go ahead, although it will go ahead dictated by the pace of change that the markets themselves will involve. You know the ground rules for that. I was talking to a multimedia producer who just tells me she can't get digital radio in her own home. Now this is still one of the big issues because, until we have got 90% coverage of the country and until we see something along the lines of 50% of people using digital, that switchover won't take place. But all the evidence is that we are still on track for that switchover to take place by 2015.

The second debate within that is how is that paid for, how far will the commercial sector – the commercial radio stations – be prepared to pay to invest in the digital networks and how far will the BBC contribute? Because what is clear is that there always will be a role for the BBC to fund because there will be parts of the country where the commercial sector simply won't take that process forward.

We know that the analogue [radio transmission] system, even if we do nothing at all to maintain it, will require investment of the order of £200m simply to keep the existing networks up and running and that money frankly is better spent on the switchover to digital and, of course, there will be consequential changes in terms of the licensing framework at the point of switchover.

[...]

This government, the Labour government, once re-elected, probably on May 6th ...."

---

[471] author's recording

94.

24 March 2010

The Digital Economy Bill: let the horse-trading begin, says Shadow Minister

Ed Vaizey MP for Wantage & Didcot
Conservative Party Shadow Minister for Culture
23 March 2010 @ Imperial War Museum North, Salford[472]

"In today's radio industry, brands have been shaped more by scarcity of analogue spectrum than necessarily by the market. Brands have been built as much on the frequencies they occupy as much as the characteristics of their content, and commercial revenues have tended to stay limited to local markets.

We very much support the move to digital switchover, both because we believe it is important obviously to upgrade the technology, but because we also think that it will encourage plurality and expand listener choice. We have got to be concerned that people will be ready before any switchover takes place and that there won't be literally millions of analogue radios which suddenly become redundant. As you know, the government has set a provisional target date of 2015 and we are sceptical about whether that target can actually be met. That is not to say that we are sceptical about digital switchover. We simply think that 2015 might be too ambitious. But we are delighted to see that Ford Ennals is now chief executive of Digital Radio UK, after having steered digital television switchover so successfully, and we hope that all hurdles can be overcome.

We hope that the advent of new digital stations will bring significant new opportunities for independent radio production and it will also free up commercial radio spend. At the moment, as I understand it, the commercial sector spends nearly 10% of its annual revenue on analogue transmission. In the battle for ratings in the new digital world, we would hope that great programming would be at the forefront and that therefore a good proportion of the £40m annual cost of analogue broadcasting will go to independent radio production.

At the moment, the BBC holds four out of the five available national FM licences, and it has the only national digital multiplex. So the aspiration as we move over to digital is as much about making more space for plurality in radio broadcasting as it is about new technology. And if new stations are broadcast, we hope there is plenty of scope for new exciting radio production.

We are also keen obviously not to switch off FM, but to maintain FM as a spectrum particularly for local radio. As you are probably aware, there has been a lot of lobbying during the passage of the Digital Economy Bill about that. I'm pleased to say, as well, that some of the new technology that seems to be coming on-stream, with radios that can switch seamlessly between digital and FM broadcasts, will ensure that there will still be a place for ultra-local FM broadcast stations.

Obviously, many of you will also be interested in what will happen with the Digital Economy Bill as we approach the dissolution of Parliament. My understanding is that the Second Reading will happen on the 6th of April, which I think is also the date that Gordon Brown drives up the Mall to see the Queen to call for the dissolution of Parliament if he wants an election on the 6th of May [...] We will have this rather surreal Second Debate in the House of Commons and then we will go straight into what is now called the 'wash up' where we horse-trade over the various clauses of the Digital Economy Bill to be passed by the 8th of April. But I can assure you that the deregulation of radio clauses in the Digital Economy Bill have strong cross-party support so, if anything is going to go through, it will be those clauses."

[...]

Q&A

Q: *It's interesting that you touch on digital radio as a platform going forward. Once we find the larger stations, commercial and the BBC, make the switch to digital, and they leave the FM spectrum, do you feel that the majority of listeners will move to digital radio when they vacate their homes, as most cars don't come with a DAB receiver, so obviously the commercial sector and the BBC are going to be losing listeners because the majority of times listeners tune in to these station is in the car? Furthermore, with DAB, it's reported and seen by some people in the media/press as being a failed format, competing with new technologies such as DRM. With these changes, do you think that, when people do make the migration to DAB, that smaller stations are going to lose out and that the money from the commercial side is going to be re-invested in programming and we're not going to lose the quality of the content...*

A: *Well, I think the problem in the last few years has been a kind of half-way house, so people weren't really sure what the future of digital radio was going to be, particularly with commercial radios stations that were having to make a double investment which was costing them a lot of money, so we supported the government in making a firm decision that we were going to move over to digital switchover. As I said in my remarks, I think that 2015 might be a bit ambitious.*

Your particular point about converting cars to digital radio is, I think, the crucial point. We have got to get to a stage where new cars are fitted – as the French have now mandated, for example – with digital radios and that it gets easy to convert to digital in the car. I think that 2015 is going to be ambitious, but that does not mean that we are sceptical about switchover.

The other point about FM, as against DAB. I think that there will be... There are radios on sale now that switch seamlessly between FM and digital as if you were simply changing channels. I think that, particularly as FM will then be, broadly speaking, a spectrum used by the local radio stations, that won't be such a problem if you've only got a digital radio in your car, as you tend to listen to a local radio station when you're at home – or you can de-construct that remark. The point you make about whether DAB is the right technology or whether we should be using DAB+, to a certain extent I slightly take the view that we have gone down this road, so let's leave it. I think the pain of trying to move to DAB+ or beyond will be too much, given how far we've come.

Q: *I also found it quite interesting that you had the idea that there were going to be more digital-only services. In the past, we have seen digital services such as Capital Life and Core which have come and now gone again because they were not commercially profitable. Do you think that is not going to have an impact when most people make the migration to DAB? Do you think that the local full-scale FM operators are going to suffer?*

A: *Er, well, er, I hope that they won't. There will be a distinction between national or big regional radio stations and local stations, and there is already a distinction between local and community which is ultra-local. As I say, we want to put in place a platform that will also enable cross-media ownership at a local level that will enable local media companies to create scale. So, what I hope is that, across the range of media. there will be opportunities for any good radio station that is likely to command a loyal audience – whether that be an ultra-local audience, a regional audience or a national audience – because, in terms of Capital Radio coming and going, I think that was frankly a symptom of that we were in a half-way house about digital. We need to drive digital, which I think is now underway.*

[...]

_____

[472] author's recording

95.

27 March 2010

When is an FM radio not a radio? When it's in a portable media player, says digital switchover group

Digital Radio UK is the new organisation funded by the BBC and commercial radio *"to ensure that the UK is ready for digital radio upgrade"*.[473] In February 2010, Digital Radio UK submitted written evidence to the House of Lords Communications Committee informing it of the latest data for UK retail sales of radio receivers. Amongst other things, the data showed that:

- Sales of digital radios in 2009 were under 2 million units, their lowest annual volume since 2006
- Sales of analogue radios **seemed** to have dropped dramatically to 5.2 million in 2009 from between 7 and 8 million during 2008
- As a proportion of the total volume of radios sold, digital radios had **apparently** leapt to 28% in 2009 from 21% only a year earlier.[474]

I was puzzled. Why had sales of analogue radios fallen so dramatically by year-end 2009 (see graph below)? There seemed to be almost no substitution effect by DAB radios, whose volume sales were also down, though not by as much as analogue radios. It appeared as if many consumers had just suddenly decided to stop purchasing radios. I wrote to *****, the company that *** Digital Radio UK, asking why the data had suddenly 'jumped' in Q4 2009.

The written response from ***** was:

"The q4 2009 drop is more about the basket of products included as areas previously included such as set top boxes and portable media players were excluded from the data at that time."[475]

***** defines a 'portable media player' as any device that plays music and has a 3.5mm headphone jack: MP3 players, iPods, portable cassette players, portable CD players, etc.[476] From Q4 2009 onwards, when any of these devices are sold in the UK and also include a radio, they are no longer counted as 'a radio'. Now, every MP3 player sold that includes a radio is simply excluded from these statistics. This is why the number of radios sold **appeared** to drop so significantly (by around 2m units per annum) in the latest Digital Radio UK data.

Why was this change in definition made? It is hard to understand the logic because a radio within an MP3 player is still used as a radio and has no other purpose. It is a real radio, not a fake radio, but to ***** it is no longer a radio.

The answer seems to be that a huge number of MP3 players are sold in the UK (value £666m in 2009) but almost none of them incorporate a DAB radio. When an MP3 player does include a radio, it is inevitably an FM radio. MP3 players are manufactured and sold globally

UK RADIO RECEIVER SALES: FOUR-QUARTER MOVING AVERAGE ('000)

DAB receivers as percentage of total receiver sales

| 8% | 9% | 10% | 12% | 13% | 14% | 14% | 17% | 18% | 18% | 19% | 20% | 20% | 21% | 21% | 22% | 27% |

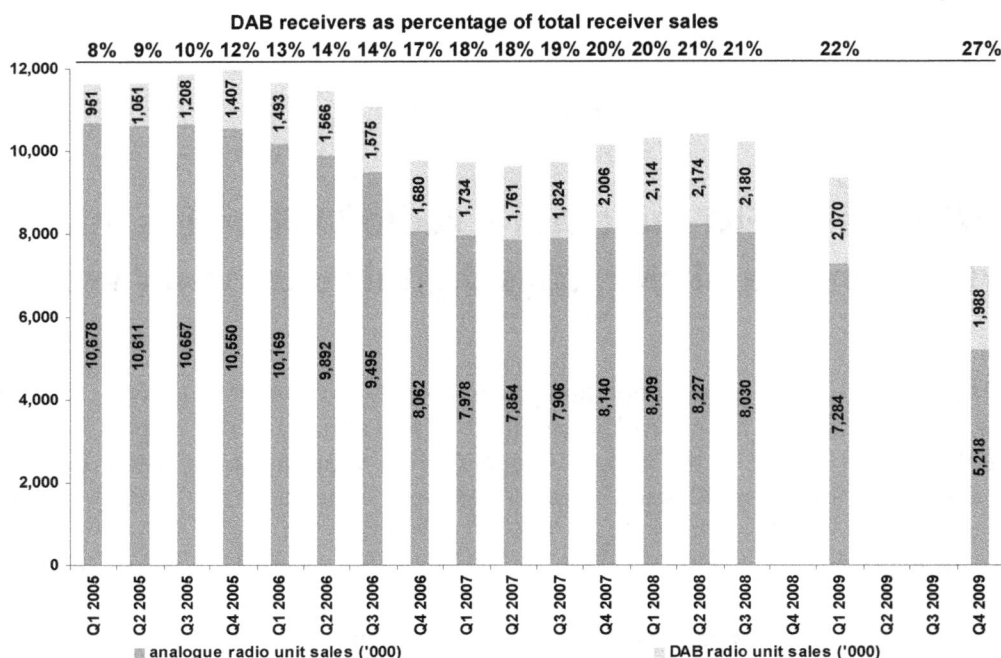

Chart data labels:

DAB radio unit sales ('000): 951, 1,051, 1,208, 1,407, 1,493, 1,566, 1,575, 1,680, 1,734, 1,761, 1,824, 2,006, 2,114, 2,174, 2,180, 2,070, 1,988

Analogue radio unit sales ('000): 10,678, 10,611, 10,657, 10,550, 10,169, 9,892, 9,495, 8,062, 7,978, 7,854, 7,906, 8,140, 8,209, 8,227, 8,030, 7,284, 5,218

Quarters: Q1 2005, Q2 2005, Q3 2005, Q4 2005, Q1 2006, Q2 2006, Q3 2006, Q4 2006, Q1 2007, Q2 2007, Q3 2007, Q4 2007, Q1 2008, Q2 2008, Q3 2008, Q4 2008, Q1 2009, Q2 2009, Q3 2009, Q4 2009

■ analogue radio unit sales ('000)   ■ DAB radio unit sales ('000)   477

by multinational electronics manufacturers who understand that FM remains the universal standard for listening to broadcast radio, while DAB is still confined to no more than a handful of countries. Global manufacturers are reluctant to mass produce an MP3 player incorporating a DAB radio because the sales market would be limited to a few, small territories.

I checked the Argos retail website this week and found it offered 82 models of MP3/MP4 player. None incorporated DAB radio, whereas there were 16 that included an FM radio and 66 that had no radio.

It seems that the last resort for Digital Radio UK to be able to demonstrate to a sceptical public (and increasingly sceptical members of the House of Lords) that DAB radio is 'taking off' with consumers is to fix the figures to make it look that way. If you cannot convince the public to stop buying analogue radios, you can 'bend' the figures to magically make it appear that the public is buying fewer analogue radios.

In an earlier chapter, I documented how Digital Radio UK had similarly fixed the same dataset from ***** to declare in its publicity that *"when buying a radio, more than 75% of people choose a digital radio".* This was not at all true. The real fact was that, in December 2009 alone (December always being the peak month for DAB radio sales), 76% of people who bought a kitchen radio bought a digital kitchen radio. That was an attempt to brazenly redefine 'a radio' as only 'a kitchen radio' so as to exclude clock radios, tuners, in-car radios, boomboxes, etc.

I can only repeat what I said then. However desperate you might be to try and make DAB radio a success, how is it justifiable to deliberately mis-state data so outrageously in print? And to Parliament?

---

[473] http://www.parliament.uk/documents/upload/DigitalRadioUKukdso.pdf

[474] http://www.parliament.uk/documents/upload/DigitalRadioUKsuppukdso.pdf

[475] the correspondent subsequently insisted that their company name not be identified

[476] http://www.mobile-ent.biz/news/33104/Nokia-claims-UKs-best-selling-music-player

[477] source: Digital Radio Development Bureau

96.

31 March 2010

French Culture Minister: launch of digital radio not "a priority"

On 29 March 2010, the French Minister of Culture & Communications, Frederic Mitterand, spoke at the monthly luncheon of the Association of Media & Communications Journalists. He was asked about the much delayed launch of digital terrestrial radio in France and replied:

"I note that the cost of the [digital radio] project is significant, that a number of the radio licensees are not at all favourable towards the project, and that it is the CSA [media regulator] that for the moment is escalating the issue. The CSA itself should still submit a report on the [digital radio] issue with recommendations, although I know roughly what will be in such a report. I have the greatest respect for the CSA, and I have the greatest feelings of respect for [CSA president] Michel Boyon, but we are not exactly on the same wavelength."

"Without organising a funeral with great pomp and ceremony, which would presume a death, I think that everything will inevitably be digital one day. And then radio will be too. Put simply, in today's economic conditions, in the general context of radio, and with the lack of consensus around this [digital radio] issue, I do not think its resolution is a priority and the launch of digital radio will not happen this year."[478]

---

[478] http://www.radioactu.com/actualites-radio/126516/radio-frederic-mitterrand-n-est-pas-favorable-au-lancement-de-la-rnt/

97.

3 April 2010

Marketing DAB radio: misleading listeners only damages the medium

The radio medium's loyalty amongst consumers derives substantially from the trust engendered between the on-air presenter and the listener. Research has demonstrated that radio is trusted more than any other medium, and that its audience feels a much greater affinity than it does with less intimate media such as television and newspapers.

In view of the importance of this 'trust' between radio and its audience, it seems a remarkable own-goal for radio to be promoting itself in a misleading way in advertisements carried on its own medium – radio. If listeners cannot trust radio people to be truthful about radio on the radio, then does it not undermine the bond that exists between a radio station and its listenership?

A recent radio advertisement placed on commercial radio stations by the Digital Radio Development Bureau, the agency tasked with persuading the public to buy and use DAB radios, stated:

"This is an advert for DAB digital radio. If you were listening to me on a conventional analogue ..." [the sound of radio interference interrupted the speaker momentarily. The voice-over then continued:] "... radio you might very well hear strange noises ..." [further sounds of radio interference followed. The voice-over continued:] "... which would ruin your enjoyment of your favourite programme ..." [more interference sounds were audible. The voice-over continued:] "... meaning you might miss out on the crucial ..." [radio interference sounds could be heard once more] "... but, with a DAB radio, you can enjoy crisp, clear digital sound. To find out more and discover loads more stations, visit getdigitalradio.com. Prices start from £24.99. Digital radio, get more from your radio".[479]

Listeners complained to the Advertising Standards Authority [ASA] that this advertisement was misleading because, when the DAB radio signal is inadequate, the audible broadcast signal is interrupted.

The Digital Radio Development Bureau responded that:
- because DAB is a digital technology which is either 'on' or 'off', the signal is always the same right up to the coverage limit
- DAB uses single-frequency networks technology where the same programme is transmitted from a number of sites, and DAB receivers add the signals from all the transmitters together, reducing gaps whereas, in an analogue radio network, gaps between transmitters cause the signal to fade in and out as the listener moves around
- a digital radio receiver is not subject to the background hiss and interference that might be audible with an analogue radio, and it is only when the listener is not in a digital station's coverage area that the signal drops out

- electrical interference from fridges, thermostats, motors or light switches can cause crackle on analogue radio, whereas digital radio is not susceptible to this
- the other interference referred to in the advert is intrusion of pirate radio broadcasters that listeners might hear on analogue radio. Because there is no low-grade, cheap equipment available for DAB, pirates are not able to broadcast on digital radio
- the advert sought to promote the fact that DAB radio was hiss- and crackle-free, which the Bureau believed was reasonable and responsible.

The ASA believed otherwise. It said it understood that *"if listening to digital radio whilst travelling, the digital signal could drop out when entering a built-up area or walking between tall buildings,"* whereas the adverts *"gave the misleading impression that listeners would never experience any interruption to a DAB signal, when that was not the case."*[480] The ASA banned future use of the advert.

This was not the first occasion on which advertisements promoting DAB radio have been found to be misleading. In 2005, the ASA had similarly banned a radio advert which had stated:

"If you're someone who thinks an iPod is something you might keep your contact lenses in you probably haven't heard about DAB digital radio. With a new digital radio costing from as little as 49.99, not only can you hear all your current favourites in crystal clear sound, you can switch on to a dial-full of digital-only stations specialising in everything from classic rock to books that talk. The future is here today with distortion free DAB digital radio: taking the hiss out of the way you listen to the radio. Message provided by TWG Emap Digital."

On this occasion, the ASA decided that *"not all DAB digital radio listeners would receive 'distortion free' and 'crystal clear' sound and concluded that the claims were misleading,"* it having *"received no evidence to show that DAB digital radio was superior to analogue radio in terms of audio quality."*

On another occasion, in 2004, Ofcom banned an advertisement broadcast on London station Jazz FM which had claimed falsely that DAB radio offers consumers *"CD-quality sound"*. Ofcom concluded that *"some listeners, in particular listening circumstances, would perceive a difference in sound quality between services using lower bit rates or broadcasting in mono compared to the quality attainable on CDs."*[481]

There is a recurring theme here of DAB radio marketing campaigns repeatedly being found to be misleading listeners. Their response: just try and try and try again. Perhaps there should be a 'three strikes, then you're out' policy. Do not pass go. Do not advertise DAB radio misleadingly on the radio. Do not continue to abuse the trust between the radio medium and its listenership.

---

[479] http://www.asa.org.uk/Complaints-and-ASA-action/Adjudications/2010/3/The-Digital-Radio-Development-Bureau/TF_ADJ_48282.aspx
[480] http://www.asa.org.uk/Complaints-and-ASA-action/Adjudications/2010/3/The-Digital-Radio-Development-Bureau/TF_ADJ_48282.aspx
[481] http://www.ofcom.org.uk/tv/obb/adv_comp/a15/a15.pdf

98.

7 April 2010

Digital radio switchover: legislation is "virtually meaningless" says Shadow Culture Secretary

House of Commons
6 April 2010 @ 1627
Digital Economy Bill, Second Reading [excerpts][482]

The Secretary of State for Culture, Media and Sport (Mr. Ben Bradshaw): *The switchover to digital radio has probably aroused more interest than any other issue in the Bill except that of unlawful file sharing. The target date of 2015, set by the Government, is an incentive not an ultimatum. We have made it clear that a decision on digital switchover will not be made until national DAB coverage is comparable to that of FM, until local DAB reaches 90 per cent of the population and all major roads and until 50 per cent of listening is through digital means. Once all those criteria have been satisfied, there will be at least two years before switchover takes place, at which point we expect coverage and listening to reach nearly universal levels — that is, about 98.5 per cent judged by television reach.*

[...]

Mr. Jeremy Hunt (South-West Surrey) (Conservative): *The Government have ducked sorting out digital radio switchover, which the Secretary of State has just talked about. They are giving Ministers the power to switch over in 2015, yes, but without taking any of the difficult measures necessary to make it practical or possible.*

[...]

Robert Key (Salisbury) (Conservative): *Is my hon. Friend content with clause 31, on the digital switchover? It is estimated that the costs to the consumer will be £800 million, and there is no sign of manufacturers of DAB radios producing cheap radios, no estimate of the cost of throwing away millions of existing FM sets, no sign that the motor car industry is going to come up with the goods — [interruption]. A Labour Back Bencher says, "Yes there is," but I have read all the papers and although there are one or two pious hopes, there is nothing more than that. This will be extremely expensive, and the 2015 deadline is unattainable. Is my hon. Friend content, therefore, or will we make some further promises?*

Mr. Hunt: *I share my hon. Friend's concerns, because I think that clause is so weak that it is virtually meaningless, as it gives the Secretary of State the power to mandate switchover in 2015 but the Government have not taken the difficult steps that would have made that possible, such as ensuring that the car industry installs digital radios as standard, as my hon. Friend suggests, and that there is proper reception on all roads and highways. As a result, a lot of people are very concerned that 110 million analogue radios will have to be junked in 2015. In particular, I would have liked the Government to find out whether we could move* *from DAB to the DAB+ technology, which most people think will be far more effective. If they had done that, this measure would not threaten smaller local radio stations.*

Mr. Siôn Simon (Birmingham, Erdington) (Labour) *rose —*

Mr. Hunt: *I will give way to the former Minister with responsibility for creative industries, and then I will make some progress.*

Mr. Simon: *Given the hon. Gentleman's desire to move to DAB+, what does he suggest the 8 million people in this country who have bought very expensive DAB radios should do?*

Mr.Hunt: *First, let me say that when the hon. Gentleman stepped down as Minister for the creative industries, it was a great shame that he was not replaced. It would have helped in the sensible framing of the Bill if we had had a Minister with that responsibility now, but there is none. The answer to the hon. Gentleman's question is simply this: when we migrate from one technology to another — whether analogue to DAB, or DAB to DAB+ — we need some kind of help scheme, as we have with TV digital switchover, but there is no mention of a help scheme in this Bill. That serves to highlight why the Government have ducked the important decisions.*

[...]

Mr. Don Foster (Bath) (Liberal Democrat): *Notwithstanding the many concerns that have been raised over the past few months about the move from analogue to digital radio, broadly speaking there is now consensus about that measure. The Secretary of State has laid down clear criteria that have to be met on listenership and coverage before the two-year starting pistol can be fired. Of course, there have been concerns. For example, some people thought that FM would be dropped, but we know that it will not be dropped; indeed, FM could become a new vibrant platform for local and micro-local radio stations and given more power. Possibly, Ofcom could start to give them even longer licences. With all the conditions that have been inserted, that is another exciting provision that we should acknowledge and accept so that everyone can have the real benefits of the digital radio era, in terms of greater interactivity and so on. The Government have done a disservice by failing to promote the real benefits of digital radio as effectively as they could. It is not surprising that the Committee in their lordships' House castigated the Government for their failure. The industry could have done more. It is a pity that it has taken so long for FM to be included in all the DAB radios now on sale. It is only very recently that we have heard of the launch of the mechanism that will ensure people can have a single tuner covering DAB and FM — a single EPG, or electronic programme guide. That is welcome, but the work could have been done sooner.*

[...]

Mr. John Whittingdale (Maldon and East Chelmsford) (Conservative): *I now turn to DAB radio. Commercial radio and the BBC have invested huge amounts in moving to DAB, and commercial radio in particular is now in real economic difficulties, as the report that my Select Committee — the Culture, Media and Sport Committee — issued this morning explains. There is no doubt that one burden on it is having to broadcast in analogue and digital simultaneously, and it would provide some help if it had a firm pathway to a future in which it need only broadcast in DAB. I believe that the 2015 date, which I know is not in the Bill, is unrealistic. It is sensible to set a date, but most people believe that that is probably too ambitious, because of the single problem of car radios. Yes, some manufacturers are* *beginning to fit DAB radios in cars, but there is a huge reservoir of cars that will not have them for a very long time. We must get to a point at which an in-car radio can easily be converted to DAB. The device that is on the market at the moment, which I have in my car, has so many wires, antennae and bits of equipment that I do not believe it will be taken up with great enthusiasm.*

[...]

Mr. Austin Mitchell (Great Grimsby) (Labour): *I cannot agree with my hon. Friend the Member for Sittingbourne and Sheppey (Derek Wyatt) in his analysis of the digital radio switchover. Clearly the industry, in the main, supports digital switchover, but of course a switchover to DAB radio by 2015 is wholly impractical and out of the question because that is too soon. It will be much more difficult to switch over to digital radio than it was to switch over to digital TV, because that process was helped by the mass subscription to Sky and by the development of Freeserve. Such provision does not exist in respect of radio, because there are 120 million radios in this country and sales of digital radio have not taken off. Digital radio is quite expensive and if we make it compulsory, that will be a heavy tax on the consumer. One of the lower prices for a digital radio is about £85, and that price has increased with devaluation. So this would be a heavy burden to impose on the consumer, and if we require switchover, it would leave about 120 stations still on FM and locked out in the cold. We do not have to switch over at this speed and we do not have to switch over to DAB because we could move to DAB+, which would allow both services to be run concurrently. I am worried about the digital switchover for radio, because the crucial factor here is car radios, for which the technology is never sold effectively. Like the hon. Member for Maldon and East Chelmsford (Mr. Whittingdale), my experience with DAB in the car has been totally unsatisfactory. Not only is it messy, but it is difficult to pick up a station, and the signal cuts in and out and fades away, so one is constantly having to switch back to FM. Digital car radio sales are crucial, but such sales have been low and there is no sign of their taking off. Only 1 per cent of cars are fitted with a digital radio, and until there is a mass fitting of digital car radios we shall not be able to have an effective switch-off. I am worried about that provision.*

[...]

Mr. John Grogan (Selby) (Labour): *Two great debates on this Bill, with commercial interests on both sides, have been referred to tonight. I will not rehearse all the arguments, but one of the debates is on digital radio. The Opposition Front-Bench team seems to be saying that it opposes the current model the Government are suggesting. The Opposition spokesman suggested that he was now in favour of DAB+. It is interesting that hundreds of radio stations listened to by our constituents throughout the land, such as Minster FM, are being offered no digital future whatever in this Bill. What they are being offered, at best, is a place on a joint FM and digital electronic programme guide that is still being developed, and even if they get on that device, they will still not have all the advantages of potential and so forth, and they will be very much second-class stations. Under the Bill as currently drafted, that is the future. Helpful amendments were tabled by the Conservatives and the Liberal Democrats in the House of Lords suggesting that before any switchover there should be full consideration of all local and community stations. I will re-table those amendments today; I hope that the hon. Member for Bath will support them, and that they might tempt the Conservative Front Bench, too, in the negotiations for the wash-up. There is another side to the debate, to do with the BBC and some other digital radio interests. This reinforces the point that we should still have a full Committee stage — and if we cannot have that, we should pass the Bill on to our successors.*

[debate ended 10pm]

---

[482] http://www.publications.parliament.uk/pa/cm200910/cmhansrd/cm100406/debtext/100406-0007.htm

99.

10 April 2010

Digital Economy Act 2010: a smokescreen for backroom radio 'deal'

On 8 April 2010 at 1732, the Digital Economy Act was given Royal Assent by Parliament. Who exactly will benefit from the radio clauses in the Act? Certainly not the consumer.

"The passing of the Digital Economy Bill into law is great news for receiver manufacturers," said Frontier Silicon CEO Anthony Sethill. As explained by Electronics Weekly: *"Much of the world DAB industry revolves around decoder chips and modules from UK companies, in particular Frontier Silicon. These firms can expect a bonanza as consumers replace FM radios with DAB receivers."*[483] Frontier Silicon says it supplies semi-conductors and modules for 70% of the global DAB receiver market.[484]

Sadly, the Bill/Act was not really about digital radio at all. For the radio sector lobbyists, it was all about securing an automatic licence extension for Global Radio's Classic FM, the most profitable station in commercial radio, so as to avoid its valuable FM slot being auctioned to allcomers. The payback on this valuable asset alone easily justified spending £100,000's on parliamentary smooching. It was interesting to see one Labour MP acknowledge the true purpose for all this parliamentary lobbying in the House of Commons debate when he congratulated *"[Classic FM managing director] Darren Henley for making a cause of the issue."*[485]

The clauses in the Digital Economy Bill on the planned expansion of DAB radio and digital radio switchover were simply promises that Lord Carter had insisted upon as the radio industry's quid pro quo for government assistance to Global Radio's most profitable asset. The existence of this 'deal' between Lord Carter and Global Radio was confirmed by Digital Radio Working Group chairman Barry Cox in his evidence to the House of Lords:

"Lord Carter did not like to do [the deal] immediately. As I understand, he wanted to get something more back from the radio industry. I think there is a deal in place on renewing these licences, yes."[486]

However, the quid pro quo promise to develop DAB radio will never come to fruition. Now that Global Radio has got what it wanted, over the coming months, the radio industry's commitment to continue with DAB will inevitably be rolled back. Every excuse under the sun will be wheeled out – the economy, the expense, the lack of industry profitability (having spent nearly £1bn on DAB to date), consumer resistance, the regulator, the Licence Fee, the government (old and new), the car industry, the French, the mobile phone manufacturers, whatever ....

The reasons that digital radio migration/switchover will never happen are no different now than they were before the Digital Economy Bill was passed into law. For the consumer, who seems increasingly unconvinced about the merits of DAB radio, this legislation changes nothing at all. Those reasons, as itemised in my written submission to the House of Lords in January 2010, are:

- The characteristics of radio make the logistics of switchover a very different proposition to the television medium
- The robustness of the existing analogue FM radio broadcasting system
- Shortcomings of the digital broadcast system, 'Digital Audio Broadcasting' [DAB], that is intended to replace analogue radio broadcasting in the UK.[487]

More specifically:

1. Existing FM radio coverage is robust with close to universal coverage
- 50 years' development and investment has resulted in FM providing robust radio coverage to 98.5% of the UK population

2. No alternative usage is proposed for FM or AM radio spectrum
- Ofcom has proposed no alternate purpose for vacated spectrum
- There is no proposed spectrum auction to benefit the Treasury

3. FM/AM radio already provides substantial consumer choice
- Unlike analogue television, consumers are already offered a wide choice of content on analogue radio
- 14 analogue radio stations are available to the average UK consumer (29 stations in London), according to Ofcom research

4. FM is a cheaper transmission system for small, local radio stations
- FM is a cheaper, more efficient broadcast technology for small, local radio stations than DAB
- A single FM transmitter can serve a coverage area of 10 to 30 miles radius

5. Consumers are very satisfied with their existing choice of radio
- 91% of UK consumers are satisfied with the choice of radio stations in their area, according to Ofcom research
- 69% of UK consumers only listen to one or two different radio stations in an average week, according to Ofcom research

6. Sales of radio receivers are in overall decline in the UK
- Consumer sales of traditional radio receivers are in long-term decline in the UK, according to GfK research
- Consumers are increasingly purchasing integrated media devices (mp3 players, mobile phones, SatNav) that include radio reception

7. 'FM' is the global standard for radio in mobile devices
- FM radio is the standard broadcast receiver in the global mobile phone market
- Not one mobile phone is on sale in the UK that incorporates DAB radio

8. The large volume of analogue radio receivers in UK households will not be quickly replaced
- Most households have one analogue television to replace, whereas the average household has more than 5 analogue radios
- The natural replacement cycle for a radio receiver is more than ten years

9. Lack of consumer awareness of DAB radio

- Ofcom said the results of its market research "highlights the continued lack of awareness among consumers of ways of accessing digital radio"

10. Low consumer interest in purchasing DAB radio receivers

- Only 16% of consumers intend to purchase a DAB radio in the next 12 months, according to Ofcom research
- 78% of radio receivers purchased by consumers in the UK (8m units per annum) are analogue (FM/AM) and do not include DAB, according to GfK data

11. Sales volumes of DAB radio receivers are in decline

- UK sales volumes of DAB radios have declined year-on-year in three consecutive quarters in 2008/9, according to GfK data

12. DAB radio offers poorer quality reception than FM radio

- The DAB transmission network was optimised to be received in-car, rather than in-buildings
- Consumer DAB reception remains poor in urban areas, in offices, in houses and in basements, compared to FM

13. No common geographical coverage delivered by DAB multiplexes

- Consumers may receive only some DAB radio stations, because geographical coverage varies by multiplex owner

14. Increased content choice for consumers is largely illusory

- The majority of content available on DAB radio duplicates stations already available on analogue radio

15. Digital radio content is not proving attractive to consumers

- Only 5% of commercial radio listening is to digital-only radio stations, according to RAJAR research
- 74% of commercial radio listening on digital platforms is to existing analogue radio stations, according to RAJAR research

16. Consumer choice of exclusive digital radio content is shrinking

- The majority of national commercial digital radio stations have closed due to lack of listening and low revenues
- After ten years of DAB in the UK, no digital radio station yet generates an operating profit

17. Minimal DAB radio listening out-of-home

- Most DAB radio listening is in-home, and DAB is not impacting the 37% of radio listening out-of-home
- Less than 1% of cars have DAB radios fitted, according to DRWG data

18. DAB radio has limited appeal to young people

- Only 18% of DAB radio receiver owners are under the age of 35, according to DRDB data
- DAB take-up in the youth market is essential to foster usage and loyalty

19. DAB multiplex roll-out timetable has been delayed

- New DAB local multiplexes licensed by Ofcom between 2007 and 2009 have yet to launch
- DAB launch delays undermine consumer confidence

20. Legacy DAB receivers cannot be upgraded

- Almost none of the 10m DAB radio receivers sold in the UK can be upgraded to the newer DAB+ transmission standard
- Neither can UK receivers be used to receive the digital radio systems implemented in other European countries (notably France)

21. DAB/FM combination radio receivers have become the norm

- 95% of DAB radio receivers on sale in the UK also incorporate FM radio
- 9m FM radios are added annually to the UK consumer stock (plus millions of FM radios in mobile devices), compared to 2m DAB radios, according to GfK data

22. DAB carriage costs are too high

- Carriage costs of the DAB platform remain too costly for content owners to offer new, commercially viable radio services, compared to FM
- Unused capacity exits on DAB multiplexes, narrowing consumer choice

23. DAB investment is proving too costly for the radio industry

- The UK radio industry is estimated to have spent more than £700m on DAB transmission costs and content in the last ten years
- The UK commercial radio sector is no longer profitable, partly as a result of having diverted its operating profits to DAB

24. DAB is not a globally implemented standard

- DAB is not the digital radio transmission standard used in the most commercially significant global markets (notably the United States)

These factors make it unlikely that a complete switchover to DAB digital terrestrial transmission will happen for radio in the UK.

With television, there existed consumer dissatisfaction with the limited choice of content available from the four or five available analogue terrestrial channels. This was evidenced by consumer willingness to pay subscriptions for exclusive content delivered by satellite. Consumer choice has been extended greatly by the Freeview digital terrestrial channels, many of which are available free, and the required hardware is low-cost.

Ofcom research demonstrates that there is little dissatisfaction with the choice of radio content available from analogue terrestrial channels, and there is no evidence of consumer willingness to pay for exclusive radio content. Consequently, the radio industry has proven unable to offer content on DAB of sufficient appeal to persuade consumers to purchase relatively high-cost DAB hardware in anywhere near as substantial numbers as they have purchased Freeview digital television boxes.

Additionally, it has taken far too long to bring DAB radio to the consumer market, and its window of opportunity for mass take-up has probably passed. Technological development of DAB was started in 1981, but the system was not demonstrated publicly in the UK until 1993 and not implemented for the consumer market until 1999. In the meantime, the internet has expanded to offer UK consumers a much wider choice of radio content than is available from DAB.

In this sense, DAB radio can be viewed as an 'interim' technology (similar to the VHS videocassette) offering consumers a bridge between a low-tech past and a relatively high-tech future. If DAB radio had been rolled out in the early 1990s, it might have gained sufficient momentum by now to replace FM radio in the UK. However, in the consumer's eyes, the appeal of DAB now represents a very marginal 'upgrade' to FM radio. Whereas, the wealth of radio content that is now available online is proving far more exciting.

The strategic mistake of the UK radio industry in deciding to invest heavily in DAB radio was its inherent belief in the mantra 'build it and they will come.' Because the radio industry has habitually offered content delivered to the consumer 'free' at the point of consumption, it failed to understand that, to motivate consumers sufficiently to purchase relatively expensive DAB radio hardware would necessitate a high-profile, integrated marketing campaign. Worse, the commercial radio sector believed that compelling digital content could be added 'later' to DAB radio, once sufficient listeners had bought the hardware, rather than content being the cornerstone of the sector's digital offerings from the outset.

In my opinion, the likely outcome is that FM radio (supplemented in the UK by AM and Long Wave) will continue to be the dominant radio broadcast technology. For those consumers who seek more specialised content or time-shifted programmes, the internet will offer them what they require, delivered to a growing range of listening opportunities integrated into all sorts of communication devices. In this way, the future will continue to be FM radio for everyday consumer purposes, with personal consumer choice extended significantly by the internet.

---

[483] http://www.electronicsweekly.com/Articles/2010/04/09/48379/government-writes-fm-death-warrant.htm
[484] http://www.frontier-silicon.com/audio/DABFlash/index.htm
[485] http://www.publications.parliament.uk/pa/cm200910/cmhansrd/cm100406/debtext/100406-0011.htm
[486] http://www.publications.parliament.uk/pa/ld/lduncorr/uccom200110ev1.pdf
[487] http://www.parliament.uk/documents/upload/GGoddard.pdf

The author can be contacted via his web site:

www.grantgoddard.co.uk

Radio Books is a London-based publisher of books about the radio broadcasting industry:

www.radiobooks.org

www.ingramcontent.com/pod-product-compliance
Lightning Source LLC
Chambersburg PA
CBHW080840270326
41926CB00018B/4098